This book is dedicated to my children
Rosemary and Jonathan

in the hope that one day they
may actually read it

Contents

SECTION DEPENDENCY DIAGRAM

I first encountered the idea of a diagram showing the dependency relation amongst the chapters of a book in W. W. Sawyer's *Prelude to Mathematics*. As a teenager I was inspired by this book to pursue a mathematical education; and the vigour and clarity of its exposition has remained as a model to me ever since.

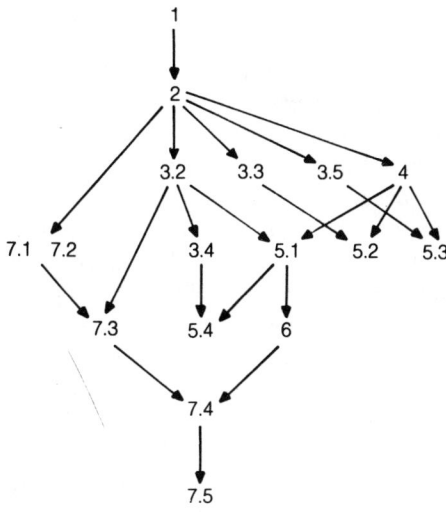

Foreword

During the past decade, the importance of mathematical logic for the computer scientist has become widely accepted. It is now recognized that logic plays a fundamental part in many areas of computer science, and that a course in logic is an essential component in the education of every computer scientist.

Today, computers are used to control military and civil aircraft, nuclear power stations and industrial robots; they run commercial and financial networks involving transfers of large funds between institutions and, in all these tasks, the consequences of malfunctioning are potentially catastrophic. New developments and applications are proceeding apace, with computers being used to monitor and control more and more of our day-to-day activities, and this, in turn, has reinforced the requirement to produce hardware and software systems which are demonstrably correct.

The value of logic techniques in circuit design has, of course, been well known for many years and the comparatively recent developments in VLSI design have merely served to emphasize this. But a thorough grounding in mathematical logic is needed for all stages of software development, especially program specification, verification and program transformation. In all these stages logic underpins the theory, thus bearing out the dictum that 'logic is the calculus of computer science'.

Dr Galton has produced an admirable book which sets out the subject with clarity and rigour. The subject is illustrated by simple examples throughout the text, and the exercises make the book useful for both undergraduates and mature students wishing to gain an understanding of this important topic.

M. H. ROGERS
University of Bristol
January 1990

Preface

In recent years, formal logic has come to occupy a central position in the repertory of technical knowledge generally admitted to be indispensable to anyone seeking more than a superficial understanding of computer science, whether the emphasis is on software engineering, artificial intelligence, or any of the flavours of the subject that come in between these extremes. In consequence, logic is being ever more widely taught at undergraduate level to students in computer science departments all over the world.

Traditionally, logic has rather been the preserve of mathematicians and philosophers, each group cultivating its own distinctive approach to the subject and developing it from its own point of view. There thus exists a wide range of accessible and in many cases excellent introductory logic textbooks, written specifically to meet the needs either of the mathematics student or of the philosophy student. Although much of the *material* in these books is highly relevant to the computer scientist, the emphasis and manner of presentation is in general not so suitable: from the point of view of the computer science student, the philosophical books contain too much philosophy, and the mathematical ones too much mathematics.

On the other hand, there *are* beginning to appear a number of books on logic specifically written for computer science students. Although some of these books are also of a high quality, none so far seems to cater for exactly the need I have in mind. Some are too advanced, and are thus suitable for graduate students and not for undergraduates. Some aim at a broad, but correspondingly less thorough coverage, quickly moving beyond the classical core of the subject to explore a wide range of further developments seem as relevant to computer science. Yet others concentrate on some specific application of logic to computing, for example logic programming, knowledge representation, or program verification.

I have nothing against these books, and indeed use some of them in my own teaching. But none of them, I find, quite fits the bill for the kind of introduction to logic that I want to give my students. That is the reason why I have written this book, which I hope will be regarded as suitably filling the role for which I have intended it.

My approach has been to stick to imparting the canonical logical theories: the propositional calculus and the first-order predicate calculus, only going beyond these with a brief excursion into modal and temporal logic in the final chapter. I have aimed to cover these classical theories as thoroughly as is compatible with the elementary nature of the text. I have tended to stress the algorithmic aspects of the subject, as something that computer science students should find particularly congenial; but I have not otherwise devoted much space to a discussion of specifically computational matters. It has also been my aim to stress the *deductive* character of logic: too often, especially in artificial intelligence, there has been a tendency to focus on its *expressive* power, as if logic were primarily a language rather than a calculus. My view is that one is not doing logic if one merely uses logical notation: one must also use that notation as a tool for drawing inferences.

In the first chapter, I introduce, informally, a wide range of general logical concepts that are applicable to any variety of logic. I introduce these in advance of any particular system of logic because otherwise the general notions come too much to be seen as inextricably bound up with the particular form of logic in connection with which they are first learnt. Thus, for example, the notion of satisfiability comes to be seen as something specifically defined in terms of the model theory of the predicate calculus, whereas in reality it is of much more general application. This approach will also help, I believe, to ease the transition from a student's pre-theoretical intuition to a more theoretically grounded understanding.

The next two chapters deal with the propositional calculus, which is fundamental in that virtually all other systems of logic include it. In Chapter 2, the propositional calculus is introduced step by step, slowly enough for the student to assimilate each new idea thoroughly before moving on to the next. The chapter ends by tying together all the ideas thus introduced into a tightly defined formal system. Chapter 3 explores the proof theory of the propositional calculus, introducing three different methods of validating propositional inferences and discussing in what respects the adequacy of such methods may be assessed.

Chapters 4 and 5 do for the predicate calculus what the two preceding chapters did for the propositional calculus. Chapter 6 goes on to discuss first-order theories, introducing in particular the problems associated with identity, and concludes with a brief look at the first-order formalization of elementary arithmetic. Finally, Chapter 7 introduces modal logic and its offshoot, temporal logic.

At various places in these chapters I touch upon some of the momentous episodes in the history of the subject: the notions of decidability and computability, as illuminated by the work of Gödel, Church, and Turing, and of complexity, with special reference to Cook's theorem and the theory of NP-completeness. These are things which all computer scientists should know about: they are part of our intellectual heritage, and helped to shape the cultural milieu in which both logic and computer science have their being. I therefore make no apologies for devoting to them the amount of space (admittedly modest) that I do.

In the course of the book I am at pains to eradicate a number of confusions and misunderstandings that I find to be widespread amongst people who have a smattering of logical knowledge but no thorough grounding. First and foremost is the confusion between the material conditional (truth-functional 'if ... then ...') and logical implication. I think this confusion has been instrumental in inducing scepticism, amongst AI researchers in particular, as to the utility of logic. For they learn, mistakenly, to read the symbol for the material conditional as 'implies', and then, reasonably enough, complain that the truth-functional relation is much to weak to do the job they want 'implies' to do. So often does one find otherwise worthy textbooks recommending 'implies' as the reading of → that I fear I am waging a losing battle; but I persevere. In this matter I am particularly indebted to the logical writings of W.V.O. Quine, who has done more than anyone else to caution the philosophical community against this misconstrual of the material conditional.

The second source of confusion which I have paid particular attention to concerns the distinction between the syntactical and semantical consequence relations: between, in other words, deducibility and entailment. (Entailment reduces to implication when the premiss set contains but a single member.) It is one thing for a conclusion to be deducible from a set of premisses by means of a given proof system: this is syntactic consequence. It is quite another thing for it to be *entailed* by that set of premisses (semantic consequence). Only by grasping this clearly can the student be got to see the real significance of soundness and completeness questions.

To the problem of confusion between use and mention my solution has been simple. Amongst logicians it is a commonplace that one must distinguish carefully between using an expression and merely mentioning it. The name which I *use* when I say that Simon plays the oboe, say, is merely *mentioned* when I say that 'Simon' is a five-letter word. My brother, the one who plays the oboe, is not a five-letter word, but his name, which I use when I refer to him, is. The last sentence but one illustrates the standard solution to use/mention confusions: enclose a mentioned expression in quotation marks. In this book, I do not *use* formal logical expressions, I only mention them; I therefore dispense with the use of quotation marks around logical expressions. Such expressions are distinguished typographically by being printed in a sanserif font. On the other hand, I have tried to be meticulous in the correct use of quotation marks with linguistic expressions. Thus I might say that the individual constant a in the schema F(a) plays a role similar to the proper name 'Simon' in the sentence 'Simon plays the oboe'.

An issue on which I take a position that is somewhat unorthodox from the point of view of mathematical logic (though not the philosophical variety) concerns the use of variables in the predicate calculus. I make no use of free variables: all my variables are bound by quantifiers. Hence, in specifying the syntax of the predicate calculus, I obtain quantified schemas by quantifying into the argument places of predicates; the predicates in turn are obtained by substituting argument places for individual constants in completed schemas, exactly as expounded by Frege when he *invented* the predicate calculus. Likewise, in the formal semantics of the predicate calculus, I assign truth values to quantified schemas by considering the extensions of the predicates from which they are obtained. These extensions are in turn obtained by the device of dummy constants. That way I avoid the cumbersome use of variable assignments, which in my experience students tend to find confusing.

Some sections of this book can be omitted without prejudicing the understanding of the rest. In particular, the material on model building and on resolution can be regarded as optional. A first course in logic can be constructed by taking any suitable combination of paths through the section-dependency diagram shown on p. ix. In some parts of the book a modest amount of standard mathematical notation is used: I assume as a prerequisite a nodding acquaintance with the elementary parts of set theory (e.g., the notions of union, intersection, complementation), the idea of a function, and the principle of mathematical induction. Some of my examples make use of the elementary arithmetic of the integers. But I do not rely heavily on any of these prerequisites, always preferring to give the reader who lacks them a sporting chance by explaining what I mean in everyday terms. But I am well aware that *parts* of the book, especially Chapter 6, will be most accessible to the more mathematically sophisticated readers.

I do not have a long list of friends and colleagues 'without whom the book would have been impossible'. My chief indebtedness is to three years' worth of second-year computer science undergraduates at Exeter who have served, albeit unwittingly, as guinea pigs for my ideas about how to teach logic. I am also grateful to Ajit Narayanan, to Mike Rogers and to my wife Carol for reading the whole manuscript or parts of it and commenting constructively on it. I should like to thank Alan Whittle, the copy editor, who picked up a number of errors which I had failed to notice. Thanks also to Rosemary Altoft and Mary Baxter at John Wiley and Sons for their help in converting my book from a dream into a reality.

ANTONY GALTON
Exeter
November 1989

1 Fundamental Concepts of Logic

1.1 INTRODUCTION

What is logic?

Making statements is a basic human activity; it is part of that complex of activities that collectively we call language. A fundamental property of statements is that they may be true or false. Which of these two a given statement is—true, or false—is called its **truth value**. When we make a statement, we are generally motivated by a belief in its truth; and when we appraise a statement made by someone else, its truth value is one of the prime objects of our appraisal.

Typically, a statement is understood as saying something about the world, and its truth value is assessed on that basis: if the world is in fact the way a statement says it is, then we call that statement **true**; if not, we call it **false**. Statements relate to the world by way of their meanings, so truth and meaning must be intimately related. Later, we shall look at some of the ways in which they are related.

But statements can also be related to each other. Sometimes two statements are so related that the first cannot be true without the second being true as well; in this case, we say that the latter is **implied** or **entailed** by the former. Again, it is possible to have a group of statements related in such a way that they cannot all be true; in this case the group of statements is said to be **inconsistent**. Later on we shall meet many concrete examples of these abstract concepts.

Logic is the systematic study of how statements can be related in ways that have repercussions for their respective truth values. Knowing such relations may allow us to determine the truth value of a statement, not by looking at the world but by looking at its relationship to other statements whose truth values are known. In such cases we are said to *reason*. Scientific progress comes about through a happy combination of observation, reasoning, and guesswork.

Logic and information technology

In the world of computing, logic has several important roles to play; here are a few of them:

Logic circuits

The simplest variety of logic is the propositional calculus, which is concerned with the logical relations amongst statements built up using words like 'and', 'or', 'if', and 'not'. This has a direct application to electronic circuit design, and hence to computer hardware. Circuits can be built up out of *gates* which regulate the flow of current in accordance with simple logical relations: an *and*-gate allows current through just so long as there is a current in *both* the input channels (i.e. in the first channel *and* the second), an *or*-gate is more lenient: it allows current through so long as there is a current in *at least one* of the input channels (i.e. in the first channel *or* the second).

Suppose we want to build a circuit which passes current just so long as the distribution of current amongst the input channels satisfies some specified condition. One can express this condition in terms of 'if', 'and', and the rest, and thus translate it into propositional calculus notation. This notation can then be manipulated using the techniques of formal logic in order to find a suitable combination of gates to do the job required. Ideally, one would like to find the *simplest* combination of gates to do the job, but no general technique is known for this!

Boolean data-types

This is another application of the propositional calculus, to software rather than hardware. Many programming languages, for example Pascal, have what is known as a *Boolean* data-type. An expression of this type can take either of two values, **true** or **false**. They may be combined using logical words such as **and**, **or**, and **not**, to yield complex expressions of the same type. The correct handling of such Boolean expressions requires an appreciation of the underlying logic, which is the propositional calculus. Thus, using logic, a program instruction of the form

if (*A* and *B*) or (*A* and *C*) then...

can be simplified to

if *A* and (*B* or *C*) then...

which not only makes for simpler code but might also result in faster execution.

Reasoning about programs

Writing programs is perhaps the central activity in information technology. But this activity does not take place in isolation. Before any code is written, some kind of *specification* is required, which describes in more or less detail how the program should behave. A specification may be an informal statement in English prose of the general requirements that the program must satisfy, or it may be a formal, technical document detailing with great precision the exact behaviour of the program. An informal specification has the advantage of being more readily intelligible to a wider range of people, but formal specifications are more useful to the programmers themselves.

Once a program has been written, it is necessary to check that it does behave in accordance with the specification. To some extent, this can be achieved by *testing* it, i.e. by actually running it on a wide range of different possible inputs to check that the behaviour of the program is as desired for these inputs. In some cases one can be reasonably confident that one's test inputs are exhaustive in the sense that between them they cover all possible types of behaviour that will be exhibited by the program. In general, though, no such confidence can be attained, and it is desirable to have more watertight ways of certifying a program's behaviour.

The desired behaviour of the program can be described by means of a set of statements: these collectively constitute a specification. If these statements truly describe the program's behaviour, then the program behaves correctly. The difficulty in checking a program against its specification arises from the fact that the relationship between the program code itself and the behaviour it generates is, especially in complex cases, very hard for human beings to understand. Merely by understanding the programming language, we can easily enough determine the behaviour generated by very short stretches of program code; the difficulty lies in extrapolating this local understanding to a global understanding of the behaviour of a large program.

This is where logic comes in. The general idea is to express in terms of a logical system the relationship between simple pieces of program code and the behaviours that they generate; by analysing the way in which a complex program is built up out of simple components it is then possible to use formal logical techniques to determine the behaviour of the whole program. The pioneering work on logic-based program verification was done by R. W. Floyd (1967) and C. A. R. Hoare (1969). Subsequently many different techniques have been developed, including the use of temporal logic for specifying and verifying parallel programs (see, e.g. Z. Manna and A. Pnueli, 1981). Here logical notation is used to express both the statements that describe the program itself and the statements that describe its desired behaviour. Logical techniques can then be used to determine whether the latter set of statements is implied by the former.

Logic programming

Conventionally, the task of programming is to convert a specification, which says *what* results are required, into executable code, which details *how* the results are to be achieved. A program thus consists of a sequence of instructions to the computer which, when executed, will (one hopes) give rise to the results described in the specification.

Logic programming departs radically from this style. Instead of instructing the computer what to do, the logic programmer sets up a data-base of statements and inference rules, and then asks questions which the computer answers by logical manipulation of the data-base. Ideally, the logic program contains no *instructions* at all, only facts. Thus, in contrast to ordinary 'imperative' programming, the programmer is here concerned only with the 'what', leaving the 'how' up to the computer itself. In effect, a logic program is a kind of specification: an *executable specification.*

In practice, this pure notion of logic programming is hard to achieve. The most widely used logic programming language, Prolog, makes so many compromises with the imperative style of programming that many people prefer not to regard it as a true logic programming language at all. None the less, it has proved to be a particularly congenial medium for certain types of programming task, finding especial favour with workers in artificial intelligence and related disciplines. And the *ideal* of a pure logic programming language is a worthy goal for future developments to aim for.

Automated reasoning

Insofar as logic programming delegates to the computer itself the task of drawing inferences from a data-base, it can be regarded as a form of automated reasoning. More generally, much research is being devoted to the task of creating automatic theorem-provers for certain branches of mathematics (e.g. number theory, abstract algebra) as well as for such special tasks as software verification and program synthesis. In all these enterprises logic plays a fundamental part.

If we could automate the process of applying logic to reasoning about computer programs, then the process of software development could be considerably streamlined. The ideal here would be to have a computer system which, when presented with a program and a specification as inputs, would decide whether or not the program correctly implements the specification. In practice it seems certain that one will have to settle for considerably less than this.

An even remoter goal would be to create a *program synthesizer*, i.e. a system which, when presented with a specification, will actually write a program that is guaranteed to meet it. Some small-scale results have been achieved in this direction, but as with automatic verification there are formidable obstacles to be overcome.

Artificial intelligence

This is the attempt to model human thought processes computationally, either for the purely practical purpose of getting computers to perform tasks which require thought, or for the more theoretical purpose of gaining insight into the mechanisms underlying human psychology. At least some human thinking involves reasoning, so automated reasoning would appear to have a role to play here.

Many researchers in artificial intelligence go further than this in believing that formal logic can provide the necessary technical underpinning for a wide variety of thought processes; in particular, logical symbolism has been widely used as a way of representing the extensive reserves of *knowledge* than an intelligent agent has to draw upon in order to think realistically about the world. It is only fair to remark that this view has plenty of detractors, too, who claim that an excessive reverence for logic-based techniques has hindered the development of artificially intelligent systems. But even if we accept this claim, it has to be admitted that the conceptual clarity which can come from an understanding of formal logic is an invaluable aid in the construction of appropriate representations, whether or not these are themselves logical in form.

In this introductory chapter, we take a broad overview of the central logical ideas of truth, inference, and validity. In subsequent chapters we apply these ideas to specific formal logics. We shall try to introduce the reader gently to the subject, in an informal, intuitively clear way. Later on, all the ideas receive a more formal treatment, with precise definitions. Inevitably, since no formally defined concept ever matches exactly any informally introduced one, there will be some discrepancies between the informal treatment and the formal treatment. I have done my best to minimize these discrepancies; but I am firmly of the conviction that they are a price worth paying for having an accessible introduction to what can otherwise become an intimidatingly technical subject.

1.2 STATEMENTS

A **statement** (or **proposition**) is a sentence which is either true or false. Perhaps the best way to get clear about what counts as a statement is to look at some sentences which are *not* statements. For a sentence not to be a statement, it must fail to be either true or false. This may happen for a variety of reasons, for example:

(1) it may be a question, like

> What is the time?

or

> Is sodium a metal?

which it would not make sense to speak of as being true or false;

(2) it may be a command, like

> Open the door!

or

> Sign on the dotted line;

(3) it may be a wish, like

> If only I hadn't missed the bus!

or

> May the Lord make us truly grateful.

All these sentences are very obviously not statements: this is already clear from their grammatical form. But there are also some more subtle reasons why a sentence may fail to express a statement. This may happen because, although it has the typical form of a statement, one or more words occurring in it are not defined. The sentence

> John is older than Mary

only expresses a statement when it occurs in a context which makes it clear exactly who John and Mary are. The present context fails to do this, as you can easily see if you try to decide whether the sentence is true or false—the question is unanswerable. But if you know someone called John and someone called Mary, and ask if the sentence is true or false as regards *them,* then there is a definite answer to this question, even if you don't happen to know what the answer is; so long as the sentence is interpreted as referring to that particular pair of people known to you then it can count as a statement.

The same issues arise in an even sharper form when we consider sentences containing pronouns, like

He is older than she is.

This has no truth value until it is settled who 'he' and 'she' are supposed to refer to. This could be done by pointing at a man and a woman while speaking the sentence; or it could be established by the context, as in

John is in the same class as Mary, but he is older than she is.

Often there will be an 'obvious' interpretation to put on a sentence to make it into a statement. Most readers of this book will understand the sentence

Margaret Thatcher lives in Downing Street

as making a statement about the Westminster residence of the present (1989) Prime Minister of the United Kingdom, notwithstanding the undoubted fact that there are other women of the same name, of each of whom it is determinately true or false that she lives in, say, Downing Street, Cambridge.

Clearly, if we are to do logic satisfactorily we must be sensitive to issues such as these; but this sensitivity may reasonably stop short of pedantry. We do not need to insist that all our sentences are proper statements, but we do need to insist that they could be made into statements given suitable interpretations of the terms they contain. In this book I shall only use pronouns in contexts where their reference is clear, but I shall make free use of sentences involving John, Mary and other names—in these cases you may assume that these names refer to certain fixed characters.

It is also important that the interpretations of names and pronouns do not change in the course of an argument. To illustrate this point, suppose you read in a music textbook that Bach, the German-born composer, never crossed the sea, and a bit later on you read that Bach lived for many years in London. You might well be puzzled by the apparent inconsistency. If the textbook is any good, though, it should have been made clear at some point that the first statement refers to Johann Sebastian Bach, while the second refers to his son Johann Christoph.

Warning: In computer science, the word 'statement' is sometimes used to refer to instructions in an imperative programming language—so that, for instance, people speak about 'the *goto* statement in Pascal'. *This use of the word 'statement' is to be very strongly discouraged.* Program components like 'goto 17' and 'if $x > 0$ then $y := x - 3$' are not statements but **instructions** or **commands**.

This is not to say that statements, in the strict logical sense, play no part in these languages. On the contrary, they form an essential part of the syntax, since the so-called *conditions* which must immediately flow the keywords 'if', 'while', and 'until' in Pascal are indeed statements— that they may be true or false is of course essential for them to be able to play the role that they do. Thus, although 'if $x > 0$ then $y := 3$' taken as a whole is not a statement but an instruction, the '$x > 0$' part *is* a statement, whose truth is the condition for the embedded instruction '$y := 3$' to be executed. Of course even '$x > 0$' is not a statement unless the identifier x is instantiated to some particular value—which in a properly written program it always will be whenever the truth-value of '$x > 0$' has to be evaluated.

Exercise 1.2   For each of the following sentences, decide whether or not it is a statement. If it is not a statement, change it so that it becomes one (e.g. change 'Shut the door' to 'The door is shut').

(1)   The moon is made of green cheese.

(2)   How many times have you been to London?

(3)   Nobody knows the answer to that question.

(4)   Nobody has ever proved Fermat's last theorem.

(5)   Name the first president of the USA.

(6)   The first president of the USA was George Washington.

(7)   Let n be the least integer with more than ten prime factors.

(8)   The least integer with more than ten prime factors is even.

(9)   Prove that the least integer with more than ten prime factors is even.

(10)   To be or not to be, that is the question.

1.3   CONSISTENCY AND INCONSISTENCY

Sometimes a set of statements is such that simply by considering what they mean we can be sure that they cannot all be true. When this happens, the set of statements is said to be **inconsistent.**

The simplest case of inconsistency arises when we have a set containing two statements which directly contradict one another, as in

$$S_1 = \{\text{John is married, John is not married}\}$$

or

$$S_2 = \{\text{pi equals } 22/7, \text{ pi does not equal } 22/7\}.$$

In each of these sets, one of the two statements must be true and the other one false. This means that they cannot *both* be true, and hence each of the sets is inconsistent.

Now consider the sets

$$S_3 = \{\text{John's wife is a doctor, John is a bachelor}\}$$

and

$$S_4 = \{\text{pi equals } 22/7, \text{ pi is greater than } 22/7\}.$$

In each of these sets we have two statements which contradict one another *indirectly*. In the case of S_3, this is because 'John's wife is a doctor' implies that John has a wife, and hence is married, whereas 'John is a bachelor' implies that John is *not* married. In S_4, 'pi is greater than 22/7' implies that pi is neither equal to nor less than 22/7, and hence contradicts 'pi equals 22/7'. Thus both S_3 and S_4 are inconsistent too.

One has to be careful in assessing whether a set of statements is inconsistent. In particular, if one of the sentences in the set is ambiguous, so that it can be understood in two ways, then it is possible that under one of the interpretations the set is inconsistent but under the other it is not. Our example S_3 is a case in point. We assumed that the word 'bachelor' here meant a man who has never been married. But another possible meaning is a person with a bachelor's degree (B.A., B.Sc., etc.). If we interpret the second sentence in S_3 so that 'bachelor' has this meaning then there is no inconsistency: many graduates are married! Here, as always when talking about language, one must be sensitive to *meanings*.

More complicated cases of inconsistency can arise in which three or more statements are collectively inconsistent, but the responsibility for this cannot be pinned on any single contradictory pair. Some examples are:

$$S_5 = \{\text{John is married to Mary,}$$
$$\text{John is older than his wife,}$$
$$\text{Mary is older than her husband}\}$$

$$S_6 = \{\text{John is taller than Marry,}$$
$$\text{Mary is taller than James,}$$
$$\text{James is taller than Judy,}$$
$$\text{Judy is taller than John}\}$$

Both the sets S_5 and S_6 are inconsistent, since in neither case could all the constituent statements be true together, but neither set contains a contradictory pair of statements. In fact, we can remove any one statement from S_5 and the remaining statements will no longer form an inconsistent set; thus, the pair of statements

$$\{\text{John is married to Mary,}$$
$$\text{John is older than his wife}\}$$

could both be true, as could the pair

$$\{\text{John is married to Mary,}$$
$$\text{Mary is older than her husband}\}$$

and the pair

$$\{\text{John is older than Mary,}$$
$$\text{Mary is older than her husband}\}.$$

It is only when we put all three statements together, as in S_5, that the inconsistency arises. The reader should do the same with S_6 to be satisfied that the same thing happens in that case too.

A set of statements is **consistent** just so long as it is not inconsistent, i.e. they *can* all be true together. Note carefully that just being consistent does not guarantee that the statements in a set are all true, only that they *could* be, given an appropriate state of the world.

For example, the set

$$\{\text{Exeter is in Yorkshire,}$$
$$\text{Yorkshire is in Scotland}\}$$

is consistent, even though neither of the statements it contains is true. It is consistent because if the facts of geography had been different, they *could* both have been true: it is easy enough to draw a map in which there is a town called Exeter in a county called Yorkshire in a country called Scotland. By contrast, the set

{Exeter is in Yorkshire,
Yorkshire is in Scotland,
Exeter is not in Scotland}

is *in*consistent: you cannot draw a map which makes all three of these statements true, however much you tamper with the facts of geography. (For more on this, see below, Section 1.5.)

Why is consistency important? If someone asserts a set of statements, and that set turns out to be inconsistent, then you have the strongest possible grounds for getting them to retract some or all of the statements they have made, because you know that at least one of them must be false. Remember that most of the time we aim to assert only true statements—and even when this is not strictly the case (e.g., in play-acting), we must surely be going astray if we do not aim to *believe* only true statements. Thus so long as truth is at stake (which it usually is), we must strive to ensure that the statements we uphold form a consistent set. This will not guarantee them to be true, of course, but at least we have a chance of attaining the truth, which we do not have if we maintain an inconsistent set of beliefs.

Exactly the same considerations apply to the 'beliefs' of a computer. Of course, computers do not really have beliefs, but when we build up a database of statements ('facts'), the computer may metaphorically be said to believe these statements. And it is just as important that they should form a consistent set in this case as it is with our own beliefs. We might regard the computer data-base as an extension of our own beliefs, and hence must apply to it exactly the same standards as we apply to ourselves. There is thus an important practical need for computer scientists to understand the logical concept of consistency.

Logical truth and falsehood

All the sets of statements we have looked at so far contain at least two members; but the concepts of consistency and inconsistency apply equally to singleton sets, that is, to sets which contain only one member. Thus the set

{Mary's husband is unmarried}

is inconsistent since the single statement 'Mary's husband is unmarried' cannot in any circumstances be true—it is self-contradictory, since being unmarried is, by definition, incompatible with being someone's husband. Self-contradictory statements are said to be **logically false**, meaning that they are false for purely logical reasons, not because of the way the world happens to be. Any inconsistent set of statements, so long as it is finite, gives rise to a logical falsehood when its members are joined together by 'and'; for example, from S_5 we obtain the logically false statement

John is married to Mary and John is older than his wife
and Mary is older than her husband.

The opposite of logical falsehood is logical truth. A statement is **logically true** if to deny it would incur logical falsehood, so that the statement itself is true for logical reasons. An example is

> If Mary is married then she has a husband.

We do not need to know anything at all about Mary in order to know that this statement is true: its truth follows directly from its meaning.

Statements which are neither logically false nor logically true, such as

> Mary's boy-friend is unmarried

are called **contingent**. The truth-value of a contingent statement cannot be determined from the meaning of the statement alone; one has also to check that it agrees with the way the world is—in this case, we have to find out whether Mary has a boy-friend, and if so, whether he is married.

A logically false statement is impossible, but an impossible statement is not necessarily logically false. For example, the statement

> Toby's car travels faster than light

is certainly impossible in the sense that we know it cannot be true; but it is not logically false. In order to know that it must be false, we need to look beyond the mere meaning of the statement: we need to know something about the laws of physics, which provide the real source of its impossibility. To put it another way, we could imagine a world in which the laws of physics were different, so that a car *could* travel faster than light. But we could not imagine a world in which husbands were unmarried, because being a husband necessarily involves, as a matter of definition, being married. (Of course, we can very easily imagine a world in which the *words* 'husband' and 'married' meant something different—but that is beside the point, since it is the *English* sentence 'Mary's husband is unmarried', with its normal English meaning, whose logical falsehood is under discussion.)

Exercise 1.3

(1) Decide which of the following sets of statements are consistent, giving a reason in each case:

*(a) {There are fewer than twenty people in the room,
      There are more than ten people in the room}
*(b) {There are more than twenty people in the room,
      There are fewer than ten people in the room}
 (c) {John lives in the same county as Mary,
      John lives in Devon,
      Mary lives in Devon}
 (d) {John does not live in the same county as Mary,
      John does not live in Devon,
      Mary does not live in Devon}
 (e) {John lives in the same county as Mary,
      John lives in Devon,
      Mary does not live in Devon}

(2) Classify each of the following statements as logically true, logically false, or contingent. For a contingent statement, describe a state of affairs which would make it true, and another which would make it false.

* *(a) It is raining and it is not raining.
* *(b) It is raining or it is not raining.
* (c) It is raining and it is snowing.
* (d) If the sun is shining and it is raining then there is a rainbow.
* (e) If the sun is shining and it is raining then it is raining.
* *(f) John is married but his wife is not.
* *(g) John is married to Mary and to Anne.
* (h) Either John's wife is married or John is unmarried.

Note: In this and all later exercises, asterisks indicate questions to which the solutions are given in the appendix (pp. 257–285).

1.4 CONSEQUENCE AND ENTAILMENT

Consider again the set of statements

$$\mathbf{S}_5 = \{\text{John is married to Mary,}$$
$$\text{John is older than his wife,}$$
$$\text{Mary is older than her husband}\}.$$

As we saw in the last section, this set is inconsistent, which means that its constituent statements cannot all be true. That means that if two of them are true, the third must be false. Thus if we are told that

$$A: \text{John is married to Mary}$$

and

$$B: \text{John is older than his wife,}$$

we can infallibly infer from these statements that

$$C: \text{Mary is not older than her husband.}$$

Note that C's being true is the same as 'Mary is older than her husband' being false.

The statements A, B, and C are related in such a way that so long as A and B are both true then C must be true as well. We say that C **follows** from A and B together, or is a **consequence** of them; alternatively, we say that A and B together **entail** or **imply** C, or that we may **infer** C from A and B together. Note that in popular usage 'infer' is often used where more careful speakers would say 'imply'. This usage is not recommended.

We thus have six different ways of expressing what is essentially the same fact:

(1) C follows from A and B

(2) C is a consequence of A and B

(3) A and B entail C

(4) *A* and *B* imply *C*

(5) We may infer *C* from *A* and *B*

(6) **S₅** is inconsistent.

The inconsistency of the same set **S₅** tells us about some other implications as well. For example, we know also that the pair of statements

B: John is older than his wife

and

D: Mary is older than her husband

together imply the statement

E: John is not married to Mary.

(Exercise: What is the third implication we can get from the inconsistency of **S₅**?)
  Suppose someone tries to infer from the two statements

F: Exeter is not in Yorkshire

and

G: Yorkshire is not in Scotland

that

H: Exeter is not in Scotland.

Have they inferred correctly? At first sight, one might imagine that since all the statements *F*, *G*, and *H* are actually true, the inference must be correct. But if this were the case, what would stop us from inferring from *F* and *G* the true statement

I: Paris is not in Wales?

After all, *F*, *G*, and *I* are all true, just as *F*, *G*, and *H* are. Since there is no plausibility whatever in the idea that we could correctly infer *I* from *F* and *G*, we must look for some better reason to justify the inference of *H* from *F* and *G*, if indeed this inference is justified.
  Going back to our equivalent formulations (1)–(6) above, we can see that the claim that *F* and *G* together imply *H* is equivalent to a certain set's being inconsistent. What set? Well, *H*'s being true is the same as the following sentence being false:

Exeter is in Scotland.

So we must show that the set

$$\mathbf{S_7} = \{\text{Exeter is not in Yorkshire,}$$
$$\text{Yorkshire is not in Scotland,}$$
$$\text{Exeter is in Scotland}\}$$

is inconsistent. Think about this before reading further.
  Remember that we are asking about whether the constituent statements *could* all be true, not whether they *are* all true. Suppose, then, that Exeter were to swap names with Edinburgh, everything else being kept the same. Then the statements in **S₇** would all be true! This means that **S₇** is actually consistent; which in turn means that *F* and *G* do not imply *H* after all. If somebody seriously tried to argue along these

lines (i.e. to convince us that Exeter is not in Scotland on the grounds that Exeter is not in Yorkshire and Yorkshire is not in Scotland), we could say: that's ridiculous, you might as well argue that *Edinburgh* is not in Scotland, since after all, Edinburgh is not in Yorkshire and Yorkshire is not in Scotland. Note that, in denying the correctness of his inference, we are not denying that Exeter is not in Scotland, we are denying that he has adduced adequate grounds for asserting this.

Exercise 1.4

*(1) For each of the questions in Exercise 1.3 for which you decided that the given set was inconsistent, write down all the entailments whose validity can be deduced from the inconsistency of that set.

(2) Look up the words 'entail', 'imply', and 'infer' in a dictionary. Do the definitions agree with the meanings introduced in this section?

1.5 INFERENCE AND VALIDITY, ENTAILMENT AND EQUIVALENCE

We have just observed that there is a relationship between implication and inference, namely that one can validly infer a statement C from a set of statements $\{P_1, \ldots, P_n\}$ just so long as C is a consequence of that set, i.e. statements $P_1, \ldots, P_n$ together imply C. To make an inference is to conclude that a certain statement C is true solely on the basis that certain other statements $P_1, P_2, \ldots, P_n$ are all true. The inferred statement C is called the **conclusion** of the inference, and the statements $P_1, \ldots, P_n$ that it is inferred from are called the **premisses** of the inference. In inferring C from $P_1, \ldots, P_n$, one is seeking to justify one's assertion of the statement C on the ground that the premisses are true statements and C is implied by them. If C is indeed implied by $P_1, \ldots, P_n$, then the inference is said to be **valid**. In order for the assertion of C to be justified on this basis, it is also necessary that $P_1, \ldots, P_n$ should all be true. In this case, the inference is said to be **sound**.

Note that sometimes we may be interested in the validity of inferences which are known not to be sound. This would be the case if, for example, we were reasoning about what *would* be the case if certain states of affairs obtained.

We shall write

$$P_1, \ldots, P_n/C$$

to denote the inference with premisses $P_1, \ldots, P_n$ and conclusion C. When the premisses and conclusion are given in full as English sentences, we usually represent the inference in vertical format thus:

$$P_1$$
$$\vdots$$
$$\frac{P_n}{C}$$

Here the '/' which separates the conclusion from the premisses is written as a horizontal rule. In either case, it may be read as 'therefore'. To assert that this inference is valid, we write

$$P_1, \ldots, P_n \vDash C.$$

The symbol $\vDash$, known as a 'turnstile', can be read as 'entail(s)'.

Sometimes we do not want to name the premises individually, but to consider them together as a set, as we did with our examples S_1–S_7 in the previous section. In this case, we can write S/C to denote the inference whose premises are the members of the set S and whose conclusion is C, and likewise $S \vDash C$ to indicate that the inference S/C is valid. Note that our notational conventions here are a little bit sloppy, because if $S = \{P_1, \ldots, P_n\}$, say, then we ought to expand S/C as $\{P_1, \ldots, P_n\}/C$ rather than $P_1, \ldots, P_n/C$. But this is harmless sloppiness, which justifies itself by its convenience.

To summarize our definitions, then, we have:

The inference $P_1, \ldots, P_n/C$ is valid so long as it is not possible for all of $P_1, \ldots, P_n$ to be true at the same time as C is false.

The inference $P_1, \ldots, P_n/C$ is sound so long as it is valid and $P_1, \ldots, P_n$ are all true.

We now apply these ideas to some examples.

Example 1.5.1   We saw this one in the last section, but repeat it here for convenience:

P_1: Exeter is not in Yorkshire
P_2: Yorkshire is not in Scotland

C_1: Exeter is not in Scotland

As we saw, it is possible for P_1 and P_2 to be true while C_1 is false (as would be the case, for example, if Exeter were situated where Edinburgh is in reality, everything else being kept the same); hence the inference is neither valid nor sound.

Example 1.5.2   Now consider the same inference with all occurrences of 'not' removed:

P_3: Exeter is in Yorkshire
P_4: Yorkshire is in Scotland

C_2: Exeter is in Scotland

Notice that the statements P_3, P_4, and C_2 are, in fact, all false. What we must consider, though, is whether it would be possible, by suitable tampering with geographical facts, to make P_3 and P_4 true at the same time as C_2 is false. A little thought will show that this cannot be done. Whatever geographical region is named 'Yorkshire', if P_3 is to be true we must make sure that the name 'Exeter' is given to a place or region inside it, thus:

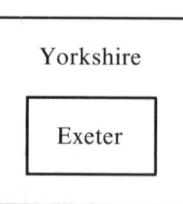

And to make P_4 true, we must make sure that the region we call 'Scotland' contains the region we have called 'Yorkshire', thus:

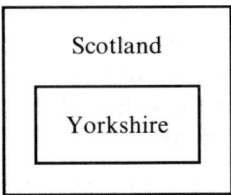

Now, putting the two pictures together, we see at once that we cannot avoid having 'Exeter' in 'Scotland':

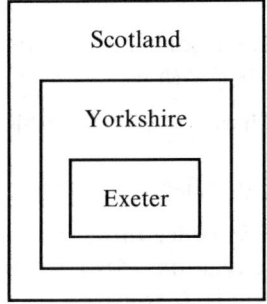

And this means that we cannot have P_3 and P_4 both true at the same time as C_2 is false.

We can conclude from all this that the inference from P_3 and P_4 to C_2 is valid. Is it sound as well? No, because P_3 and P_4 are not both true (in fact neither of them is).

Example 1.5.3   As a final example, we modify Example 1.5.2 by a judicious substitution of names to give:

P_5: Exeter is in Devon
P_6: Devon is in England

C_3: Exeter is in England

Since we have arrived at this inference from the previous one merely by substituting names ('Devon' for 'Yorkshire' and 'England' for 'Scotland'), the argument we applied to show that the previous inference was valid will apply just as well to this inference: all we need do is make the same substitution of names in the diagrams we used. Remember that validity is about what is or is not possible: and we can freely play around with what the names in an inference might refer to when we are investigating all the possibilities, so it does not matter what the names actually are, so long as they still conform to the same pattern (more on this important point later).

In addition, since P_5 and P_6 are actually true, the inference from P_5 and P_6 to C_3 is not only valid but sound as well. Notice that we can't apply the name-substitution trick when arguing for soundness, in the way that we did when arguing for validity. Soundness is concerned with what *is* the case, not just with what *could* be the case, and for this it would obviously be illegitimate to play around with what the names refer to. The best we can say is that *if* the regions we now call Devon and England were to be renamed

'Yorkshire' and 'Scotland', *then* the inference in Example 2 would become sound; but as things are, it isn't.

Just as an inconsistent set of statements $\{P_1, \ldots, P_n\}$ gives rise to a logically false statement when its members are linked together as 'P_1 and P_2 and ... and P_n', so from a valid entailment $P_1, \ldots, P_n \vDash C$ we can derive the logically *true* statement 'If P_1 and ... and P_n then C'. From the valid inference in Example 1.5.2, for example, we derive the logically true statement

> If Exeter is in Yorkshire and Yorkshire is
> in Scotland then Exeter is in Scotland.

Equivalence

The relation of entailment, when it holds between two statements, is not, in general, symmetric; that is, we can have $A \vDash B$ without having $B \vDash A$. For example, the statement

> P: Charles is older than his wife

entails the statement

> Q: Charles is married,

since a man has to *have* a wife in order to be older than her, and to have a wife is to be married. But the converse relation does not hold: Charles could be married to an older woman, which would make Q true and P false. Since Q can be true without P being true, we can be sure that Q does not entail P.

For some pairs of statements, however, entailment does work both ways. Consider, for example, the statements

> R: Charles is older than Julia
> S: Julia is younger than Charles.

It is not hard to see that each of these statements entails the other: if R is true, then S must be true too, so R entails S, and if S is true, then R must be true, so S entails R. So the relationship between R and S is stronger than that between P and Q: it is a two-way entailment. We call this **equivalence**.

Thus to say that a statement R is equivalent to a statement S is precisely to say both that R entails S and that S entails R. If R is equivalent to S, then R and S must have the same truth value: they can both be true, or they can both be false, but we cannot have one true and the other one false. Of course, two statements can have the same truth value without being equivalent (e.g., 'Exeter is in Devon' and 'Leeds is in Yorkshire' are both true, but they are not equivalent since we can easily imagine a world in which one is true and the other false). As with the other fundamental logical concepts in this chapter, equivalence involves all the possibilities, not just what happens to be true.

Notation: To say that two statements R and S are equivalent, we write

$$R \cong S.$$

This is a shorthand for '$R \vDash S$ and $S \vDash R$'.

Exercise 1.5

(1) Decide which of the following inferences are valid. Of those which are valid, which are sound?

*(a)   Norwich is in Norfolk
        Norfolk is not in Scotland

        Norwich is not in Scotland

(b)   Cardiff is in Devon
       Devon is not in England

       Cardiff is not in England

(c)   Paris is not in Yorkshire
       Yorkshire is in England

       Paris is not in England

*(d)   Leeds is not in Devon
        Devon is in Scotland

        Leeds is not in Scotland

(e)   Whenever John comes, Mary is happy
       John came yesterday

       Mary was happy yesterday

*(f)   Whenever John comes, Mary is happy
        Mary has not been·happy today

        John has not come today

(g)   Whenever John comes, Mary is happy
       John did not come on Tuesday

       Mary was not happy on Tuesday

(h)   John admires everyone who admires Jane
       Jane admires herself

       John admires himself

(2) Show that, for any statements P, Q, and R,

*(a)   if $P \vDash Q$ and $Q \vDash R$, then $P \vDash R$.
(b)   if $P \cong Q$ and $Q \cong R$ then $P \cong R$.
(c)   if $P, Q \vDash R$ and $P \vDash Q$ then $P \vDash R$.

1.6   VALIDITY AND FORM

The inferences in Example 1.5.2 and Example 1.5.3 are both valid, and one and the same argument can be used to establish the validity of either of them. Let us repeat the two inferences together:

> (2) Exeter is in Yorkshire
>   Yorkshire is in Scotland
>   ――――――――――――――――――
>   Exeter is in Scotland

> (3) Exeter is in Devon
>   Devon is in England
>   ――――――――――――――――
>   Exeter is in England

The way we argued for the validity of (2) in no way drew upon any specific properties of Exeter, Yorkshire, or Scotland: on the contrary, we deliberately avoided doing any such thing, by considering the possibility that these names could have referred, had the world been other that it actually is, to quite different places. It was this abstraction away from the actual meanings of Exeter, Yorkshire, and Scotland which allowed us to regard our argument equally as evidence for the validity of (2) and of (3).

Generalizing from this, it should be obvious that any inference which shares the same pattern as (2) and (3) will be valid, for example the inference

> (4) Eldorado is in Erehwon
>   Erehwon is in Utopia
>   ――――――――――――――――――
>   Eldorado is in Utopia

This inference is also valid (though not, of course, sound!).

In the light of this, it seems natural to say that what is really valid about (2), (3), and (4) is not so much the inferences themselves as the *pattern* which they all share, a pattern which we can represent as

> (5) A is in B
>   B is in C
>   ――――――――――
>   A is in C.

This pattern is called an **inference schema**; and its constituents 'A is in B', 'B is in C', and 'A is in C' are **statement schemas**. The letters A, B, and C occurring in (5) are called **schematic letters**. The inferences (2), (3), and (4), which can be obtained from (5) by substituting specific names for the schematic letters, are called **interpretations** of (5). In particular, (2) is what we get if we **interpret** the schematic letters A, B, and C to mean Exeter, Yorkshire, and Scotland respectively.

A given statement schema may have both true and false interpretations. For example, the schema 'A is in B' has amongst its true interpretations the statements

> Exeter is in Devon
> Leeds is in Yorkshire
> Devon is in England
> Kathmandu is in Nepal

while its false interpretations include

> Exeter is in Yorkshire
> Leeds is in Devon
> Devon is in France
> Tegucigalpa is in Nicaragua

An interpretation for a schema can be specified by giving interpretations for each of the schematic letters it contains. We write $I(A)$ to denote the meaning of the schematic letter A under the interpretation I. Thus if I is the interpretation under which 'A is in B' means 'Exeter is in Devon', then $I(A)$ is Exeter and $I(B)$ is Devon. (N.B., $I(A)$ is the city Exeter itself, not the word 'Exeter'.)

If the statement schema S comes out true under a given interpretation I, then we say that I **satisfies** S, whereas if S comes out false under I, then we say that I **falsifies** S. A *set* of statement schemas **S** is said to be **satisfiable** if there is an interpretation which satisfies every member of **S**, i.e., a way of reading the schematic letters which makes all the schemas in **S** come out true. Satisfiability is closely related to consistency: but whereas we speak of a set of *statements* as being consistent or inconsistent, it is always a set of statement *schemas* that we call satisfiable or unsatisfiable.

We are now in a position to define what we mean by the validity of an inference schema. The argument we used earlier to demonstrate the validity of (2) could equally well be applied to *any* interpretation of (5): in effect, the argument shows us that any interpretation of (5) which satisfies its premisses must also satisfy the conclusion. This suggests that we should take this to be our definition of a valid inference schema; we could then go on to define a valid inference as an inference which can be obtained as interpretation of a valid schema. We thus have

---

An inference schema is valid just so long as any interpretation which satisfies all its premisses also satisfies its conclusion

An inference is valid just so long as it is an interpretation of a valid inference schema

An inference is sound just so long as it is valid and its premisses are all true

---

An alternative, but equivalent, formulation of the first of these definitions would be that an inference schema is valid so long as the set consisting of the premisses together with the negation of the conclusion is unsatisfiable. For in that case, no interpretation of the schema can make the premisses and the negation of the conclusion all true, which means that every interpretation which makes the premisses true makes the conclusion true as well.

Note that it makes no sense to ask whether an inference schema is sound. This is because its premisses are statement schemas, and a statement schema, unlike a statement, is not the sort of thing that can be true or false. Only if its schematic letters are interpreted does the statement schema become true or false (for it is then a statement), and once this has been done to each part of an inference schema, it is no longer an inference schema but an inference.

What about inference schemas that are not valid, i.e. *invalid* inference schemas? From the definitions of validity we can say that

---

An inference schema is invalid just so long there is an interpretation which satisfies the premisses and falsifies the conclusion

---

An example of an invalid inference schema is

> (6)   A is not in B
>   B is not in C
>   _____
>   A is not in C.

To show that this is invalid, we need only find an interpretation with true premises and false conclusion. We have already seen one interpretation of this schema, namely Example 1.5.1; but this has true premises and true conclusion, so does not help us to show that the schema is invalid. But it is easy to find an interpretation of the kind we require; here is an example:

> (7)   Edinburgh is not in Yorkshire
>   Yorkshire is not in Scotland
>   _____
>   Edinburgh is not in Scotland.

Here the premisses are both true, but the conclusion is false, and this is enough to show us that the schema above is invalid.

Does the invalidity of schema (6) suffice to show that Example 1.5.1 is invalid? Let us repeat this example for ease of reference:

> (1)   Exeter is not in Yorkshire
>   Yorkshire is not in Scotland
>   _____
>   Exeter is not in Scotland.

Curiously enough, the invalidity of (6) does not *prove* that (1) is invalid, even though (1) is an interpretation of (6). The reason for this somewhat paradoxical state of affairs is as follows. An inference is valid so long as it is an interpretation of a valid schema; but an inference may be an interpretation of *more than one schema*. So even though (1) is an interpretation of (6), which is invalid, it could still be valid if we could find an example of a valid schema of which it is also an interpretation.

To illustrate, consider the schema

> (8)   A is in B
>   B is in C
>   _____
>   A is in D.

This is an invalid schema, as is clearly seen from the interpretation

> (9)   Exeter is in Devon
>   Devon is in England
>   _____
>   Exeter is in Scotland

which has true premises and false conclusion. (Note that this inference is very obviously invalid—in fact, it has no plausibility whatever, whereas it is not hard to imagine someone being taken in by (1).) Now the inference (3) is an interpretation of (8), got by interpreting A, B, C, and D to mean Exeter, Devon, England, and England respectively (there is no rule saying that different letters have to be interpreted differently). So here we have a valid inference, (3), which is an interpretation of an invalid schema, (8). It would plainly be incorrect to conclude that (3) is invalid just from the fact that it is an interpretation of the invalid schema (8); it also an interpretation of the valid schema (5), and this is enough to establish its validity.

Likewise, therefore, we cannot infer that (1) is invalid merely because it is an interpretation of (6); rather, we must satisfy ourselves somehow that it is not also an interpretation of some valid schema. In this particular case, it seems intuitively obvious that this is so, but in general we must clearly exercise caution with regard to this point. Thus, to sum up this latest idea, we have:

> An inference is invalid just so long as every inference schema of which it is an interpretation is invalid

*Exercise 1.6   For each of the valid inferences in Question 1 of Exercise 1.5, write down the schema which makes it valid. For each of the invalid ones, write down an invalid schema of which it is an interpretation and find another interpretation of this schema which has true premises and a false conclusion.

(Solutions given for (a), (d), and (f).)

1.7   SOME VARIETIES OF LOGIC

Logic as a science is concerned with the systematic investigation of the validity and invalidity of inference schemas. Different branches of logic study different kinds of schemas. In this section we briefly introduce some of the more important examples.

(1) *Propositional logic*, or the *propositional calculus*, deals with schemas like

> A or B
> Not A
> _____
> B

> If A then B
> Not A
> _____
> Not B

In these schemas, the schematic letters are meant to be interpreted as whole statements, as in the following examples:

> It is raining or it is snowing
> It is not raining
> _____
> It is snowing

> If John is coming then the party will be a success
> John is not coming
> _____
> The party will not be a success

(*Warning*: the examples given in this section are not necessarily all valid! See the exercise.)

In the propositional calculus, the statement schemas are built up out of the schematic letters by means of a rather restricted set of words called truth-functional connectives. Typical of these are 'and', 'or', and 'if'. Propositional logic is the most basic kind of logic, and almost all other logics can be regarded as extensions of it. We shall examine it in detail in the next two chapters.

(2) *Predicate logic*, or the *predicate calculus*, deals with schemas like

> Every A is a B
> a in an A
> _____
> a is a B
>
> Some As are Cs
> Some Bs are Cs
> _____
> Some As are Bs

Here there are two kinds of schematic letter, requiring two different kinds of interpretation. Roughly, the lower-case letters a and b need to be interpreted as denoting individual entities (people, things, numbers, etc.), while the upper-case letters A and B must be interpreted as referring to properties that the individuals may or may not have (also, in more complicated examples, relations that may or may not hold amongst groups of individuals). In English, properties are referred to by means of expressions having a variety of grammatical forms—common noun, adjective, or verb—but in predicate logic a uniform class of expressions, called predicates, is used to cover all these cases.

Typical instantiations of the schemas given above are:

> Every grandmother is a mother
> Anne is a grandmother
> _____
> Anne is a mother

and

> Some students are lazy
> Some of my children are students
> _____
> Some of my children are lazy

Predicate logic is the subject of Chapters 4–6.

(3) *Modal logic* deals with schemas such as

> It is necessary that if A then B
> It is possible that A
> _____
> It is possible that B
>
> It is necessary that either A or B
> _____
> Either it is necessary that A or it is necessary that B

Here, as in propositional logic, the schematic letters are to be interpreted as statements. There are many different varieties of modal logic, differing from one another with regard to exactly how the words 'necessary' and 'possible' are to be understood. For more about modal logic, see Chapter 7.

(4) *Temporal logic* is a variety of modal logic dealing with inferences involving time, as in

> It will always be the case that either A or B
> It will not always be the case that A
> _____
> It will sometimes be the case that B

> It is now the case that A
> _____
> It will always have been the case that A

Note that, in English, the conclusion of the second of these schemas is rather hard to understand—in fact, it is ambiguous. In temporal logic, a symbolic scheme for representing tenses in a clear and unambiguous way helps us to resolve these difficulties, though possibly at the cost of sacrificing any direct correspondence to the way tenses work in ordinary language.

Exercise 1.7 Examine all the schemas given in this section and try to determine which are valid and which are invalid.

2 The Propositional Calculus

2.1 TRUTH FUNCTIONS

Consider the statement

(1)   John ate fish and Mary ate steak.

We can view this as an interpretation of more than one statement schema. It is an interpretation of the schema

x ate A and y ate B

(where x and y are to be interpreted as referring to people, and A and B to substances), but equally, it is an interpretation of the schema

A and B

(where A and B are now to be interpreted as whole statements).

Each of these schemas reveals something about the structure of the original statement. This structure can be portrayed graphically as a tree:

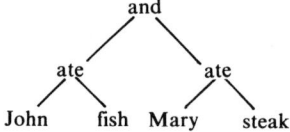

The first schema can be thought of as having been obtained by covering up the bottom layer of this structure, while the second is obtained by covering up the bottom two layers. Thus the second schema hides more than the first; more of the detailed structure is lost. What it reveals is the *high-level structure* of the statement. It is this high-level structure that forms the subject matter of the propositional calculus.

When we view our statement (1) as an interpretation of the schema 'A and B' we are seeing the statement as having a high-level structure according to which it may be described as the result of joining together two smaller statements (namely 'John ate fish' and 'Mary ate steak') by means of the link-word 'and'. There are many words in English which function in a similar way to 'and', linking together existing statements to produce new ones. In grammar, these words are generally called *conjunctions*, but because the word 'conjunction' has a special technical meaning in logic (see below, p. 27), we shall carry on calling them *link-words*.

By using different link-words to join the statements 'John ate fish' and 'Mary ate steak', we can construct a series of new statements such as:

(1)   John ate fish and Mary ate steak.
(2)   John ate fish because Mary ate steak.
(3)   John ate fish but Mary ate steak.
(4)   John ate fish while Mary ate steak.
(5)   John ate fish although Mary ate steak.
(6)   John ate fish or Mary ate steak.
(7)   John ate fish whenever Mary ate steak.
(8)   John ate fish if Mary ate steak.

All these statements differ in meaning, sometimes in quite subtle ways. Let us concentrate for now on just the first two:

(1)   John ate fish and Mary ate steak.
(2)   John ate fish because Mary ate steak.

Suppose you want to determine the truth values of these two statements. Assume for the sake of the argument that John and Mary went out for a meal last night, and it is this occasion that is being discussed. What do you have to find out in each case?

For statement (1), this is straightforward: you must find out whether John ate fish, and you must find out whether Mary ate steak. If the answer to both these questions is 'yes', then (1) is true; but if the answer to either or both of them is 'no', then (1) is false; and there the matter ends. Merely knowing the truth values of the component statements 'John ate fish' and 'Mary ate steak' is enough to determine the truth value of the compound statement (1).

Is this enough to determine the truth value of (2) as well? A little thought will show you that it certainly is not. Admittedly, if either of the component statements is false, then (2) must be false; but it is not enough, in order for (2) to be true, just for its two components to be true. The word 'because' in (2) is used to make a stronger claim, namely that Mary's eating steak is the *reason* why John ate fish; and this cannot be determined on the basis of truth values alone.

Exercise 2.1a   Before reading on, try to decide, for each of the remaining statements (3)–(8) above, whether the truth value of the whole can be determined merely by determining the truth values of the components.

You probably found the last exercise somewhat problematic. Most of the time, you should have decided that knowing the truth values of the components alone is not enough to determine the truth value of the compound. In fact, the only case where the opposite conclusion is undoubtedly warranted is statement (6). When we discussed (1) above, we concluded that (1) is true if, and only if, both its components are true. A similar examination of (6) will probably lead us to the conclusion that (6) is false if, and probably only if, both its components are false. There is a possible complication here, which is why I said 'probably only if'; for a discussion of this, see below, p. 27.

We are now ready for a bit of technical jargon. A link-word is said to be

truth-functional whenever the truth value of the statement obtained by using it to link two given statements A and B is uniquely determined by the truth values of A and B themselves.

Thus 'and' and 'or' are usually truth-functional because most of the time the truth values of statements of the form 'A and B' and 'A or B' are uniquely determined by the separate truth values of the statements A and B of which they are composed. I have to say 'usually' here because there are ways of using 'and' which are not strictly truth-functional. An example is in the statement

John felt faint and drank a glass of water,

where it seems to be an implication that John first felt faint and then as a consequence drank a glass of water—so that more is being said here than just that the two component statements 'John felt faint' and 'John drank a glass of water' are true. This is seen clearly if we reverse their order, as in

John drank a glass of water and felt faint,

which seems to suggest that there was something in the water which made John feel faint.

So is 'and' truth-functional or not? Most of the time it is, but sometimes it isn't. This is a recurring problem. Now, *the propositional calculus is only concerned with truth-functional link-words*. So, in order to avoid the problems arising from the fact that English link-words tend not to be invariably either truth-functional or not, the propositional calculus *uses its own link-words*, which are *defined* to be always truth-functional.

Instead of 'and', propositional calculus uses a special symbol $\wedge$ which can be thought of as having all the truth-functional meaning of 'and' without any of its non-truth-functional meaning. You can think of this as similar to the way that mathematics uses a special symbol '+' to capture just the arithmetical sense of 'add' while totally ignoring other uses of the word such as in 'adding insult to injury'.

The meaning of $\wedge$ is completely defined, then, by a **semantic rule**, which specifies the truth value of $A \wedge B$ in terms of the truth values of A and B individually. We want to say that $A \wedge B$ is true just so long as A and B are both true. Of course, being schematic letters, A and B are not true or false in themselves; they only acquire truth values when they are interpreted. Instead of saying that a schema is true in a given interpretation, we can say that the interpretation *satisfies* the schema. So our semantic rule tells us the precise condition for an interpretation to satisfy the schema $A \wedge B$:

An interpretation satisfies the schema $A \wedge B$ if and only if it satisfies both A and B

Another way of presenting this rule is by means of a **truth table**. Each row of this table corresponds to one of the four possible combinations of truth values which may be possessed by interpretations of the schematic letters A and B. The last entry in the row gives the truth value of the corresponding interpretations of $A \wedge B$. Since the table defines the truth value of $A \wedge B$ for each possible combination of truth values for A and B, ∧ is guaranteed to be truth-functional.

A	B	A ∧ B
true	true	true
true	false	false
false	true	false
false	false	false

Some more technical jargon: The schema A ∧ B is called the **conjunction** of the two schemas A and B; conversely, A and B are called the **conjuncts** of A ∧ B. Any symbol such as ∧ which is defined truth-functionally is called a **truth-functional connective**.

What other truth-functional connectives are there? We saw above that the link-word 'or' seems to be truth-functional, so we should be able to define a corresponding connective. I mentioned a slight complication; let's look at this now.

Consider the following two dialogues:

(1) Waitress: You may have cream or yoghurt with your apple-pie. Which would you like?
  Customer: I'd like both, please.
  Waitress: No, I said cream *or* yoghurt—you can't have both!

(2) John: What do you think Jane will do with her prize money?
  Mary: She'll buy a new car or have a holiday.
  Later—
  John: You were right! She bought a new car *and* had a holiday.

These two dialogues were designed to draw attention to the following problem: if you make a statement of the form 'A or B' (and are thus asserting that at least one of A and B is true), do you intend to be understood as *including* or *excluding* the possibility that *both* A and B might be true? In the first dialogue, this possibility is excluded, so we say that the waitress was using 'or' in the **exclusive** sense; in the second dialogue, that possibility is included (otherwise Mary would turn out not to have predicted correctly), so Mary was using 'or' in the **inclusive** sense. Both senses of 'or' occur frequently in English—though it isn't always easy to say which sense is intended on any particular occasion.[1] This need not trouble us, though, because in the propositional calculus, we can define our connectives to have whatever meaning we like, so long as that meaning is completely truth-functional. So we can introduce one symbol for the inclusive 'or' and another for the exclusive.

In practice, the sense of 'or' most often encountered in logic is the inclusive one: 'A or B or both'. The connective for this is written ∨, and its meaning is defined by the semantic rule

[1] It isn't even determinate in some cases. If I say 'John is in London or Paris', then since there is no possibility that John is in both cities, it makes no difference whether my 'or' is interpreted as having the inclusive or the exclusive sense—it is the meanings of the two component statements that determine that they cannot both be true, and I am powerless to change this by stipulating in which way my 'or' is to be understood.

> An interpretation satisfies the schema A ∨ B if and only if it satisfies at least one of A and B

or, equivalently,

> An interpretation falsifies A ∨ B if and only if it falsifies both A and B

The truth table corresponding to this rule is

A	B	A ∨ B
true	true	true
true	false	true
false	true	true
false	false	false

The statement A ∨ B is called the **disjunction of** A and B, which in turn are the **disjuncts** of A ∨ B.

For the exclusive 'or', we use the connective ⊕; its meaning is given by the rule

> An interpretation satisfies A ⊕ B if and only if it satisfies exactly one of A and B

There are other connectives in the propositional calculus, but before introducing them we shall examine the properties of ∧, ∨, and ⊕ in more detail.

Exercise 2.1b

*(1) Write down the truth table for A ⊕ B.

(2) In the following sentences, decide which occurrences of 'and' and 'or' are purely truth-functional, distinguishing between inclusive and exclusive uses of 'or'. (*Note*: There is not necessarily a single 'right' answer to each of these questions; in some cases, at least, it is a matter for interpretation.)

*(a) I grow flowers in the front garden and vegetables in the back.
 (b) I will spend the evening reading or listening to the radio.
*(c) John cooked supper, and Mary did the washing up.
 (d) Charles has two children and Toby has three.
*(e) The hedgesparrow, or dunnock, is a familiar British bird.
*(f) You must eat up your vegetables, or you shan't have any pudding.
 (g) James voted Conservative and the Conservatives won.
 (h) If $x < -1$ or $x > 1$ then $x^2 > x$.
 (i) To be or not to be, that is the question.
 (j) I had a little nut tree, and nothing would it grow.

2.2 SOME REMARKS ON STRUCTURE

Much of the later discussion will be simplified if we acquaint ourselves at this stage with some concepts and terminology relating to the structure of the schemas we shall be using. We have seen that a truth-functional connective such as ∧ can link two statements (or statement schemas) to produce a third, e.g. from A and B we get A ∧ B. We can represent this relationship in a tree structure:

Now there is nothing to stop the conjuncts from themselves having a structure. For example, instead of B we could have a disjunction, say B ∨ C; and this would give us the schema A ∧ (B ∨ C), for which the tree structure is

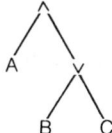

Note that B ∨ C is bracketed in order to make the structure clear: B and C are associated more closely with each other via the connective ∨ than either is to A: in fact it is the whole unit, or **subschema** B ∨ C which is linked to A by ∧. If instead we bracket together A and B, we get a different schema, (A ∧ B) ∨ C, with the structure

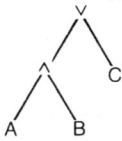

In English, we can use the word 'either' in front of the first disjunct to make clear how the bracketings go. The statement

     John will be promoted and Bill will resign or Mary will be sacked

is ambiguous in exactly the same way as the unbracketed 'schema' A ∧ B ∨ C is. We may resolve the ambiguity by inserting 'either' at a suitable position. If we say

     John will be promoted and either Bill will resign or Mary will be
     sacked

then the implied structure is that of A ∧ (B ∨ C); to get the structure of (A ∧ B) ∨ C we must say instead

     Either John will be promoted and Bill will resign or Mary will be
     sacked.

The first of these two schemas, A ∧ (B ∨ C), is *primarily* a conjunction, one of whose conjuncts is a disjunction. Although disjunction occurs in it, the schema as a whole is not a disjunction. We say that the **main functor** of this schema is the ∧, whereas the main functor of (A ∧ B) ∨ C, which is a disjunction one of whose disjuncts is a conjunction, is the ∨.

Each connective in a schema has a **scope**, which contains all the bits of the schema hanging down from that connective in the tree diagram. Thus the scope of ∧ in A ∧ (B ∨ C) is the whole schema, whereas the scope of ∨ is the subschema B ∨ C:

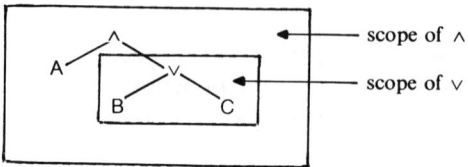

Here we do the same thing for the more complicated schema (A ∨ (B ∧ C)) ∧ (B ∨ D):

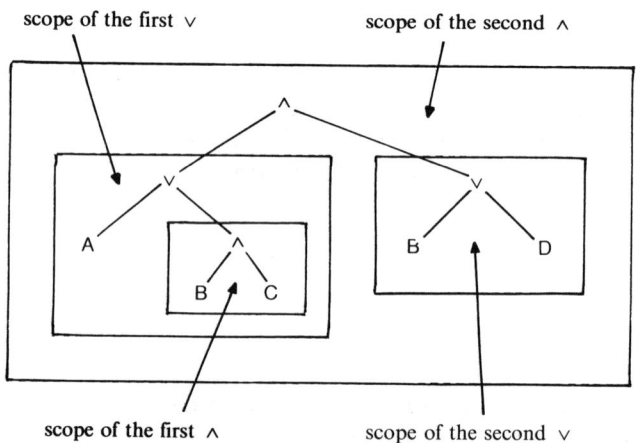

The main functor of this schema is the second occurrence of ∧. In other words, the schema is first and foremost a conjunction. Its conjuncts are the two subschemas A ∨ (B ∧ C) and B ∨ D. Each of these is first and foremost a disjunction. Since the whole schema is a conjunction, we could say that it has the same high-level structure as the simplest possible conjunction A ∧ B. It is a **substitution-instance** of that schema in the sense that we can get it from A ∧ B by making the substitutions

Substitute A ∨ (B ∧ C)   for A
Substitute B ∨ D   for B

(*Note*: the substitutions must be performed simultaneously, otherwise we would obtain different results, e.g. if we did the first substitution before the second we would end up with $(A \vee ((B \vee D) \wedge C) \wedge (B \vee D)$. The point is that $B \vee D$ must substitute for occurrences of B in the *original* schema, but not for occurrences introduced by other substitutions.)

We shall have several occasions in the rest of this book to refer to substitutions and substitution-instances, so it will be convenient at this point to introduce some special notation for them. The substitution of Y for X is written X/Y. The simultaneous substitution of $Y_1, \ldots, Y_n$ for $X_1, \ldots, X_n$ respectively is written as a *set* of individual substitutions:

$$\{X_1/Y_1, \ldots, X_n/Y_n\}.$$

Sets of substitutions are conventionally denoted, for short, by lower-case Greek letters such as θ, ϕ, and ψ (read 'theta', 'phi', and 'psi' respectively). The result of applying the set of substitutions θ to an expression E is written $E\theta$. We say that $E\theta$ is the **substitution-instance** of E obtained by applying the substitution-set θ. We may also say that $E\theta$ **instantiates** E (by substitution θ). Thus, to rephrase our example of the previous paragraph using the new notation, we can say

If E is the schema $A \wedge B$, and θ is the substitution-set
$\{A/A \vee (B \wedge C), B/B \vee D\}$, then $E\theta$ is the schema $(A \vee (B \wedge C)) \wedge (B \vee D)$.

A useful concept when discussing statement schemas is the **depth** of their structure. This may be defined as the length of the longest path through the tree structure of the formula, from top to bottom. Thus a single schematic letter

A

has depth 0, since the top level *is* the bottom level in this case. A conjunction of two schematic letters, $A \wedge B$, has depth 1:

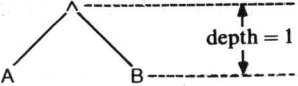

whereas $A \wedge (B \vee C)$ has depth 2:

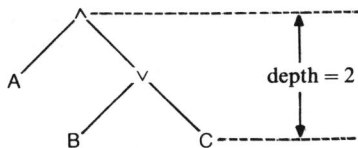

A schema of depth 4 is $A \wedge ((B \wedge (C \vee B)) \vee E)$. The reader should draw the tree structure of this schema as an exercise.

We may also speak of the depth *at which* a subschema occurs in a given schema. For example, in the schema just mentioned, the subschema $C \vee B$ occurs at depth 3, since its main functor is three steps down from the top of the tree, whereas the subschema E is at depth 2. The depth of the whole tree is then just the depth at which its deepest subschemas occur in it—in this case they are C and D, at depth 4.

Exercise 2.2

(1) Draw tree diagrams to illustrate the structure of each of the following schemas:

(a) $P \vee (Q \wedge R)$
*(b) $(P \wedge Q) \vee (P \wedge R)$
(c) $((A \wedge B) \vee C) \vee D$
*(d) $A \wedge (B \wedge (C \wedge (D \vee F)))$
(e) $(A \wedge (B \vee C)) \vee ((A \vee C) \wedge (B \vee C))$

(2) Write down the scope of each connective in the schemas in Question (1) (solutions given for (a) and (c)).

(3) Find all of the subschemas of *(a) $(P \wedge Q) \vee ((R \wedge S) \vee T)$, and (b) $P \wedge ((Q \wedge R) \vee ((S \vee T) \wedge U))$, stating the depth of each subschema in the whole schema.

*(4) What schema is obtained from the schema $(A \vee B) \wedge (C \vee (A \wedge D))$ if $A \wedge C$ is substituted for A and $B \wedge C$ is substituted for C

(a) simultaneously?
(b) successively?

(5) Apply each of the substitution-sets

$\theta = \{A/A \vee B, B/A \wedge B\}$
$\phi = \{A/C \wedge D, B/A \wedge C\}$
$\psi = \{A/B, B/B \vee A\}$

to each of the schemas $A \wedge B$, $A \vee (A \wedge B)$, $B \vee A$, $C \wedge D$.

2.3 SOME EQUIVALENCES INVOLVING CONJUNCTION AND DISJUNCTION

The main results of this section take the form of equivalences, as follows:

(1a)	$A \wedge B \cong B \wedge A$	(Conjunction is commutative)
(1b)	$A \vee B \cong B \vee A$	(Disjunction is commutative)
(2a)	$(A \wedge B) \wedge C \cong A \wedge (B \wedge C)$	(Conjunction is associative)
(2b)	$(A \vee B) \vee C \cong A \vee (B \vee C)$	(Disjunction is associative)
(3a)	$A \wedge A \cong A$	(Conjunction is idempotent)
(3b)	$A \vee A \cong A$	(Disjunction is idempotent)
(4a)	$A \wedge (B \vee C) \cong (A \wedge B) \vee (A \wedge C)$	(Conjunction distributes over disjunction)
(4b)	$A \vee (B \wedge C) \cong (A \vee B) \wedge (A \vee C)$	(Disjunction distributes over conjunction)
(5a)	$A \wedge (A \vee B) \cong A$	(Absorption)
(5b)	$A \vee (A \wedge B) \cong A$	(Absorption)

All these equivalences can be easily verified by examination of the relevant truth tables. We shall look at (3a) and (4a), and leave the rest to the reader as an exercise.

(3a) $A \wedge A \cong A$. You might imagine to begin with that the truth table for $A \wedge A$ should look like this:

A	A	A ∧ A
true	true	true
true	false	false
false	true	false
false	false	false

but with a little thought you will realise that the second and third rows make no sense since each of them provides conflicting statements about the truth value of A. So only the first and last rows are applicable, and the truth table reduces to

A	A ∧ A
true	true
false	false

Since A and A ∧ A always have the same truth value, they are equivalent, as required.

(4a) A ∧ (B ∨ C) ≅ (A ∧ B) ∨ (A ∧ C). This one requires a little more work. The truth table for A ∧ (B ∨ C) must contain a row for each of the eight possible combinations of truth values for the three schematic letters A, B, and C:

A	B	C	
T	T	T	
T	T	F	
T	F	T	
T	F	F	
F	T	T	
F	T	F	
F	F	T	
F	F	F	

(Note that from now on we abbreviate 'true' and 'false', in truth tables, to 'T' and 'F' respectively.)

The right-hand conjunct of A ∧ (B ∨ C) is itself a disjunction, B ∨ C, so we use the truth table for disjunction to fill in a column for this conjunct:

A	B	C	B ∨ C
T	T	T	T
T	T	F	T
T	F	T	T
T	F	F	F
F	T	T	T
F	T	F	T
F	F	T	T
F	F	F	F

Note that the truth values of A play no part in determining the entries in the column for B ∨ C.

Now we use the truth table for conjunction to work out the truth values of

$A \land (B \lor C)$ from the entries in the columns for the two conjuncts:

A	B	C	B ∨ C	A ∧ (B ∨ C)
T	T	T	T	T
T	T	F	T	T
T	F	T	T	T
T	F	F	F	F
F	T	T	T	F
F	T	F	T	F
F	F	T	T	F
F	F	F	F	F

For example, the 'F' entry in the fourth row is obtained as follows: in this row, A has the truth value 'T', and $B \lor C$ has the truth value 'F', and the conjunction of a 'T' with an 'F' is an 'F'.

Next we must construct the truth table for $(A \land B) \lor (A \land C)$

A	B	C	A ∧ B	A ∧ C	(A ∧ B) ∨ (A ∧ C)
T	T	T	T	T	T
T	T	F	T	F	T
T	F	T	F	T	T
T	F	F	F	F	F
F	T	T	F	F	F
F	T	F	F	F	F
F	F	T	F	F	F
F	F	F	F	F	F

Finally, we compare the truth tables for $A \land (B \lor C)$ and $(A \land B) \lor (A \land C)$ and observe that the rightmost entries are line for line the same; this means that whatever truth values A, B, and C have (and the rows of the truth tables exhaust all the possibilities), $A \land (B \lor C)$ has the same truth value as $(A \land B) \lor (A \land C)$, i.e. the two schemas are equivalent.

Exercise 2.3a

   (1) Use truth tables to verify the other equivalences given above.

   *(2) Which of the equivalences above in which $\lor$ features remain valid if $\lor$ is replaced by $\oplus$? What is the relationship between $A \land (B \oplus C)$ and $(A \land B) \oplus C$?

   (3) Write down truth tables for each of the following schemas:

      *(a) $A \lor (A \land B)$

      (b) $(A \lor B) \lor (B \lor C)$

      (c) $A \land (B \land (C \land D))$

      *(d) $(A \land B) \lor (B \lor C)$

      (e) $(A \lor B) \land (A \land (B \lor C))$

Further equivalences can be derived directly from the ones we have already got, without the need to construct more truth tables. For example, we have

$$(4c) \quad (A \land B) \lor C \cong C \lor (A \land B) \qquad (by\ 1b)$$

$$\cong (C \vee A) \wedge (C \vee B) \quad \text{(by 4b)}$$
$$\cong (A \vee C) \wedge (B \vee C) \quad \text{(by 1b)}$$

so that $\vee$ distributes over a conjunction to its left as well as over one to its right.

Notice that in this last proof, we justified the equivalence

$$(A \wedge B) \vee C \cong C \vee (A \wedge B)$$

by appealing to the equivalence

$$A \vee B \cong B \vee A.$$

The point here is that $(A \wedge B) \vee C$ is a substitution-instance of $A \vee B$, the relevant substitution being $\{A/A \wedge B, B/C\}$. Applying the same substitution to $B \vee A$ gives us $C \vee (A \wedge B)$. The underlying principle is that if the same substitution is applied to two equivalent schemas, then the resulting schemas are themselves equivalent. Thus, in an equivalence, each schematic letter does duty for any of its substitution-instances. So the equivalence

$$A \vee B \cong B \vee A$$

stands in for an unlimited set of equivalences, of which

$$A \vee (B \vee C) \cong (B \vee C) \vee A$$
$$(A \wedge B) \vee C \cong C \vee (A \wedge B)$$
$$(A \wedge B) \vee (C \wedge D) \cong (C \wedge D) \vee (A \wedge B)$$

are examples. This way of handling schematic letters is not confined to equivalences: in general, *whenever we use a schematic letter in the propositional calculus, we can think of it as standing in for any of its substitution-instances.*

Similar manoeuvres occur in the derivation:

$$(A \wedge B) \vee (C \wedge D) \cong ((A \wedge B) \vee C) \wedge ((A \wedge B) \vee D) \quad \text{(by 4b)}$$
$$\cong ((A \vee C) \wedge (B \vee C)) \wedge ((A \vee D) \wedge (B \vee D)) \quad \text{(by 4c)}$$
$$\cong (A \vee C) \wedge ((B \vee C) \wedge ((A \vee D) \wedge (B \vee D))) \quad \text{(by 2a)}$$

This equivalence displays a more complicated kind of distributivity, by which a disjunction of two conjunctions is shown to be equivalent to a conjunction of four disjunctions.

Since both $\wedge$ and $\vee$ are associative, we may omit the brackets from strings of conjuncts or strings of disjuncts. That is, since

$$(A \wedge B) \wedge C \cong A \wedge (B \wedge C),$$

both of these schemas can be unambiguously written as

$$A \wedge B \wedge C$$

(whereas, for example, since $A \wedge (B \vee C)$ is *not* equivalent to $(A \wedge B) \vee C$, we cannot omit the brackets from either of these schemas.) So our latest equivalence can be written as

$$(6) \quad (A \wedge B) \vee (C \wedge D) \cong (A \vee C) \wedge (B \vee C) \wedge (A \vee D) \wedge (B \vee D)$$

Exercise 2.3b Demonstrate the validity of each of the equivalences below by deriving them from equivalences already obtained:

(1) $(A \wedge B) \vee B \cong B$

*(2) $(A \wedge C) \vee (B \wedge C) \cong (A \vee B) \wedge C$

(3) $(A \vee B) \wedge (C \vee D) \cong (A \wedge C) \vee (B \wedge C) \vee (A \wedge D) \vee (B \wedge D)$

*(4) $(A \wedge B) \vee (B \wedge C) \vee (C \wedge A) \cong (A \vee B) \wedge (B \vee C) \wedge (C \vee A)$

(5) $(A \wedge ((B \wedge (C \vee D)) \vee E)) \vee F \cong (A \wedge B \wedge C) \vee (A \wedge B \wedge D) \vee (A \wedge E) \vee F$
$$\cong (A \vee F) \wedge (B \vee E \vee F) \wedge (C \vee D \vee E \vee F)$$

2.4 NORMAL FORMS

Look closely at Question (5) of Exercise 2.3b. It asks you to show that three schemas are all equivalent. The first is a complicated-looking schema of depth 4, whereas the second and third are much simpler in form, being a disjunction of conjunctions and a conjunction of disjunctions respectively (and thus each of depth 2). In fact, *any* schema built up out of $\wedge$ and $\vee$ connectives can be shown to be equivalent to a disjunction of conjunctions and to a conjunction of disjunctions. These two forms are called the **disjunctive normal form** (DNF) and the **conjunctive normal form** (CNF) for the schema. It is useful to be able to recast schemas in one or other of these forms, as they are particularly easy to handle (see especially below, section 3.6).

To give a precise definition of the two normal forms, we first introduce some new terms:

> A **literal** is any schematic letter

(later, in Section 2.9, we shall extend the meaning of this term slightly).

> A **minterm** is either a literal or a conjunction of distinct literals
> A **maxterm** is either a literal or a disjunction of distinct literals

Thus A is simultaneously a literal, a minterm, and a maxterm, whereas $A \wedge B$ and $A \wedge B \wedge C$ are minterms and $A \vee B$ and $A \vee B \vee C$ are maxterms; $A \wedge A$ and $A \vee A$ are neither literals, minterms nor maxterms. We can simplify the definitions of minterm and maxterm a little by adopting the convention that by a conjunction or disjunction of a single literal we mean just that literal itself. Then a minterm is just any conjunction of (one or more) distinct literals, and a maxterm is any disjunction of (one or more) distinct literals.

> A minterm (or maxterm) M_1 **absorbs** another minterm (or maxterm) M_2 if every literal in M_1 is also in M_2.

Thus any minterm (or maxterm) absorbs itself, and, for example, A absorbs A ∨ C. which absorbs A ∨ B ∨ C.

Finally, we have

An **DNF** is a disjunction of minterms none of which absorbs any of the others

A **CNF** is a conjunction of maxterms none of which absorbs any of the others

---

So

$$(A \land B) \lor (A \land D \land E) \lor C$$

is a DNF, but

$$(A \land B) \lor (A \land B \land C) \lor D$$

is not a DNF, since the first disjunct absorbs the second. Remembering our convention that a disjunction need not have more than one disjunct, we see also that any minterm, e.g. A ∧ B ∧ C, is a DNF; and any maxterm, e.g. A ∨ B ∨ C, being a disjunction of maxterms (each of which is a single literal) is also a DNF. Similarly, A, A ∧ B, A ∨ B, A ∨ (B ∧ C), (A ∧ B) ∨ (C ∧ D), and (A ∧ B ∧ C) ∨ D ∨ E are all DNFs—the first three of them, but not the rest, are also CNFs.

Exercise 2.4a   For each of the following schemas, decide which, if any, of the terms 'literal', 'minterm', 'maxterm', 'DNF', and 'CNF' apply to it (remember that more than one of the terms might apply in some cases):

*(1) A	*(2) (A ∨ E)
(3) A ∧ D	*(4) (A ∨ E) ∧ (D ∨ E)
(5) A ∧ (B ∨ C ∨ F)	*(6) (A ∧ B) ∨ (C ∧ (D ∨ E))
(7) A ∨ B ∨ (C ∨ D)	(8) A ∧ (B ∨ C) ∧ (B ∨ D ∨ F) ∧ E
(9) A ∨ B ∨ C ∨ G ∨ H	(10) B ∨ B

Algorithm   *Conversion to disjunctive normal form*

We now present an algorithm for converting any schema built up out of literals, ∧s, and ∨s into an equivalent DNF. In order to perform the conversion, we shall assume that all unnecessary bracketings have been removed in accordance with the associative laws for conjunction and disjunction, so that subschemas like A ∧ (B ∧ C) or (A ∨ B) ∨ (C ∨ (D ∨ E)) have been replaced by A ∧ B ∧ C and A ∨ B ∨ C ∨ D ∨ E respectively. The advantage of doing this is that all schemas now have the property that *either* the connectives at even depth are all disjunctions and those at odd depth all conjunctions *or* the other way round. The former case is illustrated by the schema

$$A \lor (B \land ((C \land D) \lor E) \land C)) \lor ((F \lor (G \land H)) \land I)$$

which can be represented by the tree

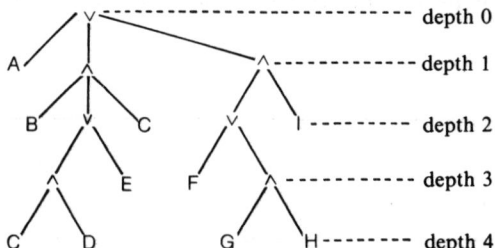

The algorithm is as follows:

(1) Look for the deepest conjunction which includes at least one disjunction amongst its conjuncts. If none is found, go to (3); if more than one is found, choose the first.
(2) In the conjunction found at (1), distribute the main functor over all the disjunctions amongst its conjuncts, removing any unnecessary bracketing. Go back to (1).
(3) Simplify each disjunct as far as possible by omitting any repeated literals it contains.
(4) Omit all but one from any set of disjuncts which contain exactly the same literals.
(5) Omit any disjunct which contains all the literals in some other disjunct.

To illustrate, we shall apply the algorithm to the schema

$$(A \lor B) \land ((B \land (C \lor ((C \lor D) \land A))) \lor C).$$

At each step of the algorithm, we shall give the tree structure of the current state of the schema in order to illustrate graphically what is going on. Initially, then, the tree is

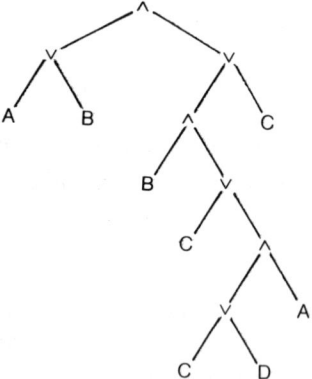

At step (1), we locate $(C \lor D) \land A$ as the deepest conjunction with a disjunctive conjunct. At step (2), we distribute this conjunction over the disjunction to give

$$(A \lor B) \land ((B \land (C \lor (C \land A) \lor (D \land A))) \lor C)$$

whose tree is

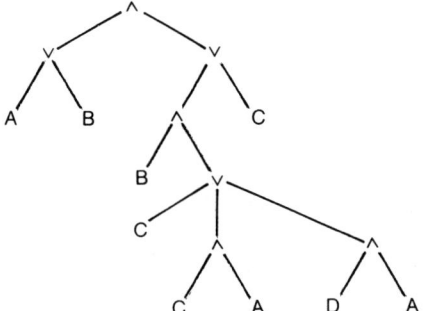

At the next repetition of step (1), we locate $B \wedge (C \vee (C \wedge A) \vee (D \vee A))$ as the conjunction to be distributed at step (2), giving us

$$(A \vee B) \wedge ((B \wedge C) \vee (B \wedge C \wedge A) \vee (B \wedge D \wedge A) \vee C),$$

with tree

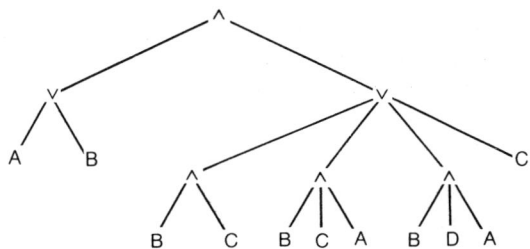

Next time round, the required deepest conjunction is the whole schema; the distribution yields

$$(A \wedge B \wedge C) \vee (A \wedge B \wedge C \wedge A) \vee (A \wedge B \wedge D \wedge A) \vee (A \wedge C) \vee (B \wedge B \wedge C) \vee$$
$$(B \wedge B \wedge C \wedge A) \vee (B \wedge B \wedge D \wedge A) \vee (B \wedge C).$$

There is now no conjunction with a disjunction amongst its conjuncts, so we move no to step (3). At this stage we have finished the iterative part of the algorithm, and all that remains is to tidy up the expression we have obtained. This is done by eliminating any unnecessary duplication. Step (3) gives us

$$(A \wedge B \wedge C) \vee (A \wedge B \wedge C) \vee (A \wedge B \wedge D) \vee (A \wedge C) \vee (B \wedge C) \vee (B \wedge C \wedge A) \vee$$
$$(B \wedge D \wedge A) \vee (B \wedge C),$$

then step (4) converts this into

$$(A \wedge B \wedge C) \vee (A \wedge B \wedge D) \vee (A \wedge C) \vee (B \wedge C)$$

which is finally simplified at step (5) to

$$(A \wedge B \wedge D) \wedge (A \wedge C) \vee (B \wedge C).$$

This last schema is thus a DNF of the one we started with; its tree is

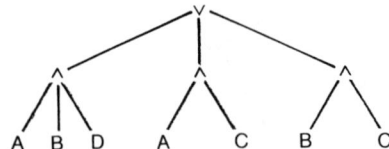

which it will readily be agreed is considerably simpler than what we began with.

The following material may be omitted on the first reading

To prove that our algorithm is correct, we must show that whatever schema we begin with, the algorithm terminates with a DNF equivalent to the initial schema. We can break this task down into three parts:

(1) show that the algorithm always terminates;
(2) show that the schema at termination is a DNF;
(3) show that the schema at termination is equivalent to the original schema.

To accomplish part (1), we note that as we progress through the algorithm, the depth of the conjunction which is being distributed decreases—and, at the same time, so does the depth of the disjunctions over which it is being distributed. Now whenever a disjunction is distributed over, it is at that time one of the deepest disjunctions in the schema—for any deeper disjunction must also be the conjunct of a conjunction and hence would be distributed over in preference to the other.

Let us use variables d and n with the meanings:

$$d = \text{the depth of the deepest disjunction(s) in the schema}$$
$$\text{at a given stage (or 0 if there are no disjunctions);}$$

$$n = \text{the number of disjunctions at depth } d.$$

Now, on execution of step (2), one or more disjunctions at depth d are distributed over, i.e. a configuration of the form

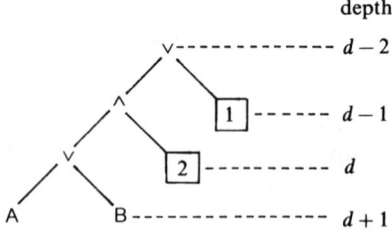

is replaced by a configuration of the form

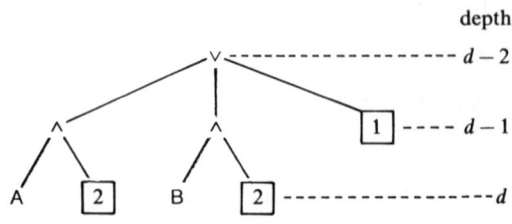

(*Note*: these diagrams are not completely general, but should suffice to give the general idea.) The net effect is that the number n of disjunctions at depth d is decreased; if it becomes 0, then the variable d is decreased since then the deepest disjunction is at a higher level than the ones just distributed over. So we have two types of occurrence at step (2), those in which the pair (d, n) is replaced by (d, n'), where $0 < n' < n$, and those in which (d, n) is replaced by (d', m), where $0 \leqslant d' < d$.

In our example above, the values taken by (d, n) are

$$(5, 1) \to (3, 1) \to (1, 2) \to (0, 1).$$

Suppose that initially the value is (d_0, n_0). Then after at most n_0 steps, the value must be (d_1, n_1), where $d_0 > d_1$; after at most n_1 further steps, the value will be (d_2, n_2), where $d_1 > d_2$; so as we go through the algorithm, we generate a decreasing sequence of values of d,

$$d_0 > d_1 > d_2 > d_3 > \cdots,$$

and this sequence will continue until the execution of the algorithm comes out of the loop consisting of steps (1) and (2), i.e. until there are no more conjunctions numbering disjunctions amongst their conjuncts. This will happen as soon as $d = 0$; and it is easy to see that this *must* happen, since otherwise we would have an infinite sequence of positive integers, each one strictly less than the one preceding it, and this is clearly impossible.

We have thus shown that the algorithm must eventually come out of the initial loop and proceed to step (3). There are no more loops, so the algorithm must terminate. At step (3), $d = 0$, i.e. either the deepest disjunction is the whole schema, or the schema contains no disjunctions. After tidying up in accordance with steps (3)–(5), such a schema must yield a DNF. It remains, then, for us to show that the DNF obtained in this way is indeed equivalent to the original schema.

To do this, we have to show that each step in the algorithm *preserves equivalence*; that is, that when a step is applied to some schema, the resulting schema is equivalent to the original one. We examine each step in turn:

(1) This step does not change the schema in any way, so naturally it preserves equivalence;
(2) This is an application of the distributive law for conjunction over disjunction. We proved this in the form

$$A \wedge (B \vee C) \cong (A \wedge B) \vee (A \wedge C),$$

but our algorithm works with a more general form of the law by which conjunction can distribute over arbitrarily many disjunctions, each of arbitrary size. To prove this is a somewhat complicated, though not really difficult, exercise in mathematical induction; it is left as an exercise for the more mathematically inclined reader. Others may prefer to restrict themselves to the following instance of the more general rule:

$$(A \vee B \vee C) \wedge (D \vee E) \wedge F \cong$$
$$(A \wedge D \wedge F) \vee (A \wedge E \wedge F) \vee (B \wedge D \wedge F) \vee (B \wedge E \wedge F) \vee (C \wedge D \wedge F) \vee (C \wedge E \wedge F).$$

(3) The omission of repeated literals is sanctioned by the idempotence of $\wedge$, which says that conjoining a schema to itself does not alter its truth value; hence, for example,

$$A \wedge B \wedge A \wedge D \wedge B \cong A \wedge A \wedge A \wedge B \wedge B \wedge D \quad \text{(by commutativity)}$$
$$\cong A \wedge B \wedge D \quad \text{(by idempotence)}$$

(4) This step is sanctioned by the idempotence of $\vee$, so that, for example.

$$\cdots \vee (A \wedge B \wedge C) \vee \cdots \vee (B \wedge C \wedge A) \vee \cdots$$
$$\cong \cdots \vee (A \wedge B \wedge C) \vee (B \wedge C \wedge A) \vee \cdots \cdots \quad \text{(commutativity of } \vee)$$

$\cong \cdots \vee (A \wedge B \wedge C) \vee (A \wedge B \wedge C) \vee \cdots \cdots$ (commutativity of $\wedge$)

$\cong \cdots \vee (A \wedge B \wedge C) \vee \cdots \cdots$ (idempotence of $\vee$)

(5)   This step is justified by the absorption laws, as for example in:

$\cdots \vee (A \wedge B \wedge C) \vee \cdots \vee (A \wedge C) \vee \cdots$

$\cong \cdots \vee (A \wedge B \wedge C) \vee (A \wedge C) \vee \cdots \cdots$ (commutativity of $\vee$)

$\cong \cdots \vee (A \wedge C \wedge B) \vee (A \wedge C) \vee \cdots \cdots$ (commutativity of $\wedge$)

$\cong \cdots \vee ((A \wedge C) \wedge B) \vee (A \wedge C) \vee \cdots \cdots$ (associativity of $\wedge$)

$\cong \cdots \vee (A \wedge C) \vee \cdots \cdots$ (absorption of $\wedge$ by $\vee$)

We thus see that each step of the algorithm preserves equivalence, and hence the algorithm as a whole does.

End of material that may be omitted at first reading

One useful feature of a DNF is the close relationship that it bears to its own truth table; in effect, the DNF is a concise *summary* of its truth table. For example, consider the schema

$$(A \wedge B \wedge D) \vee (A \wedge C) \vee (B \wedge C)$$

which we arrived at earlier as the DNF of

$$S = (A \vee B) \wedge ((B \wedge (C \vee (C \wedge A) \vee (D \wedge A))) \vee C).$$

The more energetic reader may care to construct the truth table of the latter schema before reading on.

Using the DNF, we can see that our schema is true whenever one (or more) of $A \wedge B \wedge D$, $A \wedge C$ and $B \wedge C$ is true; but it is easy to identify the lines of the truth table at which this happens:

A	B	C	D	$A \wedge B \wedge D$	$A \wedge C$	$B \wedge C$	S
T	T	T	T	T	T	T	T
T	T	T	F	F	T	T	T
T	T	F	T	T	F	F	T
T	T	F	F	F	F	F	F
T	F	T	T	F	T	F	T
T	F	T	F	F	T	F	T
T	F	F	T	F	F	F	F
T	F	F	F	F	F	F	F
F	T	T	T	F	F	T	T
F	T	T	F	F	F	T	T
F	T	F	T	F	F	F	F
F	T	F	F	F	F	F	F
F	F	T	T	F	F	F	F
F	F	T	F	F	F	F	F
F	F	F	T	F	F	F	F
F	F	F	F	F	F	F	F

Each disjunct of the DNF corresponds to a group of lines in the truth table. For example, the disjunct A ∧ C corresponds to the first, second, fifth, and sixth lines. The truth table has 'T' for each line that appears in one or more of these groups. For schemas which are already in DNF, or close to DNF, construction of the truth table can be done with much less effort than usual.

Conjunctive normal form (CNF) can be arrived at by a very similar process to the one we used for DNF: simply swap the prefixes 'con' and 'dis' wherever they occur in the algorithm. CNF does not bear quite the same straightforward relation to the truth table as does DNF; CNF is closely related to what is known as clausal form, which we introduce in Section 3.6.

The symmetry we have noted between CNF and DNF is an example of **duality**. Any valid equivalence between schemas built up out of ∧s and ∨s remains valid if ∧ is exchanged with ∨ throughout. Thus the equivalences given at the start of Section 2.3 come in dual pairs, (1a) and (1b), (2a), and (2b), etc; in all these cases, each member of the pair can be obtained from the other by exchanging ∧ with ∨.

Exercise 2.4b Convert the following formulae into (a) DNF, (b) CNF:

(1) (A ∧ B) ∨ (A ∧ C)
*(2) (A ∨ B) ∧ (A ∨ (C ∧ B))
(3) A ∧ ((B ∧ C) ∨ (B ∧ (A ∧ C)))

2.5 NEGATION

Given any schema built up out of ∧s and ∨s, we know how to construct its truth table. There is a converse to this problem: given a truth table, construct a corresponding schema. For example, suppose we want a schema X with the truth table

A	B	C	X
T	T	T	T
T	T	F	F
T	F	T	T
T	F	F	F
F	T	T	T
F	T	F	F
F	F	T	F
F	F	F	F

We observe that X is true whenever A ∧ C is true (lines 1 and 3), and also that X is true whenever B ∧ C is true (lines 1 and 5), and for no other cases. Hence

$$X \cong (A \wedge C) \vee (B \wedge C)$$

and we have our schema. The answer is not unique, of course, since any equivalent schema, such as

$$(A \vee B) \wedge C$$

would do just as well.

Now consider the truth table

A	B	X
T	T	F
T	F	F
F	T	F
F	F	T

Can we find a schema for this? Try it before reading on.

You will find that there is no schema built up out of A and B with $\wedge$s and $\vee$s which does the trick. You should quickly convince yourself that any such schema is equivalent to one of the four DNFs

$$A, B, A \wedge B, A \vee B,$$

none of which has the truth table we want. To write down a schema which has the above truth table, then, we shall have to introduce a new connective.

The connective which will solve our problem is **negation**. This has much the same meaning as the English word 'not' in a statement like

John is not married.

The relationship between 'John is not married' and 'John is married' is that at any time they cannot both be true, and they cannot both be false: exactly one of them must be true. To put it differently, 'John is not married' is true if and only if 'John is married' is false. (I am assuming here, as always, that we know exactly who we mean by 'John'. If we don't, of course, then neither statement has a definite truth value—or rather, neither sentence conveys a definite statement.)

Note that 'not' is not a link-word in the sense of joining together two statements; rather it acts on a single statement to produce a new one. Even so, it is usually truth-functional, because the truth value of a single statement with 'not' in is, usually, completely determined by the truth value of the statement obtained by leaving out the 'not'—namely, it is the opposite truth value.

A more strictly truth-functional expression is 'It is not the case that', which behaves very much as we want negation to behave in the propositional calculus. Thus instead of 'John is not married' we could say 'It is not the case that John is married', in which there is a clear separation between the negation (expressed by the first six words) and what is negated (the rest). Sometimes, it is *necessary* to use the 'It is not the case that' style to express the negation of a sentence, since simply inserting 'not' has a different effect. For example, the negation of the statement

The shop is open until midday

is not

The shop is not open until midday

since both these statements could be false (e.g. if the shop is now shut, but will open at 11 a.m.). The true negation can only readily the expressed as

It is not the case that the shop is open until midday.

As with 'and' and 'or', we replace 'not' by a special symbol, $\sim$, which is *defined* to be always truth-functional. The semantic rule for $\sim$ is

---

An interpretation satisfies $\sim A$ if and only if it falsifies A

---

or, in the form of a truth table

A	$\sim A$
T	F
F	T

Using $\sim$, we can now write down a schema for the truth table we gave earlier:

$$\sim A \wedge \sim B.$$

(Check this!) And in fact *any* truth table can be given a schema built up using $\wedge$, $\vee$, and $\sim$.

Here are some equivalences involving negation:

(1) $\sim \sim A \cong A$ (double negation)
(2a) $\sim (A \wedge B) \cong \sim A \vee \sim B$ (de Morgan's laws)
(2b) $\sim (A \vee B) \cong \sim A \wedge \sim B$
(3) $A \oplus B \cong (A \vee B) \wedge \sim (A \wedge B)$

And here is a rule enabling us to deduce one equivalence from another:

$$X \cong Y \text{ if and only if } \sim X \cong \sim Y.$$

Here, X and Y stand for any schemas; once again a schematic letter is doing duty for any of its substitution-instances. Thus, given (2a), we may deduce the equivalence

$$\sim \sim (A \wedge B) \cong \sim (\sim A \vee \sim B),$$

which reduces, using (1), to

$$A \wedge B \cong \sim (\sim A \vee \sim B).$$

Notice that this last equivalence means that we can define conjunction in terms of disjunction and negation. Similarly, we could define disjunction in terms of conjunction and negation. This means that in theory we could drop one of $\vee$ and $\wedge$, and make do with just the other and negation. In practice, though, we find it convenient to make use of all three connectives.

We can still draw tree diagrams to represent the structure of schemas containing negation, and the ideas of depth and scope apply in these cases exactly as before. For example, the schema

$$\sim (A \wedge \sim B) \vee C$$

can be represented by the diagram

The subschemas $\sim(A \wedge \sim B)$ and C occur at depth 1, $A \wedge \sim B$ occurs at depth 2, A and $\sim B$ at depth 3, and B at depth 4. The scope of the first $\sim$ is $\sim(A \wedge \sim B)$, and of the second, $\sim B$.

Exercise 2.5

(1) Verify the equivalences (1), (2a), (2b), and (3) using truth tables.

(2) For each of the following schemas, write down (i) a tree diagram to show its structure, and (ii) its truth table.

  *(a) $A \wedge \sim(B \vee C)$
  (b) $\sim P \vee (Q \wedge \sim R)$
  *(c) $(A \vee B) \wedge \sim(A \wedge C)$
  (d) $(A \wedge B) \vee (\sim A \vee \sim B)$
  (e) $(P \wedge \sim(Q \vee \sim R \vee \sim(P \vee \sim Q)$

 *(3) In this question we introduce a new connective | (read 'nand') defined so that A|B is equivalent to $\sim(A \wedge B)$.

(a) Write down the truth table for A|B.
(b) Prove the equivalences
  (i) $\sim A \cong A|A$
  (ii) $A \vee B \cong (A|A)|(B|B)$
  (iii) $A \wedge B \cong (A|B)|A\,|B)$

(H. M. Sheffer showed in 1913 that all possible truth-functions be expressed in terms of | alone.)

(4) Write a schema for the following truth table. Try to make it as simple as possible.

A	B	C	
T	T	T	T
T	T	F	T
T	F	T	F
T	F	F	F
F	T	T	F
F	T	F	T
F	F	T	F
F	F	F	T

Now write a schema using only the connective | defined in Question 3.

2.6 ENTAILMENT IN THE PROPOSITIONAL CALCULUS

So far, the only relationships between propositional calculus schemas we have considered have been equivalences. But we know from Chapter 1 that logic has to do with other relationships as well. How, for example, are entailments demonstrated in the propositional calculus?

From the definition of entailment, we know that a set **S** of statement schemas entails a statement schema A just so long as every interpretation that satisfies all the members of **S** also satisfies A. If the schemas in question are schemas of the propositional calculus, then all the possible combinations of truth values they can have will be given by the rows of their truth tables. So all we need do is to construct a truth table for all the members of **S** and for A, and check that A is true in every line of the truth table in which all the members of **S** are true.

Example 2.6.1 Show that the following inference schema is valid:

$$A \vee B$$
$$\sim A$$
$$\overline{}$$
$$B$$

For this schema to be valid, every interpretation which satisfies $A \vee B$ and $\sim A$ must satisfy B as well. In terms of truth tables, this means that any line of the truth table on which $A \vee B$ and $\sim A$ are both true must also have B true. This is easy to demonstrate

	A	B	premisses $A \vee B$	$\sim A$	conclusion B	
	T	T	T	F	T	
	T	F	T	F	F	
*	F	T	T	T	T	*
	F	F	F	T	F	

Note that only on the third line (marked with asterisks) are the two premisses true; and since on this line the conclusion is true too, the inference schema must be valid.

Example 2.6.2 Determine whether or not the following inference is valid:

$$A \vee B$$
$$A$$
$$\overline{}$$
$$\sim B$$

The truth table is

	A	B	premisses $A \vee B$	A	conclusion $\sim B$	
*	T	T	T	T	F	*
*	T	F	T	T	T	*
	F	T	T	F	F	
	F	F	F	F	T	

There are two lines on which the premisses are all true, and on one of these, the first, the conclusion is false. Hence the inference schema is invalid.

Any inference of the form A, B/C can be rewritten in the form A ∧ B/C, the two inferences being equivalent in the sense that either both are valid or both are invalid. This is because the rows of the truth table in which A ∧ B is true are exactly those rows in which A and B are both true. Hence C is true in all of the rows in which A ∧ B is true (and hence A ∧ B/C is valid) if and only if C is true in all of the rows in which A and B are both true (and hence A, B/C is valid).

This result can be extrapolated to inferences with any finite number of premisses. Thus the *n*-premiss inference

$$P_1, P_2, \ldots, P_n/C$$

is valid if and only if the one-premiss inference

$$P_1 \wedge P_2 \wedge \cdots P_n/C$$

is valid.

Exercise 2.6

*(1) Show that the inference schema

$$\begin{array}{c} \sim A \vee B \\ \underline{\sim B \vee C} \\ \sim A \vee C \end{array}$$

is valid.

(2) Determine which of the following inferences are valid, and which invalid:

*(a) A ∧ B/A
 (b) A/A ∧ B
*(c) A ∨ B/A
 (d) A/A ∨ B
*(e) A ∧ B/A ∨ B
 (f) A ∨ (B ∧ C)/A ∨ C
*(g) ∼A ∨ B, B/A
*(h) A ∨ B, ∼A ∨ C, ∼B ∨ D/C ∨ D
 (i) A ∨ ∼(B ∧ C), B ∨ C/A

(3) Show that for any schemas A, B, and C,

*(a) A ∨ B ⊨ C if and only if both A ⊨ C and B ⊨ C.
 (b) A ⊨ B ∧ C if and only if both A ⊨ B and A ⊨ C.
*(c) If either A ⊨ B or A ⊨ C then A ⊨ B ∨ C, but it is possible to have A ⊨ B ∨ C without having either A ⊨ B or A ⊨ C.
 (d) If either A ⊨ C or B ⊨ C then A ∧ B ⊨ C, but it is possible to have A ∧ B ⊨ C without having either A ⊨ B or A ⊨ C.
 (e) A ⊨ B if and only if ∼B ⊨ ∼A.

2.7 TAUTOLOGIES AND CONTRADICTIONS

Once negation has been introduced, we can construct schemas which correspond to the logically true and logically false statements discussed in Section 1.3. The simplest possibilities are the schemas:

$$(1) \quad A \lor \sim A$$
$$(2) \quad A \land \sim A$$

The truth tables for these schemas are

A	$\sim$A	A $\lor \sim$A	A $\land \sim$A
T	F	T	F
F	T	T	F

Whatever truth value A has, $A \lor \sim A$ comes out true, whereas $A \land \sim A$ comes out false. For the schema $A \lor \sim A$ is true so long as at least one of A and $\sim A$ is true, and this is bound to be the case, since if A is not true, then $\sim A$ is true. Similarly, the schema $A \land \sim A$ is true so long as *both* A and $\sim A$ are true, and this is bound *not* to be the case, since if A is true then $\sim A$ is false, and *vice versa*.

Since the truth value of $A \lor \sim A$ is independent of how the schematic letter A is interpreted, being determined purely from the logical structure of the schema, this schema can be regarded as logically true. Similarly, $A \land \sim A$ can be regarded as logically false. Logically true and logically false schemas of the propositional calculus are known as *tautologies* and *contradictions* respectively.

A **tautology** is a propositional schema which is satisfied by every interpretation

A **contradiction** is a propositional schema which is falsified by every interpretation

---

We also apply these terms to interpretations of such schemas as English statements, so that, for example, the statement 'John is married or John is not married' is a tautology (since it is an interpretation of $A \lor \sim A$), and 'John is married and John is not married' is a contradiction (interpretation of $A \land \sim A$).

Another example of a tautological schema is

$$A \lor (\sim A \land B) \lor (\sim A \land \sim B);$$

and of a contradictory one:

$$A \land B \land (\sim A \lor \sim B).$$

Tautologies and contradictions stand in a special relation to inference. Consider an inference of the form

$$P_1, \ldots, P_n / C$$

in which the conclusion C is a tautology. For this inference to be invalid, there would have to be a line in the truth table in which $P_1, \ldots, P_n$ are all true, and C is false. But since C is a tautology, C *cannot* be false; and this means that the inference must be *valid*. We have just succeeded in showing that

---

Any inference whose conclusion is a tautology is valid

This may seem paradoxical, especially when we are given actual inferences and not just inference schemas:

> Leeds is in Yorkshire
> Exeter is in Devon
> _____
> Either Cambridge is in Kent or Cambridge is not in Kent

Since the conclusion of this inference is a tautology, the inference must be a valid one. But no one in their right mind would argue in this way: there is no point in citing the premisses as grounds for believing the conclusion when the conclusion is bound to be true anyway. What this shows us is that the class of valid inferences is rather wider than the class of *useful* valid inferences. Our definition of validity seems quite natural, and cannot really be improved upon; but definitions have a habit of surprising us, and we must always be prepared for this. At least we will not go wrong in our reasoning if we admit inferences with tautological conclusions—it's just that, usually, such inferences do not get us very far.

Since premisses are not really necessary to support the inference of a tautology, we can slightly broaden our idea of what an inference is by including the possibility of an inference with no premisses, just a conclusion. We can represent the fact that there are no premisses by writing the set of premisses as $\emptyset$, the symbol for the empty set. Thus the inference with no premisses and conclusion C is $\emptyset/C$. To say that this inference is valid, and hence that C is a tautology, we could write $\emptyset \vDash C$; but it is usual to leave out the symbol for the empty set and just write $\vDash C$. To say that C is a contradiction, we write $\vDash {\sim} C$.

An even more paradoxical consequence of our definition of validity emerges if we consider inferences with contradictory premisses. Suppose that in the inference

$$P_1, \ldots, P_n/C$$

one of the premisses P_i is a contradiction. For the inference to be invalid, there must be a line in the truth table with $P_1, \ldots, P_n$ all true and C false. But if one of the P_i is a contradiction, $P_1, \ldots, P_n$ *can't* all be true; and hence the inference must be valid.

What we have shown here is that

> Any inference which includes a contradiction amongst its premisses is valid

This is paradoxical, because the validity of the inference has nothing to do with what the conclusion is. An example is

> Exeter is in Devon and Exeter is not in Devon
> _____
> Cambridge is in Kent

The point is that as soon as you start founding your reasoning on contradictions, it loses all grip on reality and can be led in any direction you please. As logicians say, 'From a contradiction, everything follows.'

It is not necessary for any one of the premisses to be a contradiction in order to obtain this effect, in which an inference is automatically valid, whatever the conclusion is. Suppose the set $\{P_1, \ldots, P_n\}$ is unsatisfiable, so that its members cannot all be true together. Then the conjunction $P_1 \wedge \cdots \wedge P_n$ must be a contradiction, so the inference $P_1 \wedge \cdots \wedge P_n / C$ is valid, for any C. But as we saw in the last section, if this inference is valid, so is the inference $P_1, \ldots, P_n / C$. We thus have the result

> Any inference whose premisses form an unsatisfiable set is valid

A propositional calculus schema which is neither a contradiction nor a tautology is **contingent**, because its truth value is contingent (i.e. depends) on the truth values of its constituent schematic letters.

Tautologies and contradictions give rise to a new set of absorption rules. Suppose some schema, which we abbreviate **T**, is a tautology, and hence is true for every possible assignment of truth values to its schematic letters; and let S stand for any schema whatever. Then we have the following truth tables.

S	T	S $\wedge$ T	S $\vee$ T
T	T	T	T
F	T	F	T

Note that the truth table has only two lines since the schema **T** cannot take the value 'F'. From this table, we see that, if **T** is a tautology, then S $\wedge$ **T** always has the same truth value as S, whereas S $\vee$ **T** always has the same value as **T**; so we have

> if **T** is a tautology, then
>
> $$S \wedge T \cong S$$
>
> and
>
> $$S \vee T \cong T$$

Now suppose that **F** is any contradiction, and S any schema as before. The reader should verify that

> If **F** is a contradiction, then
>
> $$S \wedge F \cong F$$
>
> and
>
> $$S \vee F \cong S$$

(This can be done either using truth tables or by derivation from the previous two equivalences.)

We remarked earlier that any truth table corresponds to a schema built up from schematic letters using $\sim$ and either $\lor$ or $\land$ (or both). We now illustrate this with an example:

A	B	C	X
T	T	T	F
T	T	F	F
T	F	T	T
T	F	F	F
F	T	T	T
F	T	F	T
F	F	T	T
F	F	F	F

In this truth table, the truth values in the X column were chosen at random. The problem is to find a schema involving just A, B and C which can stand in place of X to give the same set of truth values. We can do this by considering the truth table line by line. Note first that X is true in lines 3, 5, 6 and 7 and no others. Line 3 corresponds to A and C being true and B false, i.e. to A, $\sim$B and C all being true, i.e. to $A \land \sim B \land C$ being true. Similarly, line 5 corresponds to $\sim A \land B \land C$, line 6 to $\sim A \land B \land \sim C$, and line 7 to $\sim A \land \sim B \land C$. Since X is true so long as one of these lines holds, the schema corresponding to X is

$$(A \land \sim B \land C) \lor (\sim A \land B \land C) \lor (\sim A \land B \land \sim C) \lor (\sim A \land \sim B \land C).$$

This schema can be simplified using the equivalence you are asked to prove in Question (2a) of Exercise 2.7. First we can collapse together the two middle disjuncts to give

$$(A \land \sim B \land C) \lor (\sim A \land B) \lor (\sim A \land \sim B \land C);$$

then we can collapse together the two outer disjuncts to give

$$(\sim B \land C) \lor (\sim A \land B).$$

Looking back at the truth table, we see that the first of the two disjuncts in this simplified schema corresponds to lines 3 and 7, while the second disjunct corresponds to lines 5 and 6.

You have probably spotted that the schema

$$(\sim B \land C) \lor (\sim A \land B)$$

looks suspiciously like a DNF; and indeed, if we extend the definition of DNFs to take account of negation then that is just what it is. The extended definitions of DNF and CNF are to be found in Section 2.9.

Exercise 2.7

(1) Classify the following schemas as contingent, contradictory, or tautologous:

    *(a) $A \land \sim B$

    (b) $(A \land \sim B) \lor \sim A$

    *(c) $(A \land \sim B) \lor \sim A \lor B$

    *(d) $A \land B \land \sim A \land C$

(e) (A ∧ B) ∨ (A ∧ ∼ B)
(f) (A ∧ B) ∨ (A ∧ ∼ B) ∨ ∼ A
(g) (A ∨ B) ∧ (A ∨ ∼ B)
(h) (A ∨ B) ∧ (∨ ∼ B) ∧ ∼ A

(2) Derive the following equivalences:
   (a) (A ∧ B) ∨ (A ∧ ∼ B) ≅ A
   (b) (A ∨ B) ∧ (A ∨ ∼ B) ≅ A

(3) Show that for any schemas A and B,
   (a)   A is a tautology if and only if ∼ A is a contradiction.
   *(b) If A is a tautology, and A⊨B, then B is a tautology.
   (c)   If B is a contradiction, and A⊨B, then A is a contradiction.
   *(d) If A is a tautology, then so is A ∨ B.
   (e)   If A is a contradiction, then so is A ∧ B.
   *(f) A ∧ B is a tautology if and only if both A and B are tautologies.
   (g)   A ∨ B is a contradiction if and only if both A and B are contradictions.

2.8 *MATERIAL IMPLICATION AND MATERIAL EQUIVALENCE*

So far, we have introduced two different kinds of special logical symbol. On the one hand, we have the symbols

$$\vDash \quad \cong$$

which represent two of the fundamental logical concepts introduced in the first chapter. These symbols can be defined without reference to the propositional calculus, and in particular, *they are not truth-functional connectives*. On the other hand, we have the symbols

$$\sim \quad \wedge \quad \vee$$

which are the truth-functional connectives introduced so far in this chapter, and these of course do belong to the propositional calculus.

Note that symbols of the latter group occur *within* schemas of the propositional calculus, whereas those of the former group occur *outside* them, as in

$$(1) \quad \sim (A \vee B) \vDash \sim A.$$

Here, ∼ (A ∨ B) and ∼ A are propositional calculus schemas, but the whole entailment is a statement *about* propositional calculus schemas, it is not itself a propositional calculus schema.

Is there any way in which a statement that one schema logically implies another can be expressed by a schema of the propositional calculus? In other words, can any single propositional calculus schema say the same as (1)? It turns out that there is no exact equivalent of (1) within the propositional calculus, but that there is a schema which is related to (1), and which can do duty for (1) in many cases.

Consider first the schema

$$\sim X \vee Y,$$

where X and Y are any schemas you choose. This has the following truth table:

X	Y	$\sim X \vee Y$
T	T	T
T	F	F
F	T	T
F	F	T

The only way that $\sim X \vee Y$ can be false is if X is true and Y is false. Now suppose X logically implies Y (e.g. X could be $\sim (A \vee B)$ while Y is $\sim A$.) Then the combination of X true and Y false cannot occur, so that in effect the truth table is really

X	Y	$\sim X \vee Y$
T	T	T
F	T	T
F	F	T

You can see this by making the substitutions suggested above:

A	B	$\sim (A \vee B)$	$\sim A$	$\sim \sim (A \vee B) \vee \sim A$
T	T	F	F	T
T	F	F	F	T
F	T	F	T	T
F	F	T	T	T

Notice that the truth values of $\sim (A \vee B)$ and $\sim A$ are TT, FT, and FF, just as in the second truth table we gave for $\sim X \vee Y$. And in both these cases, the right-hand column is all Ts, which means we are dealing with a tautology. We have shown, in fact, that

> if X logically implies Y then $\sim X \vee Y$ is a tautology

(In symbols, if $X \vDash Y$ then $\vDash \sim X \vee Y$.)

Now suppose we do not know whether X logically implies Y, but we find that $\sim X \vee Y$ is a tautology. We can then reason as follows: suppose there were a combination of truth values which made X true at the same time as Y were false; then $\sim X \vee Y$ would come out false for that combination of truth values, which means that it *cannot* be a tautology. So if $\sim X \vee Y$ is a tautology then it is not possible for X to be true at the same time as Y is false, i.e. X implies Y. So we have now shown that

> if $\sim X \vee Y$ is a tautology then X logically implies Y

Putting our last two results together, we see that

> X logically implies Y if and only if $\sim X \vee Y$ is a tautology

We have not succeeded in translating the statement 'X logically implies Y' into a schema of the propositional calculus: we have shown that $X \vDash Y$ is equivalent to $\vDash \sim X \vee Y$, but the latter is no more a schema of the propositional calculus than is $X \vDash Y$, since it contains the nontruth-functional symbol $\vDash$. But we have translated a statement about *two* propositional calculus schemas (X and Y) into a statement about just *one* ($\sim X \vee Y$), so that that part of the meaning of $\vDash$ which involves linking two schemas together has been taken inside the single resulting schema.

Because of its important relation to logical implication, the combination $\sim \cdots \vee \cdots$ is replaced by a single truth-functional connective $\rightarrow$, called **material implication**. The definition of material implication is thus given by the equivalence

$$A \rightarrow B \cong \sim A \vee B,$$

and its truth table is

A	B	A → B
T	T	T
T	F	F
F	T	T
F	F	T

We thus have two kinds of implication, logical implication (i.e. entailment) and material implication, which must be carefully distinguished. The schema $A \rightarrow B$ can be read 'if A then B'; but this can be a little misleading because the range of meanings of the English link-word (or rather 'link-phrase') 'if ... then ...' far exceeds the simple truth-functional connection expressed by $\rightarrow$ (for a discussion of this, see below). Another possible reading of $A \rightarrow B$ is 'A only if B'; or, more cumbersomely, 'A materially implies B'.

IMPORTANT WARNINGS
(1)   The schema $A \rightarrow B$ is quite often read simply as 'A implies B'. *This is to be strongly discouraged.* The schema $A \rightarrow B$ only says that it is not the case that A is true and B is false; but to say that A implies B is to say something much stronger than this, namely that it is not *possible* for A to be true while B is false. Thus the word 'implies', on its own, suggests a meaning much closer to logical implication than to material implication. The connection between $\rightarrow$ and $\vDash$ is, as we have seen, that $A \vDash B$ holds if and only if $A \rightarrow B$ is a tautology. It is easy to give instances of A and B to illustrate the distinction. For example, since it is true that Exeter is in Devon and Leeds is in Yorkshire, the statement

Exeter is in Devon $\rightarrow$ Leeds is in Yorkshire

is true; but the statement

Exeter is in Devon $\vDash$ Leeds is in Yorkshire

is false, since we can easily imagine a world in which Exeter is in Devon but Leeds is not in Yorkshire. Similarly, since 'Exeter is in Yorkshire' is false, the statement

$$\text{Exeter is in Yorkshire} \rightarrow \text{Leeds is in Yorkshire}$$

is true; but again, the corresponding logical implication is false. Material implication is but a pale reflection of logical implication: whereas the latter takes into account all the possibilities, the former takes into account only what is actually the case, the material facts.

(2) Be careful to distinguish the logical use of 'if ... then ...', as a connective which forms a statement out of two given statements, from the use of 'if ... then ...' in instructions and commands. Compare, for example, the statement

$$\text{If if rains, we shall get wet}$$

with the instruction

$$\text{If it rains, take an umbrella.}$$

It is reasonable to paraphrase the former as 'Either it won't rain or we shall get wet', but we cannot paraphrase the latter as 'Either it won't rain or take an umbrella', which doesn't even make sense! It is this second use of 'if ... then ...' which is used in programming languages such as Basic and Pascal to form conditional commands, e.g.

$$\text{if } x > 0 \text{ then } x := x - 1.$$

Of course, these two uses of 'if' are not entirely unrelated; what is important to realize is that they are nonetheless distinct.

How well does the truth-functional connective $\rightarrow$ correspond to the English 'if ... then...'? Consider the statement

$$\text{(1)   If it rains we will get wet.}$$

(Note that the word 'then' is often omitted, as here.) If we understand 'if' in this statement truth-functionally, we are regarding the statement as equivalent to

$$\text{(2)   Either it will not rain or we will get wet.}$$

For this to be false, it must rain without our getting wet. With our empirical knowledge of the world, we know that this can happen, e.g. if we stay indoors or take an umbrella. To assert (1) is to imply that these falsifying circumstances will not hold. It is tempting to suppose that (1) also implies something extra, such as

$$\text{(3)   Under such-and-such circumstances, rain makes people wet.}$$

This temptation should be resisted: the truth of (3) provides us with a *reason for believing* both (1) and (2), but it would be a mistake to conclude from this that either (1) or (2) actually *implies* (3). Thus we can consistently maintain that (1) is a purely truth-functional compound of 'It will rain' and 'We shall get wet', while recognizing that the grounds we might have for asserting this compound are supplied by some causal law such as (3).

Whereas the material implication $A \rightarrow B$ merely asserts a relation between the actual truth values of A and B (namely that it is *not the case* that A is true and B false), the logical implication

A⊨B asserts a relation between all *possible* truth values of A and B (namely, that it is *not possible* for A to be true and B false). It is tempting to say that the corresponding English statement 'if A then B' generally asserts something in between these two: something stronger than A→B, because it is not merely a matter of the actual truth values, but rather it asserts that there is some connection between A and B which results in the appropriate relation between the truth values (e.g. it is not just that either it won't rain or we will get wet; but the rain, if it occurs, will *make* us wet); and something weaker than A⊨B, because the connection asserted between A and B is not so strong that under some possible circumstances it may not obtain (in other circumstances we might be carrying an umbrella).

Arguably, though, 'if A then B' only *says* the same as A→B; but to have good *grounds* for asserting it may be one should find a connection between A and B that goes deeper than just the actual truth values. Indeed, as far as the purely logical properties of 'if' are concerned, → is as good a translation as any. Thus, given an inference in English, we might test it for validity by translating the corresponding schema into the propositional calculus, and then testing that for validity.

Example 2.8.1   Consider the inference

> If it rains we will get wet
> It will rain
> ——————————
> We will get wet

To test this argument for validity, we translate it into the propositional calculus schema

$$R \to W$$
$$R$$
$$\overline{}$$
$$W$$

and then test whether *this* is valid (it is!).

Even though → may not capture everything that is implied by 'if ... then ...', at least we can say that a statement of the form A→B will be true whenever (if not more often than) 'if A then B' is true, so that the English inference with statements of this form amongst its premises will be valid so long as the propositional calculus translation is.

The cases where this doesn't seem to work well are those in which the *conclusion* of the inference is of the form 'if A then B':

> Exeter is in Devon
> Leeds is in Yorkshire
> ————————————————————
> If Exeter is in Devon then Leeds is in Yorkshire

If we translate this inference into the propositional calculus, we get

$$A$$
$$B$$
$$\overline{}$$
$$A \to B$$

and this is *valid*; yet it seems implausible to claim validity for the inference as originally stated. We raise this difficulty here, not to solve it, but because we think it important, even at this introductory stage, to draw the reader's attention to the sort of problems that can arise from an uncritical application of formal logic to everyday reasoning. Formal logic is a powerful tool, and like all powerful tools it must be used with great care!

As remarked above, a possible reading of A→B is 'A only if B'. Be careful not to confuse 'if' and 'only if' with 'if and only if'. The relationship between these three link-phrases is as follows:

'If A then B', 'B if A', and 'A only if B' are all readings of A→B
'A if and only if B' is equivalent to 'A if B, and A only if B', and hence to 'If A then B, and if B then A'.

Another reading makes use of 'unless': logically 'A unless B' means the same as 'A or B'; so 'if A then B', which is equivalent to 'not-A or B', can also be read as 'not-A unless B'.

Some of these readings will generally be more natural than others, depending on the interpretations of A and B. Thus both

$$\text{I shan't go out unless it has stopped raining}$$

and

$$\text{I shall only go out if it has stopped raining}$$

seem quite natural, whereas the logically equivalent statement

$$\text{If I go out, then it has stopped raining}$$

seems strange, possibly because there is a faint suggestion that my going out caused it to stop raining (whereas in reality it is the fact that the rain has stopped that allows me to go out).

All three types of statement,

$$A \rightarrow B$$
$$A \vDash B$$
$$\text{if A then B}$$

are called **conditional** statements. In a conditional statement, the first part (A in our examples) is called the **antecedent**, and the second part (B) is called the **consequent**.

Just as → is the nearest we can get to expressing ⊨ by means of a truth-functional connective, so we might ask how we may similarly express ≅ (equivalence). Remember that A ≅ B is true so long as A and B always agree in truth value. No propositional calculus schema can say *this*, of course; but on the other hand the schema

$$(A \wedge B) \vee (\sim A \wedge \sim B)$$

is true just when A and B do agree in truth value, so that we have a rule

$$A \cong B \text{ if and only if } \vDash (A \wedge B) \vee (\sim A \wedge \sim B),$$

i.e., in words,

> A is equivalent to B if and only if (A ∧ B) ∨ (∼ A ∧ ∼ B) is a tautology

So (A ∧ B) ∨ (∼ A ∧ ∼ B) is related to A ≅ B in exactly the same way as A→B is related to A⊨B. For this reason, we abbreviate the schema $(A \land B) \lor (\sim A \land \sim B)$ as

$$A \leftrightarrow B.$$

The new truth-functional connective $\leftrightarrow$ is called **material equivalence**. Its truth table is

A	B	A↔B
T	T	T
T	F	F
F	T	F
F	F	T

The nearest we can come to a satisfactory English reading of A↔B is 'A if and only if B'. As with $\rightarrow$ and 'if ... then ...', the correspondence is not exact and gives rise to many difficult philosophical questions about the relationship between formal logic and natural language. We shall not pursue these issues further here.

All three types of statement

$$A \leftrightarrow B$$
$$A \cong B$$
A if and only if B

are called **biconditional** statements.

To sum up the connectives introduced in this section:

An interpretation satisfies A→B if and only if it does not both satisfy A and falsify B
A⊨B if and only if ⊨A→B

An interpretation satisfies A↔B if and only if it either satisfies both A and B or falsifies both A and B

A≅B if and only if ⊨A↔B

Exercise 2.8

(1) Verify the following equivalences:

*(a) $A \leftrightarrow B \cong (A \rightarrow B) \land (B \rightarrow A)$ (b) $\sim (A \rightarrow B) \cong A \land \sim B$

(c) $A \rightarrow B \cong \sim B \rightarrow \sim A$ (d) $\sim (A \leftrightarrow B) \cong A \oplus B$

(e) $A \leftrightarrow B \cong \sim A \leftrightarrow \sim B$ *(f) $(A \land B) \rightarrow C \cong A \rightarrow (B \rightarrow C)$

(g) $A \rightarrow (B \land C) \cong (A \rightarrow B) \land (A \rightarrow C)$ (h) $(A \lor B) \rightarrow C \cong (A \rightarrow C) \land (B \rightarrow C)$

(i) $A \rightarrow (B \lor C) \cong (A \rightarrow B) \lor C$ (j) $\sim (A \land B) \cong A \rightarrow \sim B$

(2) Show that the following inferences are all valid:

*(a) A→B   (b) A→B   (c) A→B
     A   B→C   ~B
     ___   ___   ___
     B   A→C   ~A

*(d) A∨B   (e) A→B   (f) A
     A→C   A→~B   ~A
     B→C   ___   ___
     ___   ~A   B
     C

(3) Classify the following schemas as contingent, contradictory, or tautologous:

   *(a) $A \rightarrow (A \rightarrow B)$ (b) $A \rightarrow (B \rightarrow A)$

   (c) $A \rightarrow A$ *(d) $\sim A \rightarrow A$

   (e) $\sim A \leftrightarrow A$ * (f) $(\sim A \rightarrow A) \rightarrow A$

   (g) $(\sim A \rightarrow A) \rightarrow \sim A$ *(h) $(\sim A \rightarrow A) \leftrightarrow (A \rightarrow \sim A)$

   (i) $A \vee (A \rightarrow B)$ *(j) $(A \rightarrow B) \vee (B \rightarrow A)$

   (k) $(A \leftrightarrow B) \vee (A \leftrightarrow \sim B)$ (l) $A \leftrightarrow (\sim A \leftrightarrow A)$

(4) For each of the statements below, discuss to what extent it is reasonable to regard 'if ... (then) ...' as purely truth-functional.

(a) If my telephone number is a multiple of four then it is even.

(b) If that car is for sale, then I shall buy it.

(c) If that's not stealing then I don't know what stealing is.

(d) If you get a first, I'll eat my hat.

(e) If you pay by cash you can have a discount.

(f) If a number if even then all its multiples are even.

(g) If you believe that, you'll believe anything.

(5) Translate each of the following inferences into propositional calculus schemas, and determine whether or not it is valid:

   *(a) If it rains and I do not take an umbrella, I shall get wet

       I shall take an umbrella

      —————————————————————

       I shall not get wet

   *(b) It will not rain

       I shall fall into the canal

       If I fall into the canal I shall get wet

      —————————————————————

       If it doesn't rain I shall get wet

   (c) If you eat wild fungi and feel ill, it is necessary to call a doctor

      —————————————————————

       If you eat wild fungi and do not feel ill, it is not necessary to call a doctor

   (d) If God exists then He is both loving and omnipotent.

       If God is unable to eliminate suffering then He is not omnipotent.

       If God is unwilling to eliminate suffering then He is not loving.

       If God were both able and willing to eliminate suffering then there would be no suffering.

       There is suffering.

      —————————————————————

       God does not exist

       (I got this example from Peter Millican; I don't know where he got it from)

   (e) Either Anne or Barbara will go.

       If Anne goes, either Charles or David will go.

       If Charles goes, Elizabeth and Fiona will go too.

       If Fiona goes but David doesn't, Elizabeth won't either.

      —————————————————————

       If David doesn't go, Barbara will.

(f) You will only get a degree if you have satisfied the residence requirements of the university

You will not get a degree unless you pass your exams

You have satisfied the residence requirements

You have passed your exams

You will get a degree

*(g) If this inference is valid and the premises are all true, then the conclusion is true too.

The premises of this inference are all true.

The conclusion of this inference is not true.

This inference is not valid.

2.9 MORE ON NORMAL FORMS

Our original definitions of DNF and CNF were for schemas built up from schematic letters using conjunction and disjunction only. Now that we have introduced negation as well, the definition must be extended. We shall insist, in order for a schema containing negations to be in normal form, that all occurrences of $\sim$ must have as small a scope as possible. Thus

$$\sim (A \wedge B \wedge C)$$

may not occur as part of a normal form, but the equivalent

$$\sim A \vee \sim B \vee \sim C$$

may.

For DNFs, the extended definition runs as follows:

A **literal** is any schematic letter or negated schematic letter.

A **minterm** is either a literal, a conjunction of distinct literals none of which is the negation of any other, or one of the symbols **F** and **T**.

A minterm M_1 **absorbs** a minterm M_2 if every literal in M_1 is also in M_2; and also every minterm absorbs **F** and is absorbed by **T**.

Two minterms M_1 and M_2 **coalesce** to a new minterm M_3 if M_1 and M_2 each contains all the literals in M_3 plus one other, the extra literal in one being the negation of the extra literal in the other; and also any literal and its negation coalesce to yield the minterm **T**.

A **DNF** is any disjunction of minterms none of which absorbs or coalesces with any of the others.

Examples (and some non-examples) of each of these terms are:

Literals: A, $\sim$ A, B, $\sim$ B.

Minterms: A, $\sim$ A, A $\wedge \sim$ B, $\sim$ A $\wedge$ B $\wedge$ C, **F**, **T**, but not A $\wedge$ B $\wedge \sim$ B.

Absorption: Each minterm in the following list absorbs all the minterms which follow it: **T**, A, A $\wedge$ C, A $\wedge \sim$ B $\wedge$ C, **F**.

Coalescence: $A \wedge \sim B \wedge \sim C \wedge D$ and $A \wedge \sim B \wedge C \wedge D$ coalesce to give $A \wedge \sim B \wedge D$, and A and $\sim A$ coalesce to give **T**.

DNF: $(A \wedge \sim B) \vee (B \wedge C \wedge \sim D) \vee (A \wedge D)$, **T**, and **F** are all DNFs, but $(A \wedge \sim B) \vee (A \wedge \sim B \wedge C) \vee (B \wedge C \wedge D)$, $(A \wedge \sim B \wedge C) \vee (A \wedge \sim B \wedge \sim C) \vee (A \wedge B \wedge C)$, $A \vee F$ and $\sim B \vee$ **T** are not.

It is possible to have equivalent DNFs built out of distinct sets of minterms; if we take our second non-example above,

$$(A \wedge \sim B \wedge C) \vee (A \wedge \sim B \wedge \sim C) \vee (A \wedge B \wedge C)$$

and coalesce the first two minterms, we get the DNF

$$(A \wedge \sim B) \vee (A \wedge B \wedge C),$$

whereas if we coalesce the first and the last minterms we get

$$(A \wedge C) \vee (A \wedge \sim B \wedge \sim C).$$

These two DNFs are equivalent, and correspond to two different groupings of the T lines in the truth table.

You may have been puzzled by references to **T** and **F** as minterms in the new definition of DNFs. These terms are needed so that tautologies and contradictions can have DNFs. Without admitting **T** and **F** as minterms, $A \vee \sim A$, $B \vee \sim B$, $A \vee B \vee \sim B$, $A \vee B \vee C \vee \sim C$, $A \vee D \vee \sim D$, etc would all be DNFs, and they would all be equivalent since they are all tautologies; replacing them all by the single term **T** makes everything much simpler. Again, if we did not allow **F** as a minterm, there would be no DNF equivalent to the contradiction $A \wedge \sim A$, which itself fails to be a DNF because it is not even a minterm (a minterm cannot contain a literal together with its negation): so **F** plays the role of the DNF of a contradiction.

The algorithm for converting any schema into DNF consists of three main phases: first, all connectives apart from $\wedge$, $\vee$ and $\sim$ must be eliminated by means of the equivalences

(1) $A \rightarrow B \cong \sim A \vee B$

(2) $A \leftrightarrow B \cong (A \wedge B) \vee (\sim A \wedge \sim B)$.

Next, the negation signs must be pushed inwards so that their scopes are made as small as possible; this is achieved by means of the De Morgan laws:

(3) $\sim (A \wedge \cdots \wedge B) \cong \sim A \vee \cdots \vee \sim B$

(4) $\sim (A \vee \cdots \vee B) \cong \sim A \wedge \cdots \wedge \sim B$.

If at any stage two consecutive negation signs appear, they may be eliminated using the equivalence

(5) $\sim \sim A \cong A$

Our schema will now be built up from literals using $\wedge$ and $\vee$ alone, and hence we can apply the algorithm we had before for the restricted form of DNF (the only difference being that now some of the literals will consist of a negation sign and a schematic letter, instead of just being a schematic letter pure and simple). We thus have:

Algorithm *To convert any propositional calculus schema into disjunctive normal form*

(1) Eliminate all occurrences of → and ↔ by means of the equivalences (1) and (2) above.
(2) Use equivalences (3), (4) and (5) to reduce the scope of each occurrence of ~ to a single literal.
(3) Eliminate any disjunctions coming within the scope of a conjunction by distributing the conjunction over them.
(4) Tidy up the resulting schema by eliminating any duplicated literals within any minterm, eliminating duplicated minterms, replacing any contradictory minterm by **F**, and absorbing or coalescing minterms wherever possible.

We have not given this algorithm to the same degree of refinement as we did in the previous case; many different strategies are possible for the detailed execution of the algorithm, and we leave this as an exercise to the reader. Try refining the algorithm to the point where you can convert it into a computer program which you can actually run. Here we shall merely illustrate the working of the algorithm by means of an example.

We shall use the algorithm to convert into DNF the schema

$$((A \rightarrow B) \rightarrow C) \leftrightarrow (A \rightarrow C).$$

Step (1): We eliminate the deepest occurrences of → to give

$$((\sim A \vee B) \rightarrow C) \leftrightarrow (\sim A \vee C),$$

and then eliminate the remaining → to give

$$(\sim (\sim A \vee B) \vee C) \leftrightarrow (\sim A \vee C).$$

Finally, elimination of ↔ results in

$$((\sim (\sim A \vee B) \vee C) \wedge (\sim A \vee C)) \vee (\sim (\sim (\sim A \vee B) \vee C) \wedge \sim (\sim A \vee C)).$$

Step (2): Pushing the outermost negations inwards first gives us

$$(((\sim \sim A \wedge \sim B) \vee C) \wedge (\sim A \vee C)) \vee (\sim \sim (\sim A \vee B) \wedge \sim C \wedge \sim \sim A \wedge \sim C)$$

which, after elimination of the double negations is

$$(((A \wedge \sim B) \vee C) \wedge (\sim A \vee C)) \vee ((\sim A \vee B) \wedge \sim C \wedge A \wedge \sim C).$$

Step (3): We now distribute conjunction over disjunction wherever possible. This gives us

$$(A \wedge \sim B \wedge \sim A) \vee (A \wedge \sim B \wedge C) \vee (C \wedge \sim A) \vee (C \wedge C)$$
$$\vee (\sim A \wedge \sim C \wedge A \wedge \sim C) \vee (B \wedge \sim C \wedge A \wedge \sim C).$$

Step (4): Eliminating duplicated literals gives us

$$(A \wedge \sim B \wedge \sim A) \vee (A \wedge \sim B \wedge C) \vee (\sim A \wedge C) \vee C \vee (\sim A \wedge A \wedge \sim C) \vee (B \wedge \sim C \wedge A).$$

Several of the minterms contain contradictory literals, and hence collapse to **F**:

$$\textbf{F} \vee (A \wedge \sim B \wedge C) \vee (\sim A \wedge C) \vee C \vee \textbf{F} \vee (B \wedge \sim C \wedge A).$$

We can drop the **F**s, since they are absorbed by the other minterms; also, the two minterms $(A \wedge \sim B \wedge C)$ and $(\sim A \wedge C)$ are absorbed by the minterm C, so we have

$$C \vee (B \wedge \sim C \wedge A)$$

as our final DNF. The reader should verify by means of a truth table that this DNF is indeed equivalent to the schema we started with.

It should be noticed that although our algorithm always delivers a DNF for the input schema, this is not always the simplest possible DNF. In particular, if the input schema is already a DNF, it will pass through the algorithm unchanged; so even if there exists a simpler DNF that is equivalent to it, this will not be found by the algorithm. An example is the DNF $(A \wedge \sim B) \vee B$, which is equivalent to the simpler DNF $A \vee B$, although our algorithm won't reduce the former to the latter. Because of this, the algorithm might fail to reduce a tautological schema to **T**: for example, the tautology $(A \rightarrow B) \rightarrow (A \rightarrow B)$ gets reduced to the DNF $(A \wedge \sim B) \vee \sim A \vee B$, but this is not then reduced to **T**, although it is equivalent to it.

We could try to remedy this by 'patching up' the algorithm, for example by adding to the final tidying up stage a rule allowing us to eliminate from a minterm any literal whose complement appears as a minterm elsewhere in the schema, the justifying equivalences being

$$(A \wedge \sim B) \vee B \cong A \vee B$$
$$(A \wedge B) \vee \sim B \cong A \vee \sim B$$

Unfortunately, even this will not work in every case, and in fact no method is known which is guaranteed to deliver the *simplest* DNF.

Exercise 2.9

(1) Use truth tables to verify that the DNF $C \vee (A \wedge B \wedge \sim C)$ is equivalent to the schema $((A \rightarrow B) \rightarrow C) \leftrightarrow (A \rightarrow C)$.

(2) Convert the following schemas into DNF:

    *(a) $(A \rightarrow B) \rightarrow C$
    *(b) $(A \vee B) \rightarrow (C \wedge D)$
    *(c) $\sim ((A \rightarrow B) \wedge A \wedge \sim B)$
    (d) $(A \leftrightarrow B) \wedge (A \vee C \rightarrow B)$
    (e) $(A \rightarrow B) \leftrightarrow (B \rightarrow A)$
    (f) $(A \rightarrow B) \vee (B \rightarrow A)$
    (g) $A \wedge \sim (B \rightarrow A)$

(3) Define conjunctive normal form (CNF) for general propositional calculus schemas, and construct an algorithm for converting any schema into CNF.

(4) *Project*: Write a computer program to convert propositional calculus schemas into DNF. Use arguments similar to the ones given in Section 1.4 to verify that your program is correct.

2.10 THE PROPOSITIONAL CALCULUS AS A FORMAL SYSTEM

We have now introduced all the main ingredients of the propositional calculus. We have done so in a somewhat piecemeal fashion, introducing concepts one at a time, and pausing for fairly detailed explanations to allow the reader to absorb the material thoroughly. Now it is time to bring it all together into a single coherent system, which we shall call **Prop**. This is a *formal* system, which means that it is defined in a mathematically precise and rigorous fashion, in which intuition may aid our understanding but is never the final arbiter.

The fundamental units of **Prop** are the *schemas*, and the first thing we must do in setting up the formal system is to provide a precise definition of what counts as a schema. In **Prop**, schemas are constructed out of a stock of primitive symbols collectively known as the *lexicon*:

> *Lexicon of* **Prop**
>
> Schematic letters: $P_0 P_1 P_2 P_3 \cdots$
>
> Connectives: $\sim \ \wedge \ \vee \ \rightarrow \leftrightarrow$
>
> Parentheses: $(\quad)$

There is no limit on the number of schematic letters that are allowed: there may even be infinitely many of them, one for each natural number. Schemas are now defined recursively from the lexicon by means of the following *rules of formation* or *syntactic rules*:

> *Rules of formation for* **Prop**
>
> (1) If P_i and P_j are schematic letters, then P_i, $\sim P_i$, $(P_i \wedge P_j)$, $(P_i \vee P_j)$, $(P_i \rightarrow P_j)$, $(P_i \leftrightarrow P_j)$ are all schemas.
>
> (2) If S is a schema, then so is any substitution-instance of S obtained by substituting schemas for the schematic letters in S.
>
> (3) Nothing is a schema unless it can be generated by the rules 1 and 2 above.

To illustrate the rules of formation, we shall build up the schema

$$((P_0 \wedge (P_1 \vee P_2)) \rightarrow \sim P_3).$$

By rule 1, there is a schema $(P_0 \rightarrow P_3)$. By rule 2, any substitution instance of this schema is a schema; in particular, the substitution $P_0/(P_0 \wedge P_1)$ gives us the schema $((P_0 \wedge P_1) \rightarrow P_3)$. Finally, we apply to *this* schema the substitutions $P_1/(P_1 \vee P_2)$ and $P_3/\sim P_3$ to give the desired schema.

Schemas formed according to these rules tend to be little unreadable. For the sake of readability, it is customary to relax the rules slightly, in two ways.

(1) We allow different schematic letters such as P, Q, R, S, ... This allows us to avoid using subscripts to a large extent.

(2) We adopt two conventions allowing us to leave out all but a few of the parentheses:

(a)   If a schema begins with an opening parenthesis and ends with the corresponding closing parenthesis, then these parentheses may be dropped. Thus for example, $(P \wedge Q)$ and $((P \wedge Q) \vee R)$ may be written as $P \wedge Q$ and $(P \wedge Q) \vee R$ respectively. Note that we cannot drop the outermost parentheses from $(P \wedge Q) \vee (P \wedge R)$, since these do not form a pair.

(b)   We shall regard the connectives $\vee$ and $\wedge$ as binding their arguments more strongly than $\to$ and $\leftrightarrow$ do, so that, for example,

$$P \to Q \vee R$$

is to be understood to mean the same as $P \to (Q \vee R)$, and not $(P \to Q) \vee R$.

With these conventions, the schema we constructed above can be written much more clearly as

$$P \wedge (Q \vee R) \to\, \sim S.$$

Readers familiar with the Backus–Naur form (BNF) used for specifying the syntax of programming languages may prefer to see the rules of formation for PC written as follows:

$$\langle \text{schema} \rangle ::= \langle \text{schematic letter} \rangle |\sim \langle \text{schema} \rangle | (\langle \text{schema} \rangle \wedge \langle \text{schema} \rangle)|$$
$$(\langle \text{schema} \rangle \vee \langle \text{schema} \rangle)|(\langle \text{schema} \rangle \to \langle \text{schema} \rangle)|$$
$$(\langle \text{schema} \rangle \leftrightarrow \langle \text{schema} \rangle)$$
$$\langle \text{schematic letter} \rangle ::= P_0 |P_1 |P_2 |P_3 | \cdots$$

This way of presenting the rules is more suitable for checking whether or not a given string of symbols is a schema, whereas the way we used before is more suitable for actually constructing schemas.

We now define the *formal semantics* of **Prop** as follows. An **interpretation** I is *any subset of the schematic letters*. The members of I are the schematic letters which are to count as satisfied by the interpretation. Corresponding to an interpretation, we can define an **interpretation function**, which maps each schematic letter onto one of the truth-values T and F. The interpretation function corresponding to the interpretation I is the function which maps the members of I onto T and all other schematic letters onto F.

We next define that what it means for an interpretation to **satisfy** a schema. We write $\vDash_I S$ to mean that the interpretation I satisfies the schema S, and $\nvDash_I S$ to mean that it does not. In the table, 'iff' is used as the customary abbreviation of 'if and only if':

---

(1)   For a schematic letter A, $\nvDash_I A$ iff $A \in I$.

(2)   For a negation $\sim S$, $\vDash_I \sim S$ iff $\nvDash_I S$.

(3)   For a conjunction $S \wedge T$, $\vDash_I S \wedge T$ iff both $\vDash_I S$ and $\vDash_I T$

(4)   For a disjunction $S \vee T$, $\vDash_I S \vee T$ unless both $\nvDash_I S$ and $\nvDash_I T$

(5)   For a conditional $S \to T$, $\vDash_I S \to T$ unless both $\vDash_I S$ and $\nvDash_I T$

(6)   For a biconditional $S \leftrightarrow T$, $\vDash_I S \leftrightarrow T$ iff both $\vDash_I S \to T$ and $\vDash_I T \to S$

These rules encapsulate in a rather terser form the meanings of the connectives given by the truth tables. To illustrate them, consider the interpretation $I = \{P, Q\}$. We shall determine whether or not I satisfies the schema $P \vee R \to \sim Q \wedge S$. By Rule (1), we have $\vDash_I P$, so by Rule (4), $\vDash_I P \vee R$. Again, since by Rule (1) we have $\vDash_I Q$, by Rule (2) we have $\nvDash_I \sim Q$ and so by Rule (3), $\nvDash_I \sim Q \wedge S$. Finally, since $\vDash_I P \vee R$ and $\nvDash_I \sim Q \wedge S$, it follows by Rule (5) that $\nvDash_I P \vee R \to \sim Q \wedge S$.

We conclude this chapter by summarizing some of the logical terms introduced so far, as they apply to **Prop**. In what follows, A stands for any schema, **S** for any set of schemas, I for any interpretation.

(1) A is **satisfied** by I if $\vDash_I A$.
(2) A is **falsified** by I if $\nvDash_I A$.
(3) **S** is satisfied by I if every member of **S** is satisfied by I.
(4) **S** is falsified by I if at least one member of **S** is falsified by I.
(5) **S** (or A) is **satisfiable** if there is at least one interpretation that satisfies it.
(6) **S** (or A) is **unsatisfiable** if there are no interpretations that satisfy it.

Satisfaction and falsification are the formal equivalents of *truth* and *falsity* respectively: sometimes we speak of a schema being *true* in an interpretation instead of satisfied by it; and likewise *false* instead of falsified. Similarly, satisfiability and unsatisfiability are the formal analogues of consistency and inconsistency.

(7) **S entails** A, written $S \vDash A$, if every interpretation which satisfies **S** also satisfies A.
(8) Two schemas A and B are **equivalent**, written $A \cong B$, if every interpretation either satisfies them both or falsifies them both.
(9) A is a **tautology**, written $\vDash A$, so long as it is satisfied by every interpretation; or equivalently, if $\sim A$ is unsatisfiable.
(10) A is a **contradiction** so long as it is falsified by every interpretation; or equivalently, if A is unsatisfiable.

The reader should verify that all these formal definitions agree, in substance, with what has been presented more informally in earlier sections. The point of formalizing everything in this way is that it makes propositional logic amenable to all the rigour of a mathematical style of reasoning. Since this is only an introductory textbook, we shall continue to temper the severity of the formal notations with a good deal of expository material in a more informal style. But having got this far, the reader should expect to see rather more formal notations in the following chapters than we have made use of up to now. For further information, the reader should consult a more advanced textbook such as Gallier (1987).

Exercise 2.10

(1) Decide which of the following is a well-formed schema of **Prop**, assuming that none of the bracket-dropping conventions is to be used. Which ones become acceptable if bracket-dropping is allowed?

*(a) $((P_0 \wedge P_1) \vee \sim P_2)$.   *(b) $(P_1 \vee P_2 \vee P_3)$.
(c) $\sim (P_1 \wedge P_4) \leftrightarrow P_6$.   (d) $(\sim P_1 \wedge P_2)$.
*(e) $(P_1 \to (\sim (P_2 \vee P_0)))$.   *(f) $\sim \sim \sim \sim P_3$.
(g) $\sim (\sim P_1 \wedge \sim P_8)$.   (h) $((P_1 \vee P_2) \vee P_3))$.
(i) $(P_1 \to P_2 \vee P_3)$.   *(j) $P_1 \vee P_2 \to P_3 \wedge P_4$.
*(k) $P_1 \to P_2 \to P_3$.   (l) $P_1 \vee P_2 \wedge P_3$.

(2) The following schemas are all well-formed, given the bracket-dropping conventions. In each case, restore all the dropped brackets.

   (a) P ∨ Q → R
*(b) P ∨ Q ↔ Q ∨ P
   (c) ~ P ∨ Q ∨ ~ (R → S)
*(d) P ∨ (Q ∧ ~ R) → P ∨ Q
   (e) P ∧ (~ Q → R) → ~ P ∨ Q ∨ S.

(3) The interpretation *I* satisfies the schemas P, Q and S, but falsifies R. Determine which of the following schemas are satisfied by *I*:

*(a) P ∧ (Q ∨ R)   *(b) P ∨ (Q ∧ R)
   (c) P → Q ∧ R   *(d) P ∨ Q → R ∧ S
   (e) P ∧ Q → R ∨ S   *(f) P ∧ ((Q → R) ∨ S)
   (g) ~ S ∨ ~ (P → Q)   (h) (P → R) → (Q → S)
*(i) (Q → S) → (P → R)   (j) (P ↔ ~ Q) ∨ (R ↔ ~ S)

(4) In this question, **S** stands for a set of schemas, while A, B, C,... stand for individual schemas. Show that

*(a) **S**⊨A if and only if **S**∪{ ~ A} is unsatisfiable.
*(b) **S**⊨A → B if and only if **S**∪{A}⊨B.
   (c) if **S**⊨A → B and **S**⊨A then **S**⊨B.
   (d) If **S**⊨A and **S**⊨ ~ A then **S** is unsatisfiable.
   (e) If **S**∪{B}⊨A and **S**∪{B}⊨ ~ A then **S**⊨ ~ B.
   (f) For any schemas A, B, and C, {A → C, B → C}⊨A ∨ B → C and {A → B, A → C}⊨A → B ∧ C.

*(5) A set of schemas is said to be **minimally unsatisfiable** so long as it is itself unsatisfiable, but all of its proper subsets are satisfiable. Determine which of the following sets are minimally unsatisfiable:

   (a) {P → ~ Q, P → ~ R, P ∧ Q, R}
   (b) {P → ~ Q, P → ~ R, Q ∨ R, P}
   (c) {P → Q, P → R, Q ∧ R, ~ P}

Show that any finite unsatisfiable set has a minimally unsatisfiable subset. (This result is also true for infinite sets, but the proof is harder: interested readers might like to investigate this.)

3 Proof Systems for the Propositional Calculus

3.1 BEYOND TRUTH TABLES: DEDUCTION AND REFUTATION

Given an inference in **Prop**, how can we tell whether it is valid or not? So far, the only method available to us is to draw up a truth table for the constituent schemas, and check that there is no line of the table in which the premisses of the inference are true, but the conclusion is false. We have here a simple *inference-checking algorithm*. This algorithm is correct in the sense that we know that if it is performed correctly then it will give us a correct answer; but it is rather inefficient. To see how inefficient it is, we note that if the number of distinct schematic letters occurring in the inference is n, then the truth table will have 2^n rows. Since, in general, the size of the inference is roughly proportional to the number of distinct schematic letters it contains, this means that the time taken to write out the complete truth table will be an exponential function of the size of the inference; and tasks which take an exponential time are not usually considered practicable for any but the simplest instances.

In order to increase the efficiency of our inference-checking algorithm, we must somehow avoid writing out the full truth table. There are a number of ways of doing this. In testing the validity of the inference

$$P_1, \ldots, P_n / C,$$

we are only interested in whether there are any rows of the truth table in which $P_1, \ldots, P_n$ are all true but C is false. In particular, therefore, we are only interested in rows in which C is false. So as a first step towards improving the efficiency of the method, we could write out the full truth table just for C and then strike out all the rows in C is true before going any further (if C is true in every row then C is a tautology, so the inference is automatically valid). We then have to check whether all the P_i are true in any of the remaining rows.

For example, consider the inference

$$P \wedge Q \rightarrow R, S \wedge \sim Q \rightarrow P, R \rightarrow Q \wedge S / R \rightarrow P.$$

The truth table for $R \rightarrow P$ is

P	R	R→P
T	T	T
T	F	T
F	T	F
F	F	T

From what we said above, we need only consider the third row of this table, i.e., the case of P false and R true. So for the whole inference the relevant part of the complete truth table is

P	Q	R	S	P∧Q→R	S∧~Q→P	R→Q∧S
F	T	T	T	T	T	T
F	T	T	F	T	T	F
F	F	T	T	T	F	F
F	F	T	F	T	T	F

The first line has all the premisses of the inference true and the conclusion false, and hence the inference is invalid. And we have discovered this by writing down two truth tables with a total of 16 entries, as opposed to the 64 entries we should have written down if we had constructed the complete truth table. (By 'entries' here I mean *computed* entries, not the ones in the columns headed by schematic letters.)

We could simplify matters still further by cutting out some of the lines of the second truth table. Observe, for example, that R → Q ∧ S is only true in the first row of this table, so we need only consider this row when computing the truth values of the other schemas. At a stroke, that eliminates six entries, bringing the total we have to compute down to ten.

Of course, we would be lucky to hit on R → Q ∧ S as the premiss to evaluate first. If we had started with P ∧ Q → R, no lines would be eliminated. So we cannot be sure, beforehand, of eliminating the greatest number of lines. Thus although we can, with care and a little luck, reduce the amount of work involved in validating inferences by means of truth tables, there is a limit to how much reduction can be *guaranteed*. Moreover, the method of truth tables is somewhat unnatural in the sense that it does not relate very directly to our intuitive understanding of what the schemas mean. It would be much better if we could find some other more congenial way of validating inferences. In fact, there are a number of such methods in existence: they are called **proof systems** for **Prop**. In this chapter we shall examine three such systems.

It is only fair to inform the reader that although the proof systems we shall examine below are in general more efficient than using truth tables for validating inferences, even for them there will always be *some* inferences whose validation (or invalidation) takes exponential time. No proof system for **Prop** is known which is guaranteed to determine the validity of an arbitrary inference in less than exponential time. On the other hand, neither is it known for sure that such a proof system cannot exist! For more on this, see Section 3.4.

Exercise 3.1 Return to the inferences in Exercise 2.8, Question 2, and see how far it is possible to cut down on the number of computed truth table entries that are necessary for determining validity, in the way suggested in this section.

3.2 NATURAL DEDUCTION

Proof systems may be broadly classified into **deduction systems** and **refutation systems**. In a deduction system, the inference $P_1, \ldots, P_n/C$ is validated by *deducing* the conclusion C directly from the premiss-set $\{P_1, \ldots, P_n\}$ by means of a set of *inference rules*. In a refutation system, on the other hand, we make use of the connection between validity and unsatisfiability noted in Chapter 1: C is a logical consequence of $\{P_1, \ldots, P_n\}$ just so long as $\{P_1, \ldots, P_n, \sim C\}$ is unsatisfiable. So a refutation system works by trying to *refute* the set $\{P_1, \ldots, P_n, \sim C\}$, i.e. to show that it is unsatisfiable. (Note that in popular usage the word 'refute' is often used with the meaning of 'deny'. This is quite incorrect. To refute a claim is not merely to assert that it is false; it is to *show* that it is false.) The proof system we introduce in this section is an example of a deduction system; refutation systems are treated in Sections 3.3 and 3.6.

In **natural deduction**, the validation of an argument is presented in the form of a **derivation** of the conclusion from the premises. Each step in the derivation involves either asserting one of the premises or using an inference rule to derive a new schema from earlier ones already established earlier in the proof. In the system we shall study, there are two kinds of inference rules, known as **introduction rules** and **elimination rules**, one (or sometimes two) of each for each connective of **Prop**.

The simplest rules allow us to pass directly from given schemas to new schemas. Consider, for example, the rules for handling conjunction ($\wedge$). The $\wedge$-**introduction** rule says that if at earlier stages of the derivation you have been able to assert the schemas A and B (either as premises or by derivation from still earlier schemas), then you may now derive the schema $A \wedge B$. Symbolically, this rule can be written:

$$(\wedge\text{-intro}) \quad \begin{array}{|l} A \\ B \\ \hline A \wedge B \end{array}$$

The $\wedge$-**elimination** rule says that if at an earlier step you have asserted the schema $A \wedge B$, then you may now assert either A or B. In fact, we have here two rules, one for A and one for B:

$$(\wedge\text{-elim-left}) \quad \begin{array}{|l} A \wedge B \\ \hline A \end{array}$$

$$(\wedge\text{-elim-right}) \quad \begin{array}{|l} A \wedge B \\ \hline B \end{array}$$

The next two examples illustrate how these rules are used in constructing a derivation.

Example 3.2.1 Validate the inference $P \wedge Q/Q \wedge P$.
At Step (1), we assert $P \wedge Q$, the justification for doing so being that $P \wedge Q$ is a premiss:

$$(1) \quad P \wedge Q \quad \text{(premiss)}$$

Then we invoke the $\wedge$-elimination rules to separate the conjuncts:

$$(2) \quad P \quad (1, \wedge\text{-elim-left})$$
$$(3) \quad Q \quad (1, \wedge\text{-elim-right})$$

The annotation at the right of each step gives the justification for the assertion made at that step. For example, the assertion of P at step 2 is justified by the application of the inference rule ∧-elim-left to the schema P ∧ Q already asserted at step 1.

Finally we use ∧-introduction to recombine the two schemas in the reverse order:

$$\text{(4)} \quad Q \wedge P \qquad (3, 2, \wedge\text{-intro})$$

Note that, in the annotation, we cite the earlier steps 3 and 2 in the order corresponding to the ordering of the conjuncts in the new schema.

Collecting together the steps, we have

(1)	P ∧ Q	(premiss)
(2)	P	(1, ∧-elim-left)
(3)	Q	(2, ∧-elim-right)
(4)	Q ∧ P	(3, 2, ∧-intro)

Example 3.2.2 Validate the inference P ∧ (Q ∧ R)/(P ∧ Q) ∧ R.

(1)	P ∧ (Q ∧ R)	(premiss)
(2)	P	(1, ∧-elim-left)
(3)	Q ∧ R	(1, ∧-elim-right)
(4)	Q	(3, ∧-elim-left)
(5)	R	(3, ∧-elim-right)
(6)	P ∧ Q	(2, 4, ∧-intro)
(7)	(P ∧ Q) ∧ R	(6, 5, ∧-intro)

The other rules of this simple kind are

$$(\rightarrow\text{-elim}) \qquad \begin{array}{|l} A \\ A \rightarrow B \\ \hline B \end{array}$$

$$(\vee\text{-intro-right}) \qquad \begin{array}{|l} A \\ \hline A \vee B \end{array}$$

$$(\vee\text{-intro-left}) \qquad \begin{array}{|l} B \\ \hline A \vee B \end{array}$$

$$(\leftrightarrow\text{-elim-left}) \qquad \begin{array}{|l} A \leftrightarrow B \\ A \\ \hline B \end{array}$$

$$(\leftrightarrow\text{-elim-right}) \qquad \begin{array}{|l} A \leftrightarrow B \\ B \\ \hline A \end{array}$$

$$(\sim\text{-elim}) \qquad \begin{array}{|l} \sim \sim A \\ \hline A \end{array}$$

More complicated inference rules make use of **subderivations**. An example is the →-**introduction** rule, which says that if you can derive B from a set of schemas

$\{A, S_1, \ldots, S_n\}$, then you are entitled to derive $A \rightarrow B$ from the set $\{S_1, \ldots, S_n\}$. This is a sensible rule because if B is a logical consequence of $\{A, S_1, \ldots, S_n\}$ then given the truth of $A, S_1, \ldots, S_n$, we cannot have B false; another way of saying this is that given the truth of $\{S_1, \ldots, S_n\}$, we cannot have A true and B false, i.e. we cannot have $A \rightarrow B$ false—and this means that $A \rightarrow B$ is a logical consequence of $\{S_1, \ldots, S_n\}$.

The following example illustrates the application of $\rightarrow$-introduction.

Example 3.2.3   Validate the inference

$$P \rightarrow Q, Q \rightarrow R / P \rightarrow R.$$

The rule of $\rightarrow$-introduction says that I can derive $P \rightarrow R$ from $P \rightarrow Q$ and $Q \rightarrow R$ so long as I can derive R from P, $P \rightarrow Q$, and $Q \rightarrow R$. But this is easy:

(1)	P	(premiss)
(2)	$P \rightarrow Q$	(premiss)
(3)	Q	$(1, 2, \rightarrow\text{-elim})$
(4)	$Q \rightarrow R$	(premiss)
(5)	R	$(3, 4, \rightarrow\text{-elim})$

This done, we can now invoke the rule $\rightarrow$-intro to derive $P \rightarrow R$ from $P \rightarrow Q$ and $Q \rightarrow R$. The derivation consisting of lines 1 to 5 above is called a **subderivation** of the overall derivation. A neat way of presenting the derivation is:

(1)	**SUBDERIVATION**	
	(1.1) P	(assumption)
	(1.2) $P \rightarrow Q$	(premiss)
	(1.3) Q	$(1.1, 1.2, \rightarrow\text{-elim})$
	(1.4) $Q \rightarrow R$	(premiss)
	(1.5) R	$(1.3, 1.4, \rightarrow\text{-elim})$
(2)	$P \rightarrow R$	$(1, \rightarrow\text{-intro})$

Note that the extra premiss of the subderivation (P in the example) is annotated as an **assumption** in order to distinguish it from the premisses of the derivation as a whole. When we come out of the subderivation at step 2, we are said to **discharge** the assumption: it has done its work for us, and is now past history.

Another way of presenting the derivation just given is as follows:

(1)	$P \rightarrow Q$	(premiss)
(2)	$Q \rightarrow R$	(premiss)
(3)	**SUBDERIVATION:**	
	(3.1) P	(assumption)
	(3.2) Q	$(1, 3.1, \rightarrow\text{-elim})$
	(3.3) R	$(2, 3.2, \rightarrow\text{-elim})$
(4)	$P \rightarrow R$	$(3, \rightarrow\text{-intro})$

This shows that it is permissible, within a subderivation, to make use of schemas already asserted *outside* the subderivation. On the other hand, it is not possible to bring schemas asserted inside the subderivation out into the main body of the derivation. Once an assumption has been discharged, neither it nor anything deduced from it can be cited in support of any further step in the derivation.

The formal statement of →-intro is

$$(\rightarrow\text{-intro}) \qquad \begin{array}{|l} \begin{array}{|l} A \\ \hline B \end{array} \\ \hline A \rightarrow B \end{array}$$

Here $\begin{array}{|l} A \\ \hline B \end{array}$ represents the subderivation that is required in order for →-elimination to be applicable.

Other rules which make use of subderivations are

$$(\vee\text{-elim}) \qquad \begin{array}{|l} A \vee B \\ \begin{array}{|l} A \\ \hline C \end{array} \\ \begin{array}{|l} B \\ \hline C \end{array} \\ \hline C \end{array}$$

$$(\leftrightarrow\text{intro}) \qquad \begin{array}{|l} \begin{array}{|l} A \\ \hline B \end{array} \\ \begin{array}{|l} B \\ \hline A \end{array} \\ \hline A \leftrightarrow B \end{array}$$

$$(\sim\text{-intro}) \qquad \begin{array}{|l} \begin{array}{|l} A \\ \hline B \\ \sim B \end{array} \\ \hline \sim A \end{array}$$

The ∨-elimination rule formalizes the procedure known as **reasoning by cases**. This is the kind of reasoning involved in an argument of the following kind:

The Prime Minister is either a criminal or insane. Suppose she is a criminal; then she ought to be locked up. Suppose on the other hand she is insane; then again she ought to be locked up. So in any case she ought to be locked up.

This argument is certainly valid; whether or not it is sound is a matter for debate! Note that this argument involves issues about which many people have strong feelings one way or the other: the merits of the Prime Minister, and the proper treatment of the criminal or the insane. It is of the utmost importance to realize that any feelings you may have on these matters are *in no way* relevant to the logical question of whether or not the argument is valid. They do relate, on the other hand, to the question of whether or not the argument it sound; the argument cannot be sound unless it is also valid, but in addition it is necessary for the premises to be true, and this is precisely what the 'strong feelings' you may have are concerned with. On this issue, cf. also Exercise 2.8, Question (5d), where the *validity*, as opposed to the soundness, of the inference is a matter of logic, not of theology.

The $\sim$-introduction rule formalizes the principle of **proof by contradiction** which occurs very frequently in mathematics. This principle says that if an assumption (A) leads to a contradiction (B and $\sim$ B), then that assumption must be false (i.e. we can infer $\sim$ A). An example of its use in mathematics is the famous proof, discovered by the ancient Greeks and handed down to us by Euclid, that no perfect square is exactly twice another perfect square (i.e. that the square root of 2 is irrational). The proof, in modern notation, can be presented as follows:

Suppose there exist integers m and n such that

$$m^2 = 2n^2.$$

Let the highest common factor of m and n be k, and let $M = m/k$ and $N = n/k$. Then we have

$$k^2 M^2 = 2k^2 N^2$$

so

$$M^2 = 2N^2$$

where *M and N have no common factor*. From the last equation, M must be even (since its square is), say $M = 2L$, which gives us

$$4L^2 = 2N^2.$$

This simplifies to

$$2L^2 = N^2$$

which implies that N must be even too. Since M and N are both even, they are both divisible by 2, i.e., *M and N have a common factor*. Now the two italicized assertions contradict one another, so our original assumption must be false. Hence no perfect square is exactly twice another perfect square.

One final rule, neither an introduction rule nor an elimination rule, is the **repetition** rule, which allows a schema already asserted to be asserted again later on in the derivation:

$$(\text{rep}) \quad \begin{array}{|l} A \\ \hline A \end{array}$$

Here are some more examples of derivations to illustrate the new rules:

Example 3.2.4 Show that $\{P \lor Q, \sim P\} \vdash Q$.

(1)	$P \lor Q$	(premiss)
(2)	SUBDERIVATION	
	(2.1) P	(assumption)
	(2.2) SUBDERIVATION	
	(2.2.1) $\sim Q$	(assumption)

(2.2.2)	P	(2.1, rep)
(2.2.3)	~ P	(premiss)
(2.3)	~ ~ Q	(2.2, ~-intro)
(2.4)	Q	(2.3, ~-elim)
(3)	SUBDERIVATION	
(3.1)	Q	(assumption)
(3.2)	Q	(3.1, rep)
(4)	Q	(1, 2, 3, ∨-elim)

This derivation illustrates a number of points. First, it shows that subderivations can be *nested*: the subderivation at (2.2) occurs *within* the subderivation at 2. It also shows why we need the repetition rule: in order to apply ~-intro to derive ~ ~ Q at (2.3), we need to have collected together, within a subderivation from the assumption ~ Q, some schema and its negation, in this case P and ~ P. We can get ~ P as a premiss; but to get P we notice that we have already asserted P at (2.1) and hence must use the repetition rule to bring it inside the current subderivation.

You may wonder what is the point of the rather trivial subderivation (3) in this derivation. It is needed for a correct application of ∨-elimination. This rule requires three inputs: a disjunction A ∨ B, together with a subderivation of the desired conclusion C from *each* of the two disjuncts. In our example, the derivation of Q from the first disjunct P is given by the subderivation 2, while the derivation of Q from the second disjunct Q is given by subderivation 3.

Example 3.2.5   Show that $\{P \rightarrow Q, ~R \rightarrow ~Q, R \rightarrow ~S\} \vDash S \rightarrow ~P$.

(1)	SUBDERIVATION	
(1.1)	S	(assumption)
(1.2)	SUBDERIVATION	
(1.2.1)	P	(assumption)
(1.2.2)	P → Q	(premiss)
(1.2.3)	Q	(1.2.1, 1.2.2, →-elim)
(1.2.4)	SUBDERIVATION	
(1.2.4.1)	~ R	(assumption)
(1.2.4.2)	~ R → ~ Q	(premiss)
(1.2.4.3)	~ Q	(1.2.4.1, 1.2.4.2, →-elim)
(1.2.4.4)	Q	(1.2.3, rep)
(1.2.5)	~ ~ R	(1.2.4, ~-intro)
(1.2.6)	R	(1.2.5, ~-elim)
(1.2.7)	R → ~ S	(premiss)
(1.2.8)	~ S	(1.2.6, 1.2.7, →-elim)
(1.2.9)	S	(1.1, rep)
(1.3)	~ P	(1.2, ~-intro)
(2)	S → ~ P	(1, →-intro)

Before reading on, try your hand at validating the next inference:

Example 3.2.6 $\sim P \vee Q / P \rightarrow Q$.

You may well have followed the derivations given above perfectly well, without really knowing how to begin when faced with the task of performing a derivation by yourself. The point is that this natural deduction proof system we are using is highly *non-algorithmic*. It tells us what moves are allowed, but gives no guidance as to how to proceed.

Nevertheless, there are a number of useful rules of thumb (or *heuristics*, to use the fashionable term in computer circles) which can help here. A couple of very general rules are:

(I)   If the conclusion to be derived has a connective * as its main functor, try using the *-introduction rule.

(II) If a premiss has * as its main functor, try using the *-elimination rule.

To illustrate, Example 3.2.4 has a premiss containing $\lor$. This suggests that the overall structure of the derivation should be an application of the $\lor$-elimination rule, as indeed it is. In Example 3.2.5, the conclusion has $\rightarrow$ as its main functor, so the overall structure of the derivation is an application of $\rightarrow$-introduction; but the three premisses each contain a $\rightarrow$ too, so we can expect three applications of $\rightarrow$-elimination, and this expectation can be verified by examining the derivation.

These rules are only rules of thumb, they are not to be followed in every case. For example, it would be a mistake to try using $\rightarrow$-introduction to validate the inference $P, P \rightarrow (Q \rightarrow R)/Q \rightarrow R$. The reason is that the conclusion $Q \rightarrow R$ is already present as a subschema of one of the premisses, so there is no need to construct it. Similarly, we would not use $\lor$-elimination to validate the inference $P \lor Q, R/(P \lor Q) \land R$: since we want the premiss $P \lor Q$ to survive as a subschema of the conclusion, it would be pointless trying to eliminate it.

The considerations of the last paragraph do not apply to Example 3.2.6, so let us now try applying the rules in this case. Since the conclusion has $\rightarrow$ as its main (and only) functor, we shall aim for a derivation using $\rightarrow$-introduction. So an outline sketch of the derivation should look like this:

(1)	SUBDERIVATION	
	(1.1) P	(assumption)
	$\vdots$	
	(1.*n*) Q	
(2)	P $\rightarrow$ Q	(1, $\rightarrow$-intro)

Next, we notice that the premiss has $\lor$ as its main functor, so we shall go for an application of $\lor$-elimination within the subderivation at 1; this expands our outline derivation to:

(1)	SUBDERIVATION	
	(1.1) P	(assumption)
	(1.2) $\sim P \lor Q$	(premiss)
	(1.3) SUBDERIVATION	
	(1.3.1) $\sim P$	(assumption)
	$\vdots$	
	(1.3.*m*) Q	

    (1.4)   SUBDERIVATION
           (1.4.1)   Q   (assumption)
             ⋮
           (1.4.*n*)   Q
       (1.5)   Q   (1.2, 1.3, 1.4, ∨ -elim)
   (2)   P → Q   (1, → -intro)

The subderivation at (1.4) is trivial, just like the one at (3) in Example 3.2.4. What about the subderivation which precedes it? In order to find this, we can make use of a third heuristic:

(III)   When all else fails, try using ∼ -intro.

In order to get a conclusion Q using ∼ -introduction, we have first to get ∼ ∼ Q and then reduce that to Q using ∼ -elimination (as was done at (2.2)–(2.4) in Example 3.2.4 and at (1.2.4)–(1.2.6) in Example 3.2.5). The resulting complete derivation is

   (1)   SUBDERIVATION
        (1.1)   P   (assumption)
        (1.2)   ∼ P ∨ Q   (premiss)
        (1.3)   SUBDERIVATION
               (1.3.1)   ∼ P   (assumption)
               (1.3.2)   SUBDERIVATION
                       (1.3.2.1)   ∼ Q   (assumption)
                       (1.3.2.2)   P   (1.1, rep)
                       (1.3.2.3)   ∼ P   (1.3.1, rep)
               (1.3.3)   ∼ ∼ Q   (1.3.2, ∼ -intro)
               (1.3.4)   Q   (1.3.3, ∼ -elim)
        (1.4)   SUBDERIVATION
               (1.4.1)   Q   (assumption)
               (1.4.2)   Q   (1.4.1, rep)
        (1.5)   Q   (1.2, 1.3, 1.4, ∨ -elim)
   (2)   P → Q   (1, → -intro)

Corresponding to each correct derivation in our natural deduction system, there is a valid inference rule which allows us to pass immediately from the premisses to the conclusion without having to spell out the intervening steps. In the case of Example 3.2.4, the corresponding rule is

$$(\vee\text{-elim2}) \quad \left| \begin{array}{c} A \vee B \\ \sim A \\ \hline B \end{array} \right.$$

This is just as valid a rule of inference as the introduction and elimination rules we started with, but unlike them, this rule does not have to be postulated as a basic element of our system, since we can prove it from the other rules. It is called a **derived rule of inference**. Having validated it, we can store it in a 'library' of such derived rules, ready to use, should the occasion demand, in just the same way as we use the basic rules.

Using derived rules can make our work quite a lot easier. Example 3.2.6, for instance, becomes much more straightforward if we are allowed to use the rule ($\lor$-elim2):

(1)	$\sim$P $\lor$ Q	(premiss)
(2)	SUBDERIVATION	
	(2.1) P	(assumption)
	(2.2) SUBDERIVATION	
	(2.2.1) $\sim$P	(assumption)
	(2.2.2) P	(2.1, rep)
	(2.3) $\sim$ $\sim$P	(2.2, $\sim$-intro)
	(2.4) Q	(1, 2.3, $\lor$-elim2)
(3)	P $\rightarrow$ Q	(2, $\rightarrow$-intro)

In addition to derivations, natural deduction can be used to perform **proofs**. A proof is a derivation from no premisses; the conclusion of a proof should therefore be a tautology. The simplest example is

Example 3.2.7   Prove the tautology P $\rightarrow$ P.

(1)	SUBDERIVATION	
	(1.1) P	(assumption)
	(1.2) P	(1.1, rep)
(2)	P $\rightarrow$ P	(1, $\rightarrow$-intro)

A somewhat more complicated example is

Example 3.2.8   Prove the schema P $\lor$ $\sim$ P. The proof starts off with our third heuristic ('When all else fails...'):

(1)	SUBDERIVATION	
	(1.1) $\sim$ (P $\lor$ $\sim$P)	(assumption)
	(1.2) SUBDERIVATION	
	(1.2.1) P	(assumption)
	(1.2.2) P $\lor$ $\sim$P	(1.2.1, $\lor$-intro-right)
	(1.2.3) $\sim$ (P $\lor$ $\sim$P)	(1.1, rep)
	(1.3) $\sim$P	(1.2, $\sim$-intro)
	(1.4) P $\lor$ $\sim$P	(1.3, $\lor$-intro-left)
	(1.5) $\sim$ (P $\lor$ $\sim$P)	(1.1 rep)
(2)	$\sim$ $\sim$ (P $\lor$ $\sim$P)	(1, $\sim$-intro)
(3)	P $\lor$ $\sim$P	(2, $\sim$-elim)

A schema derivable from the empty set of premisses is called a **theorem** of the proof system. Corresponding to a theorem, we can write down a special kind of derived rule of inference in which the premiss set is empty. Thus the theorem P $\lor$ $\sim$ P which we have just proved corresponds to the rule

(LEM) $\dfrac{}{\text{P} \lor \sim\text{P}}$

The title LEM given to this rule stands for 'law of excluded middle', the name traditionally given to this theorem (a 'middle' would be a third possibility distinct from P and $\sim$ P and as it were 'between' them: LEM explicitly rules out such a possibility).

Example 3.2.9   Validate the inference $\sim (P \wedge Q)/\sim P \vee \sim Q$

(1)	$Q \vee \sim Q$		(LEM)
(2)	SUBDERIVATION		
	(2.1)	Q	(assumption)
	(2.2)	SUBDERIVATION	
		(2.2.1) P	(assumption)
		(2.2.2) $P \wedge Q$	(2.2.1, 2.1, $\wedge$-intro)
		(2.2.3) $\sim (P \wedge Q)$	(premiss)
	(2.3)	$\sim P$	(2.2, $\sim$-intro)
	(2.4)	$\sim P \vee \sim Q$	(2.3, $\vee$-intro-right)
(3)	SUBDERIVATION		
	(3.1)	$\sim Q$	(assumption)
	(3.2)	$\sim P \vee \sim Q$	(3.1, $\vee$-intro-left)
(4)	$\sim P \vee \sim Q$		(1, 2, 3, $\vee$-elim)

Despite its name, natural deduction is often not a particularly natural way of going about validating inferences. Consider, for example, the valid inference

$$P \rightarrow Q, \sim (P \rightarrow R)/Q.$$

How should we set about constructing a natural deduction derivation for this? At first sight, it may seem impossible. Our heuristics don't seem to be of much help here, and it is not obvious how we should proceed.

The clue is to observe that if $P \rightarrow R$ is false (as implied by the second premiss), then P must be true. Once we have shown that, the rest is straightforward. But how can we use natural deduction to deduce P from $\sim (P \rightarrow R)$? Our first two heuristics are of no use here, and there does not seem to be anything else we can do. In desperation, we try the third heuristic: assume $\sim P$ and try to derive a contradiction. What contradiction? Since we have $\sim (P \rightarrow R)$ as a premiss, the obvious strategy is to try to derive $P \rightarrow R$, for which we must begin another subderivation, assuming P, in order to use $\rightarrow$-introduction. We may seem to be getting into deep water, but eventually, if we persist, we shall come up with something like the following derivation:

(1)	SUBDERIVATION		
	(1.1)	$\sim P$	(assumption)
	(1.2)	SUBDERIVATION	
		(1.2.1) P	(assumption)
		(1.2.2) SUBDERIVATION	
		(1.2.2.1) $\sim R$	(assumption)
		(1.2.2.2) P	(1.2.1, rep)
		(1.2.2.3) $\sim P$	(1.1, rep)
		(1.2.3) $\sim \sim R$	(1.2.2, $\sim$-intro)
		(1.2.4) R	(1.2.3, $\sim$-elim)
	(1.3)	$P \rightarrow R$	(1.2, $\rightarrow$-intro)

(1.4)	$\sim(P \rightarrow R)$	(premiss)
(2)	$\sim \sim P$	(1.1, $\sim$-intro)
(3)	P	(2, $\sim$-elim)
(4)	$P \rightarrow Q$	(3, premiss)
(5)	Q	(3, 4, $\rightarrow$-elim)

Study this derivation carefully, making sure that you understand how each move contributes to the final outcome. The derivation undoubtedly has a certain bizarre, even sophisticated, beauty, but it is hardly 'natural'. Notice how the premisses are not introduced until over half-way through: to the uninitiated, the first part of the derivation looks aimless. Imagine trying to program a computer to conduct derivations like this! A program would have to be very ingenious indeed to be able to handle the subtle twists and turns of reasoning necessary to validate arguments like the one above using natural deduction. (Of course, other derivations for this inference are possible, but I have not found one that seems any more 'natural' than the one given above).

And there is another problem with natural deduction: it can only tell us when an inference is valid, not when it is invalid. If we try very hard to come up with a derivation of P from $\{P \rightarrow Q, Q\}$, say, and fail, we may conclude that the reason for our failure is that P really doesn't follow from $\{P \rightarrow Q, Q\}$. But what if we had been just a little bit more ingenious? How can we be sure that then we should not find the desired derivation? In view of the difficulty of finding a derivation in some cases, it may seem rather rash to assert categorically that no such derivation exists. Yet this is what we should need to be sure of if we are to be confident that the inference is invalid.[1]

For all these reasons, natural deduction is not really a suitable method to use if we want to use a computer to do our inference-checking for us; and this is true of deduction systems generally. Instead, we shall now turn to an examination of two different refutation systems, which are much more readily automatable.

Exercise 3.2

(1) Use natural deduction to validate the following inferences:

*(a)	$P \wedge Q, R \wedge S / P \wedge S$	*(b)	$P \rightarrow Q, P \rightarrow R / P \rightarrow Q \wedge R$
(c)	$\sim(P \vee Q) / \sim P$	*(d)	$\sim P / \sim(P \wedge Q)$
(e)	$P \rightarrow (Q \rightarrow R), P \wedge Q / R$	(f)	$P \rightarrow Q, \sim Q / \sim P$
*(g)	$P, Q \vee R / (P \wedge Q) \vee (P \wedge R)$	(h)	$P \wedge Q \rightarrow R, P / Q \rightarrow R$
*(i)	$P \rightarrow Q, P, \sim Q / R$	(j)	$P / Q \rightarrow P$
*(k)	$(P \vee Q) \vee R / P \vee (Q \vee R)$	(l)	$P \rightarrow Q \vee R, Q \rightarrow S, R \rightarrow S / P \rightarrow S$
(m)	$P \vee Q, \sim Q \vee R / P \vee R$		

(2) Use natural deduction to prove the following theorems:

*(a)	$\sim(P \wedge \sim P)$
(b)	$P \vee (Q \wedge R) \leftrightarrow (P \vee Q) \wedge (P \vee R)$
*(c)	$R \rightarrow (P \rightarrow (Q \rightarrow R))$
(d)	$(Q \rightarrow R) \rightarrow ((P \rightarrow Q) \rightarrow (P \rightarrow R))$
*(e)	$(P \rightarrow Q) \vee (Q \rightarrow P)$

---

[1] I am assuming here that all and only those inferences which are valid have natural deduction derivations. This is in fact the case for **Prop**, though I haven't proved it yet. For more on this, see Section 3.4.

(3) Show that if we replace the rule ($\vee$-elim) by ($\vee$-elim2), the resulting system of natural deduction is equivalent to the original system (to do this you will have to *derive* ($\vee$-elim) from ($\vee$-elim2), making sure that you don't use any results which depend for their proof on ($\vee$-elim)).

(4) Similarly show that we can replace ($\sim$-elim) by the rule

$$
(\sim\text{-elim2}) \qquad
\begin{array}{|l}
\begin{array}{|l}
\sim A \\
\hline
B \\
\sim B
\end{array} \\
\hline
A
\end{array}
$$

(Since ($\sim$-elim2) was not proved in the text, you have two things to do, i.e., derive each of the two rules from the other.)

3.3 MODEL BUILDING

We mentioned above the distinction between deduction systems and refutation systems, and presented natural deduction as a typical example of the first kind. We now introduce model building as an example of a refutation system. This system is essentially a method for determining whether or not a set of schemas is satisfiable. In order to show that a conclusion C follows from a set of premises $\{P_1, \ldots, P_n\}$, we show that $\{P_1, \ldots, P_n, \sim C\}$ is unsatisfiable.

The method works as follows. In order to determine whether a set **S** is satisfiable, we systematically search for an interpretation which satisfies **S**, i.e. an assignment of truth values to the schematic letters with respect to which every member of **S** comes out true. If we find such an interpretation, then **S** is satisfiable; if not, not. An interpretation satisfying a set of schemas **S** is called a **model** for **S**; hence the name we give to the method.

Suppose that we are looking for a model for some set which contains a conjunction $A \wedge B$, say the set $\mathbf{S} \cup \{A \wedge B\}$. We know that an interpretation satisfies $A \wedge B$ if and only if it satisfies both A and B; hence it satisfies $\mathbf{S} \cup \{A \wedge B\}$ if and only if it satisfies $\mathbf{S} \cup \{A, B\}$. We therefore replace the former set by the latter and try to construct a model for that.

Similarly, an interpretation satisfies a set $\mathbf{S} \cup \{A \vee B\}$ if and only if either it satisfies $\mathbf{S} \cup \{A\}$ or it satisfies $\mathbf{S} \cup \{B\}$; so in order to find a model for $\mathbf{S} \cup \{A \vee B\}$, we try to construct a model for one of $\mathbf{S} \cup \{A\}$ and $\mathbf{S} \cup \{B\}$. Similar simplifications can be effected for the other connectives.

If we keep on simplifying in this way, we will either find that all the sets we are trying to satisfy contain contradictions, in which case the original set was unsatisfiable, or we will find that we can satisfy the original set by satisfying a set containing only schematic letters or their negations; in this case the schematic letters of the latter set provide a model for the original set.

The procedure can be presented in various ways. It was originally formulated as the method of **semantic tableaux** by E. Beth in 1955 (see also Smullyan, 1968). A different, but equivalent presentation, is the **truth-trees** method expounded by, for example, R. Jeffrey (1965) and more recently by Bergmann, et al. (1980).

In the presentation adopted here, the procedure involves maintaining a **current set** **C** of sets of schemas (i.e. the members of **C** are themselves sets of schemas). Initially, **C** = {**S**}, where **S** is the set whose satisfiability or otherwise it is desired to determine. The current set is successively updated by applying certain **replacement rules**, which are designed so that complex schemas always get broken down into simpler ones. Whenever one of the sets in the current set contains both A and $\sim$ A, for some schema A, it is deleted. Eventually, no more of the replacement rules can be applied, and at this stage there are two possibilities: *either* **C** is empty, in which case all attempts at satisfying **S** have failed, so **S** is unsatisfiable, *or* **C** is not empty, in which case each member of **C** corresponds to a model for **S**.

If all we are interested in is whether **S** is satisfiable, and are not interested in finding a model, we can stop the process as soon as **C** contains a member to which none of the replacement rules applies and which cannot be deleted: for then we know that this member will be present in the final value of **C**, making the original set **S** satisfiable.

The replacement rules all have the form 'If **C** has a member of the form X, replace that member by Y', where Y may be one set, two sets, or nothing at all. The following table gives the form of X and Y for each replacement rule.

Rule	X	Y
[$\wedge$]	$S \cup \{A \wedge B\}$	$S \cup \{A, B\}$
[$\vee$]	$S \cup \{A \vee B\}$	$S \cup \{A\}, S \cup \{B\}$
[$\rightarrow$]	$S \cup \{A \rightarrow B\}$	$S \cup \{\sim A\}, S \cup \{B\}$
[$\leftrightarrow$]	$S \cup \{A \leftrightarrow B\}$	$S \cup \{A, B\}, S \cup \{\sim A, \sim B\}$
[$\sim \wedge$]	$S \cup \{\sim (A \wedge B)\}$	$S \cup \{\sim A\}, S \cup \{\sim B\}$
[$\sim \vee$]	$S \cup \{\sim (A \vee B)\}$	$S \cup \{\sim A, \sim B\}$
[$\sim \rightarrow$]	$S \cup \{\sim (A \rightarrow B)\}$	$S \cup \{A, \sim B\}$
[$\sim \leftrightarrow$]	$S \cup \{\sim (A \leftrightarrow B)\}$	$S \cup \{A, \sim B\}, S \cup \{\sim A, B\}$
[$\sim \sim$]	$S \cup \{\sim \sim A\}$	$S \cup \{A\}$
[del]	$S \cup \{A, \sim A\}$	(delete)

The rules for which the replacement Y consists of two sets are called **branching rules**; the others are **non-branching rules**. To avoid undue proliferation of sets, it is advisable to delay the application of branching rules as far as possible, using the heuristic

> Never apply a branching rule if a non-branching rule can be applied first.

Another good principle is

> If [del] can be applied, apply it before any other rule.

We shall illustrate the method by validating again three of the examples from Section 3.2.

Example 3.3.1   Show that $\{\sim P \vee Q, P\} \vDash Q$.
We must show that $S = \{\sim P \vee Q, P, \sim Q\}$ is unsatisfiable. The initial value of the current set, C_0, is just $\{S\}$:

$$C_0 = \{\{\sim P \vee Q, P, \sim Q\}\}.$$

The only schema to which a replacement rule applies is $\sim P \vee Q$; the appropriate rule is $[\vee]$. It tells us to replace the set $\{\sim P \vee Q, P, \sim Q\}$ by the pair of sets $\{\sim P, P, \sim Q\}$ and $\{Q, P, \sim Q\}$, giving us

$$C_1 = \{\{\sim P, P, \sim Q\}, \{Q, P, \sim Q\}\}.$$

Intuitively, what is going on here is that $\{\sim P \vee Q, P, Q\}$ is satisfiable if and only if one or other of $\{\sim P, P, \sim Q\}$ and $\{Q, P, \sim Q\}$ is satisfiable. Next, we apply [del] to $\{\sim P, P, \sim Q\}$, giving

$$C_2 = \{\{Q, P, \sim Q\}\}$$

which after a further application of [del] gives

$$C_3 = \{\quad\}.$$

Since the current set is now empty, we can conclude that S is unsatisfiable. Hence the original inference is valid.

Example 3.3.2   Show that $\{P \to Q, \sim R \to \sim Q, R \to \sim S\} \vDash S \to \sim P$.
This time we simply display successive values of C, at each step giving the rule by which the current value of C was obtained from the previous one:

$$
\begin{aligned}
C_0 &= \{\{P \to Q, \sim R \to \sim Q, R \to \sim S, \sim (S \to \sim P)\}\} \\
C_1 &= \{\{P \to Q, \sim R \to \sim Q, R \to \sim S, S, \sim \sim P\}\} && [\sim \to] \\
C_2 &= \{\{P \to Q, \sim R \to \sim Q, R \to \sim S, S, P\}\} && [\sim \sim] \\
C_3 &= \{\{\sim P, \sim R \to \sim Q, R \to \sim S, S, P\}, \{Q, \sim R \to \sim Q, R \to \sim S, S, P\}\} && [\to] \\
C_4 &= \{\{Q, \sim R \to \sim Q, R \to \sim S, S, P\}\} && [\text{del}] \\
C_5 &= \{\{Q, \sim \sim R, R \to \sim S, S, P\}, \{Q, \sim Q, R \to \sim S, S, P\}\} && [\to] \\
C_6 &= \{\{Q, \sim \sim R, R \to \sim S, S, P\}\} && [\text{del}] \\
C_7 &= \{\{Q, R, R \to \sim S, S, P\}\} && [\sim \sim] \\
C_8 &= \{\{Q, R, \sim R, S, P\}, \{Q, R, \sim S, S, P\}\} && [\to] \\
C_9 &= \{\{Q, R, \sim S, S, P\}\} && [\text{del}] \\
C_{10} &= \{\quad\} && [\text{del}]
\end{aligned}
$$

Note that in passing from C_0 to C_1, we chose the rule $[\sim \to]$ because this was the only applicable non-branching rule; similarly, in going from C_1 to C_2, $[\sim \sim]$ was chosen for the same reason. At C_2, no non-branching rules were applicable, so we had to apply $[\to]$. If we had delayed applying $[\sim \to]$ until after the first two applications of $[\to]$, our C_3 would have contained four sets instead of only two. This was what I meant by 'undue proliferation of sets' above. However, our heuristics do not always lead to a shorter proof: for example, if we apply $[\to]$ at C_6 instead of $[\sim \sim]$, then we can go straight to C_9 by an application of [del], avoiding the use of $[\sim \sim]$ altogether (since $\sim \sim R$ and $\sim R$ form a contradictory pair just as well as R and $\sim R$ do).

Example 3.3.3 Show that $\{\sim P \vee Q\} \vDash P \rightarrow Q$.

$$
\begin{array}{lll}
\mathbf{C}_0 = \{\{\sim P \vee Q, \sim (P \rightarrow Q)\}\} & & \\
\mathbf{C}_1 = \{\{\sim P \vee Q, P, \sim Q\}\} & & [\sim \rightarrow] \\
\mathbf{C}_2 = \{\{\sim P, P, \sim Q\}, \{Q, P, \sim Q\}\} & & [\vee] \\
\mathbf{C}_3 = \{\{Q, P, \sim Q\}\} & & [\text{del}] \\
\mathbf{C}_4 = \{\ \ \} & & [\text{del}]
\end{array}
$$

In order to avoid having to write each schema several times over, we can present our proofs in the form of a tree structure in which each schema appears once only, and the components of the current set correspond to paths running down from the root to the current set of leaves (this is the *truth-trees* presentation mentioned earlier). We illustrate the growth of such a tree in the case of Example 3.3.3.

Initially, we write out the elements of **S**, the set whose satisfiability is to be determined:

$$
\begin{array}{ll}
(1) & \sim P \vee Q \\
(2) & \sim (P \rightarrow Q)
\end{array}
$$

Next, we indicate the application of a rule by means of a horizontal line annotated with the number of the schema the rule is being applied to; this is placed at the bottom of the structure obtained so far, and the result(s) of applying the rule are written below that. In addition, the schema to which the rule is applied is marked as having been 'used up':

$$
\begin{array}{ll}
(1) & \sim P \vee Q \\
\times (2) & \sim (P \rightarrow Q) \\
\hline
\multicolumn{2}{r}{\qquad\qquad\quad 2} \\
(3) & P \\
(4) & \sim Q
\end{array}
$$

Note that it is not necessary to state which rule is being applied: only one rule can apply to a given schema. The next rule is a branching rule, so the replacement schemas are written in different 'branches', i.e. side by side:

$$
\begin{array}{ll}
\times (1) & \sim P \vee Q \\
\times (2) & \sim (P \rightarrow Q) \\
\hline
\multicolumn{2}{r}{\qquad\qquad\quad 2} \\
(3) & P \\
(4) & \sim Q \\
\hline
\end{array}
$$

$$
\begin{array}{ll}
(5) \quad \sim P & \qquad (6) \quad Q
\end{array}
$$

Finally, we indicate an application of [del] by drawing a double horizontal bar, annotated with the number of the schema which, together with the schema immediately above the double bar, activates the rule. This **closes** the branch of the tree to which it is appended:

$$
\begin{array}{ll}
\times (1) & \sim P \vee Q \\
\times (2) & \sim (P \rightarrow Q) \\
\hline
\multicolumn{2}{r}{\qquad\qquad\quad 2} \\
(3) & P \\
(4) & \sim Q \\
\hline
\end{array}
$$

$$
(5) \quad \sim P \qquad\qquad (6) \quad Q
$$
$$
\overline{\overline{\quad}}3 \qquad\qquad \overline{\overline{\quad}}4
$$

When all the schemas apart from the literals (to which no replacement rules apply) have been used up, the tree is **complete**. If every branch is closed, then the original set of schemas (above the uppermost horizontal bar) is unsatisfiable. This corresponds to each member of the current set having been deleted. Note that to 'use up' a complex schema, it is necessary to apply it to each branch of the tree that remains open at the time that it is being applied (cf. Exercise 3.3, question (1i)).

What happens if we try to validate an inference which is *not* valid? We illustrate this with the invalid inference

Example 3.3.4 $\{(P \to Q) \vee R\} \nvdash P \to (Q \to R)$
The tree is as follows:

$$
\begin{array}{cll}
\times (1) & (P \to Q) \vee R \\
\times (2) & \sim (P \to (Q \to R)) \\
\hline
& & \quad 2 \\
(3) & P \\
\times (4) & \sim (Q \to R) \\
\hline
& & \quad 4 \\
(5) & Q \\
(6) & \sim R \\
\hline
& & \quad 1
\end{array}
$$

$$
\begin{array}{cll}
\times (7) \quad P \to Q & \quad (8) \quad R \\
\hline
\qquad\qquad 7 & \qquad\qquad = 6 \\
(9) \quad \sim P \quad (10) \quad Q \\
= 3
\end{array}
$$

The tree is complete, since no more replacement rules are applicable, but there remains one branch, the one ending at (10), which is not closed. Reading up the branch, collecting literals, we can construct the set $\{Q, \sim R, P\}$. If we had used the 'current set' presentation, we would have ended up with $\mathbf{C}_6 = \{\{Q, \sim R, P\}\}$. It tells us that the original set $\mathbf{S} = \{(P \to Q) \vee R, \sim (P \to (Q \to R))\}$ is satisfied by the interpretation $\{P, Q\}$ which makes P and Q true and R false. This in turn tells us that the corresponding inference is invalid.

Thus the model building method has the advantage over natural deduction that it makes it just as easy to show that an inference in **Prop** is invalid as to show that one is valid. In addition, it is much more nearly algorithmic. To make it fully so, all we need do is to establish a hard and fast procedure for determining which element should be replaced at each step. We have already suggested that the use of branching rules should be delayed as long as possible, and that [del] should take the highest priority. Both of these principles aim for maximum efficiency. To secure complete determinism, we need only add a third principle to the effect that where the other two principles still leave more than one schema to act on, we should choose the earliest schema, i.e. the one highest up the tree. Alternatively, we could choose the one leftmost in the listing of the current set (in general this will not be the same as the one highest up the tree!).

Exercise 3.3

(1) Use model building to test the following inferences for validity:

*(a) $P \wedge Q, R \wedge S / P \wedge S$ *(b) $P \vee Q, R \vee S / P \vee S$

*(c) $P \to Q, P \to R / P \to Q \wedge R$ (d) $P \to (Q \to R), P \wedge Q / R$

(e) $P \wedge Q, \sim (Q \to R) / \sim (P \vee R)$ *(f) $P \to Q, P, \sim Q/R$
(g) $P \to Q \wedge R, Q \to S, R \to S/P \to S$ (h) $P \wedge Q \to R, R \wedge S \to T/P \wedge S \to R \wedge T$
*(i) $P \vee Q, R \vee S/P \to S$

(2) *Project.* Implement the model building method as a computer program which accepts as input a set of schemes and delivers as output an indication of whether or not the input set is satisfiable.

3.4 SOUNDNESS AND COMPLETENESS

In the last two sections, we gave two proof systems for **Prop**. These systems have a quite definite purpose: to enable us to discover which inferences in **Prop** are valid. How far do they fulfil this purpose?

There are several ways in which a proof system might fail. One way is if it tells us that some inference is valid when in fact that inference is *not* valid. For example, a proof system which told us that the inference

$$P \to Q$$
$$\underline{Q}$$
$$P$$

was valid would not be much use, since this inference is actually invalid. A proof system which does *not* fail us in this way is said to be **sound**: that is, a sound proof never wrongly identifies an invalid inference as valid (or, to put the same thing differently, a proof system is sound if, whenever it tells us that an inference is valid, then that inference really *is* valid).

Another possibility is that the proof system fails to tell us that some valid inference *is* valid; in this case we say that it is **incomplete**. A proof system for the propositional calculus which could not tell us that the inference

$$P \vee Q$$
$$\underline{\sim P}$$
$$Q$$

is valid would be incomplete. The opposite of this, **completeness**, is when a proof system will correctly identify *every* valid inference as valid.

Let us say that a schema A is **deducible** from a set of schemas **S** in a given proof system PS if PS identifies A as a logical consequence of **S**. So in natural deduction, A is deducible from **S** if there is a derivation of A from **S** using the rules of inference of the system. In the model building method, *A* is deducible from **S** if the set $\{S \cup \{ \sim A\}\}$ can be reduced to the empty set using the replacement rules. We shall write

$$S \vdash_{PS} A$$

to mean that A is deducible from **S** in the proof system PS.

We can now formally define soundness and completeness as follows. A proof system

PS for the propositional calculus is **sound** so long as

$$S \vDash A \text{ whenever } S \vdash_{PS} A$$

and it is **complete** so long as

$$S \vdash_{PS} A \text{ whenever } S \vDash A.$$

So a **sound and complete** system is one for which

$$S \vdash_{PS} A \text{ if and only if } S \vDash A$$

and this is obviously the most desirable state of affairs.

Before we consider whether the two proof systems we have met so far are sound and complete, let us look at two extreme examples to help fix these concepts in our minds.

(1)   Consider the proof system PS1 which says that

$$S \vdash_{PS1} A \text{ if and only if } A \in S.$$

This system correctly identifies the inference

$$\frac{\begin{array}{c} P \\ Q \end{array}}{P}$$

as valid, but cannot recognize the validity of

$$\frac{\begin{array}{c} P \to Q \\ P \end{array}}{Q}$$

From the definition of validity, it is obvious that if $A \in S$ then $S \vDash A$, so PS1 is sound. But since it misses many (indeed most) valid inferences, it is not complete. In fact it falls so far short of being complete that in practical terms it is completely useless.

(2)   Now consider the proof system PS2 which says that

$$S \vdash_{PS2} A \text{ for every schema } A$$

(so PS2 identifies *every* inference as valid). Since it does not miss any valid inference, PS2 is complete; but it is no more useful than PS1, since it drastically fails to be sound: most of the inferences it identifies as valid are actually invalid.

We can see that a proof system like PS1, which is sound but not complete, is unsatisfactory because it tells us too little, whereas a proof system like PS2, which is complete but not sound, is unsatisfactory because it tells us too much. PS1 itself tells us *much* too little, and PS2 much too much. A system which classified almost all inferences correctly would be much better than these, but best of all would be a proof system which is sound and complete, for such a system tells us *precisely* which inferences are valid.

It turns out that both of the proof systems presented so far, natural deduction and model building, are sound and complete. This means that either of these systems can be used for testing inferences for validity, and any results they produce are bound to be correct. In this book, we restrict ourselves to demonstrating the soundness and completeness of natural deduction.

Soundness of natural deduction

We shall demonstrate the soundness of our natural deduction system by first showing that the system is equivalent to a simpler system with fewer rules, and then demonstrating the soundness of *that*. The simpler system contains only two connectives, namely conjunction and negation. Its rules are $\wedge$-intro, $\wedge$-elim-left, $\wedge$-elim-right, $\sim$-intro, $\sim$-elim, and rep. We use the equivalences

$$\text{(1)} \quad A \vee B \cong \sim (\sim A \wedge \sim B)$$
$$\text{(2)} \quad A \rightarrow B \cong \sim (A \wedge \sim B)$$
$$\text{(3)} \quad A \leftrightarrow B \cong \sim (\sim (A \wedge B) \wedge \sim (\sim A \wedge \sim B))$$

to convert any schema of the full system into an schema in the restricted system. For ease of reference, we shall call the full system ND and the restricted system ND⁻.

Let I be any inference. We must show that I can be validated in ND if and only if the inference I^- obtained from I by replacing each connective apart from $\wedge$ and $\sim$ in accordance with the equivalences (1)–(3) above can be validated in ND⁻: this is what is meant by calling the two systems equivalent.

The 'if' part of this is straightforward. Since the equivalences (1)–(3) can be validated in ND (See Exercise 3.4, Question (1)), it follows that if I^- can be validated in ND⁻ then I can be validated in I. For the derivation by which I^- is validated in ND⁻ is also a derivation of ND; moreover, each premiss of I^- can be derived in ND from the corresponding premiss of I (in accordance with the derivation of the equivalences (1)–(3) governing the replacement of connectives), and the conclusion of I can likewise be derived in ND from the conclusion of I^-.

Example 3.4.1 We illustrate this point by showing how a derivation in ND of $\sim B \rightarrow \sim A$ from $A \rightarrow B$ can be obtained from a derivation in ND⁻ of $\sim (\sim B \wedge \sim \sim A)$ from $\sim (A \wedge \sim B)$. The derivation in ND⁻ is:

(1)	SUBDERIVATION	
	(1.1) $\quad \sim B \wedge \sim \sim A$	(assumption)
	(1.2) $\quad \sim B$	(1.1, $\wedge$-elim-left)
	(1.3) $\quad \sim \sim A$	(1.1, $\wedge$-elim-right)
	(1.4) $\quad A$	(1.3, $\sim$-elim)
	(1.5) $\quad A \wedge \sim B$	(1.4, 1.2, $\wedge$-intro)
	(1.5) $\quad \sim (A \wedge \sim B)$	(premiss)
(2)	$\sim (\sim B \wedge \sim \sim A)$	(1, $\sim$-intro)

We need to supplement this with a derivation in ND of $\sim (A \wedge \sim B)$ from $A \rightarrow B$ and of $\sim B \rightarrow \sim A$ from $\sim (\sim B \wedge \sim \sim A)$. We obtain the following ND derivation:

(1)	$A \rightarrow B$	(premiss)

(2)   SUBDERIVATION
    (2.1)   A ∧ ~B   (assumption)
    (2.2)   A   (2.1, ∧-elim-left)
    (2.3)   B   (1, 2.2, →-elim)
    (2.4)   ~B   (2.1, ∧-elim-right)
(3)   SUBDERIVATION
    (3.1)   ~B ∧ ~ ~A   (assumption)
    (3.2)   ~B   (3.1, ∧-elim-left)
    (3.3)   ~ ~A   (3.1, ∧-elim-right)
    (3.4)   A   (3.3, ~-elim)
    (3.5)   A ∧ ~B   (3.4, 3.2, ∧-intro)
    (3.6)   ~(A ∧ ~B)   (2, ~-intro)
(4)   SUBDERIVATION
    (4.1)   ~B   (assumption)
    (4.2)   SUBDERIVATION
        (4.2.1)   ~ ~A   (assumption)
        (4.2.2)   ~B ∧ ~ ~A   (4.1, 4.2.1, ∧-intro)
        (4.2.3)   ~(~B ∧ ~ ~A)   (3, ~-intro)
    (4.3)   ~ ~ ~A   (4.2, ~-intro)
    (4.4)   ~A   (4.3, ~-elim)
(5)   ~B → ~A   (4, →-intro)

Note carefully how the three phases of this derivation 'dovetail' together. The first phase consists of a derivation of ~(A ∧ ~B) from A→B; its conclusion is fed directly into the line of the second phase (derivation of ~(~B ∧ ~ ~A) from ~(A ∧ ~B)) where the original ND⁻ derivation used that same conclusion as a premiss. Similarly, the conclusion of the second phase acts as the premiss for the third phase, in which the conclusion is ~B → ~A.

Of course, there exist much shorter derivations of ~B → ~A from A→B in ND! The point of this example is to illustrate that there is a *systematic method* for producing derivations in ND from 'equivalent' derivations in ND⁻.

Turning now to the 'only if' part of our claim, we must show that if the inference *I* can be validated in ND, then *I*⁻ can be validated in ND⁻. To do this, it suffices to show that this is the case for each of the rules of inference governing the connectives ∨, →, and ↔ in ND; if we can validate these, or rather their paraphrases in terms of ∧ and ~, in ND⁻, then *any* derivation in ND can be converted into a corresponding derivation in ND⁻.

We shall illustrate the method by validating in ND⁻ the paraphrases of ∨-intro and ∨-elim. The corresponding validations for the other two connectives are left as an exercise for the reader (Exercise 3.4, Question (2)).

The rule ∨-intro becomes, after elimination of disjunction, the rule

$$\frac{A}{\sim(\sim A \land \sim B)}.$$

The derivation of this rule in ND⁻ is as follows:

(1)   SUBDERIVATION
    (1.1)   ~A ∧ ~B   (assumption)

(1.2) ∼A   (1.1, ∼-elim-left)
(1.3) A   (premiss)
(2) ∼(∼A ∧ ∼B)   (1, ∼-intro)

The rule ∨-elim is a little more complicated. With disjunction eliminated, the rule can be stated as

```
│ ∼(∼A ∧ ∼B)
│┌──────────
││ A
│├──────────
││ C
│├──────────
││ B
│├──────────
││ C
│└──────────
│ C
```

To derive this rule in ND⁻, we assume that we can derive C from A and that we can derive C from B, and show that we can derive C from ∼(∼A ∧ ∼B). Let us call the derivability of C from A 'rule 1' and the derivability of C from B 'rule 2'. Then we have:

(1)   SUBDERIVATION
   (1.1)   ∼C   (assumption)
   (1.2)   SUBDERIVATION
      (1.2.1)   A   (assumption)
      (1.2.2)   C   (1.2.1, rule 1)
      (1.2.3)   ∼C   (1.1, rep)
   (1.3)   ∼A   (1.2, ∼-intro)
   (1.4)   SUBDERIVATION
      (1.4.1)   B   (assumption)
      (1.4.2)   C   (1.4.1, rule 2)
      (1.4.3)   ∼C   (1.1, rep)
   (1.5)   ∼B   (1.4, ∼-intro)
   (1.6)   ∼A ∧ ∼B   (1.3, 1.5, ∧-intro)
   (1.7)   ∼(∼A ∧ ∼B)   (premiss)
(2)   ∼∼C   (1, ∼-intro)
(3)   C   (2, ∼-elim)

Having established that paraphrases of all the inference rules of ND can be derived in ND⁻, we know that the two systems are equivalent in the sense mentioned above. It follows that ND is sound and/or complete if and only if ND⁻ is sound and/or complete. We therefore shall work with the simpler system ND⁻.

A proof system is sound so long as every inference that it identifies as valid really is valid. In ND⁻, a conclusion is identified as a logical consequence of a set of premisses when it can be derived from those premisses using the rules of inference of the system. What we must check, then, is that the rules of inference are all valid.

This is straightforward in the case of the simple inference rules which do not make use of subderivations. We can dispose of these quickly as follows.

(1)   ∧-intro is valid. Since the schema A ∧ B is satisfied by exactly those interpretations which satisfy both A and B, A ∧ B is a logical consequence of {A, B}, i.e., the inference A, B/A ∧ B is valid.

(2)   The ∧-elim rules are valid. Both A and B are satisfied by any interpretation which satisfies A ∧ B. Hence both the inferences A ∧ B/A and A ∧ B/B are valid.

(3)   ∼-elim is valid. An interpretation satisfies ∼∼A if and only if it falsifies ∼A, and *this* happens if and only if the interpretation satisfies A. So any interpretation which satisfies ∼∼A also satisfies A, i.e., the inference ∼∼A/A is valid.

This leaves us with the rule ∼-intro, which involves a subderivation:

$$
\begin{array}{l}
A \\
\hline
B \\
\sim B \\
\hline
\sim A
\end{array}
$$

To show that this rule is valid we must show that if we can derive both B and ∼B from A and some set **S** of premises, then ∼A is a logical consequence of **S**. It is easy to see that if B and ∼B are logical consequences of **S** ∪ {A} then **S** ∪ {A}, and hence **S** ∪ {∼∼A}, must be unsatisfiable, which means that ∼A is a logical consequence of **S**. But this is not what we need to prove, since in order to pass from 'we can derive B and ∼B from **S** ∪ {A}' to 'B and ∼B follow from **S** ∪ {A}' we need to know that all our derivations are in fact valid—in other words, we appear to need to know already that our proof system is sound! If we assumed this we would be arguing round in circles.

The key to the problem is to prove the soundness of our method *cumulatively*, i.e., by mathematical induction. For this purpose we introduce the idea of the **subderivation depth** of a derivation. If a derivation contains no subderivations then it has subderivation depth 0. Otherwise, its subderivation depth is one more than that of its deepest subderivation—i.e. if we write down the subderivation depth of each subderivation occurring in the derivation, the largest number we have written down is one less than the subderivation depth of the whole derivation. For example, the derivation in Example 3.2.4 has subderivation depth 2, whereas the derivations in Example 3.2.5 and Example 3.2.6 both have subderivation depth 3.

We also make use of the idea of the depth *at which* a subderivation occurs within a derivation. A subderivation which is not within another subderivation is at depth 1. A subderivation at depth 1 is a **primary** subderivation of the overall derivation. Then a primary subderivation of a subderivation at depth n occurs at depth $n + 1$. The subderivation depth of a derivation is then just the depth at which its deepest subderivation occurs (or 0 if it contains no subderivations). In abstract terms, we are here dealing with exactly the same concept as the depth at which a subschema occurs in a schema, which we defined in Section 2.2.

Now since we have proved that those rules of inference in ND⁻ which do not use subderivations are all valid, we know that any derivation with no subderivations must be valid. This means that the system is sound for derivations of subderivation depth 0.

To prove that it is sound for *all* subderivations, of whatever subderivation depth, all we now need to do is to show, for any positive integer n, that if the system is sound for all derivations of subderivation depth *less than* n, then it is also sound for all derivations of subderivation depth exactly equal to n.

Suppose, then, that all derivations of subderivation depth less than n are valid, and let

Derivation 1

(1)   A

$\vdots$

(m)   B

be a derivation of subderivation depth n. We may assume that it has *exactly one* primary subderivation of subderivation depth $n - 1$: if not, we can split it up into sections each of which does, and consider the validity of the sections individually. Let this primary subderivation occur at line l, so we have

(1)   A

$\vdots$

(l)   SUBDERIVATION
    (l.1)   C   (assumption)
                                $\vdots$
    ($l.k - 1$)   D
    ($l.k$) $\sim$D
($l + 1$) $\sim$C   (l, $\sim$-intro)

$\vdots$

(m)   B

The derivation spanning lines 1 to ($l - 1$) has subderivation depth less than n, since we assumed that the subderivation at l was the only primary subderivation of the whole derivation having subderivation depth $n - 1$. So lines 1 to ($l - 1$) can be brought inside the subderivation at l without altering its subderivation depth. In this case line l becomes line 1, k becomes $k + l - 1$ (which we shall call r), and m becomes $m - l + 1$ (which we call s). So Derivation 1 can be replaced by

Derivation 2

(1)   SUBDERIVATION
    (1.1)   C   (assumption)
                                $\vdots$
    (1.$r - 1$)   D
    (1.r) $\sim$D
(2) $\sim$C   (1, $\sim$-intro)

$\vdots$

(s)   B

The derivation spanning lines 2 to s is of subderivation depth less than n, so by the induction hypothesis is valid. We are therefore only concerned to demonstrate the validity of the derivation spanning lines 1 to 2.

Consider, then, the derivation

> *Derivation 3*
>
> (1)   C   (premiss)
>
>   ⋮
>
> $(r-1)$ D
>
> (r) $\sim$ D

Since this has subderivation depth $n-1$, it is valid, by the induction hypothesis. This means that if **S** is the set of premisses of the original derivation (which includes, of course, all the premisses of *this* derivation apart from C), then $S \cup \{C\}$ logically implies both D and $\sim$ D. This means that any interpretation which satisfies $S \cup \{C\}$ also satisfies both D and $\sim$ D. But no interpretation can satisfy both D and $\sim$ D, hence no interpretation satisfies $S \cup \{C\}$, i.e., $S \cup \{C\}$ is unsatisfiable. And this means that the inference $S/\sim C$ is valid. But this is exactly the inference derived in lines 1 to 2 of Derivation 2; and since the rest of that derivation is, as we have seen, valid, the whole derivation must be valid. And this means, finally, that our original derivation, Derivation 1, is valid too.

Now *any* derivation of subderivation depth n can be split up into sections of the form of Derivation 1, and hence into sections all of which are individually valid. It follows that the whole derivation is valid. So we have proved that if all derivations of subderivation depth less than n are valid, then so are all derivations of subderivation depth exactly n. Since, as we saw, derivations of subderivation depth 0 are all valid, this now finally gives us the result that *all* derivations of any subderivation depth are valid. And this in turn means that the system ND⁻, and hence also ND, is sound.

Completeness of natural deduction

As with the proof of soundness, we shall work with the restricted system ND⁻. We begin with a definition:

A set of schemas **S** is **inconsistent** in ND⁻ if some schema and its negation can both be derived from **S** in ND⁻, i.e. if there is a schema A such that

$$S \vdash_{\text{ND}^-} A \wedge \sim A.$$

If **S** is not inconsistent in ND⁻, we call it **consistent** in ND⁻. If **S** is consistent in ND⁻ but such that no schemas can be added to **S** without making it inconsistent in ND⁻, then we say that **S** is **maximally consistent**. In the rest of this section, we shall drop the qualification 'in ND⁻' and simply speak of consistent, maximally consistent, and inconsistent sets of schemas.

The first step in the demonstration of completeness of ND⁻ is to show that *every consistent set of schemas can be extended to a maximally consistent set*. To do this, we first *order* the schemas of the language, that is, we assign each schema to a unique positive integer so that we can talk about the first schema, the second schema, the third schema, and so on. Thus ordered, the schemas are

$$S_1, S_2, S_3, \ldots$$

and every schema is S_n for some positive integer n. We shall call n the **index** of S_n.

Now let S be any consistent set of schemas. We define a sequence

$$S_0, S_1, S_2, S_3, \ldots$$

of consistent sets as follows:

$$S_0 = S;$$

$$S_{n+1} = \begin{cases} S_n \cup \{S_{n+1}\} & \text{(if this is consistent)} \\ S_n & \text{(otherwise)} \end{cases}$$

In other words, we consider each schema in turn, and if it is consistent with the set we have constructed so far, we add it to the set, otherwise we discard it. Since S_0 is consistent, and each set in the sequence is consistent so long as the one before it is, it follows (mathematical induction again!) that *all* the sets in the sequence are consistent.

Now we collect all these sets together into one big set S^+: that is, the members of S^+ are just those schemas which occur in at least one of the sets in the sequence $S_0, S_1, S_2, \ldots$. (Technically, S^+ is the **union** of all the S_i.) Then it is easy to see that S^+ is also consistent. For suppose it were not. Then there would be a derivation in ND$^-$ of some contradictory schema of the form $P \wedge \sim P$. Suppose the number of lines in the derivation is k; then at most k members of S^+ can appear as premises in the derivation. Of these premises, let S_m be the one with the highest index. In that case all the premises used in the derivation are already in S_m—so the contradictory conclusion can be derived from S_m. But this would mean that S_m is inconsistent, which it is not. We conclude that S^+ is, after all, consistent.

It is, moreover, *maximally* consistent. For suppose $A \notin S^+$, and let the index of A be n, so $A = S_n$. Then we know that $S_{n-1} \cup \{S_n\}$ must be inconsistent, otherwise S_n would have been included in S_n, and hence would be in S^+. But if $S_{n-1} \cup \{S_n\}$ is inconsistent, $S^+ \cup \{S_n\}$ must be inconsistent too, since a derivation of a contradiction from the former set is also a derivation of that contradiction from the latter. Thus no schema not already in S^+ can be added to that set without incurring inconsistency—which is just to say that S^+ is maximally consistent.

By virtue of being maximally consistent, S^+ enjoys a number of properties which will be important to us later on. In particular, we have

(a) A schema is in S^+ if and only if its negation is not in S^+.
(b) Two schemas are in S^+ if and only if their conjunction is in S^+

The proof of these assertions is straightforward. For the first one, note first that A and $\sim A$ cannot both be in S^+, for that would make the set inconsistent. On the other hand they cannot both *not* be in S^+, for if A is not in S^+ then $S^+ \cup \{A\}$ is inconsistent (since S^+ is maximally consistent), so there is a derivation of a contradiction from $S^+ \cup \{A\}$. But this derivation immediately yields, using $\sim$-intro, a derivation of $\sim$ A from $\mathbf{S}^+$, which means that $\sim$ A must already be in $\mathbf{S}^+$ (otherwise that set would not be *maximally* consistent).

To prove (b), note that if A and B are both in $\mathbf{S}^+$, then A $\wedge$ B can be derived from $\mathbf{S}^+$ by $\wedge$-intro and hence must already be in $\mathbf{S}^+$, by maximal consistency. And conversely, if A $\wedge$ B is in $\mathbf{S}^+$, then both A and B can be derived from that set, using the $\wedge$-elim rules, and hence again must already be in it.

We now show that the set $\mathbf{S}^+$ is satisfiable. We do this by exhibiting an interpretation which satisfies it. Specifically, define the interpretation I by the rule that for each schematic letter P,

$$P \in I \text{ if and only if } P \in \mathbf{S}^+$$

(so the interpretation simply consists of the schematic letters in $\mathbf{S}^+$). We shall show that I satisfies a schema A if and only if A is in $\mathbf{S}^+$.

We use induction on the depth of the schema. A schema of depth 0 is just a schematic letter, for which we have

$$I \vDash P \text{ if and only if } P \in I,$$

and hence

$$I \vDash P \text{ if and only if } P \in \mathbf{S}^+.$$

Now suppose the result holds for every schema of depth n or less, and let C be a schema of depth $n + 1$. Then there are two cases to consider:

(i)   C is $\sim$ A, where A has depth n;
(ii)   C is A $\wedge$ B, where A and B have depth at most n.

By the induction hypothesis, if A and B have depth at most n, then

$$I \vDash A \text{ if and only if } A \in \mathbf{S}^+$$

and

$$I \vDash B \text{ if and only if } B \in \mathbf{S}^+$$

We thus have that I satisfies $\sim$ A if and only if I does not satisfy A, i.e. if and only if A is not in $\mathbf{S}^+$, i.e. if and only if $\sim$ A *is* in $\mathbf{S}^+$ (by property (a) above). That takes care of case (i).

For case (ii), we note that I satisfies A $\wedge$ B if and only if it satisfies both A and B, i.e. if and only if both A and B are in $\mathbf{S}^+$, i.e. if and only if A $\wedge$ B is in $\mathbf{S}^+$.

We conclude, then, that for every schema A,

$$I \text{ satisfies A if and only if A is in } \mathbf{S}^+.$$

This means in particular that I satisfies $\mathbf{S}^+$, and since $\mathbf{S} \subseteq \mathbf{S}^+$, that I satisfies $\mathbf{S}$. But $\mathbf{S}$ was *any consistent set of schemas*; and we have shown that $\mathbf{S}$ is satisfiable. In other words, we have shown that *any consistent set of schemas is satisfiable*.

From here it is, but a short step to completeness. To demonstrate completeness, we must show that whenever $\mathbf{S}$ entails A, there is a derivation of A from $\mathbf{S}$. Suppose, then, that $\mathbf{S}$ entails A. This means that $\mathbf{S} \cup \{\sim A\}$ is unsatisfiable. By the result we have just proved, this implies that $\mathbf{S} \cup \{\sim A\}$ is inconsistent, i.e., that there is a derivation of some contradiction, say $B \wedge \sim B$, from $S \cup \{\sim A\}$. We use this derivation as the subderivation 1 in the following larger derivation:

(1) SUBDERIVATION

    (1.1) $\sim A$ (assumption)

    $\vdots$

    $(1.k-1)$ B

    $(1.k)$ $\sim B$

(2) $\sim \sim A$ $(1, \sim\text{-intro})$

(3) A $(2, \sim\text{-elim})$

We thus have a derivation of A from **S**, as required.

The method of proof used here was first discovered by L. Henkin (1949).

Exercise 3.4

(1) Use **ND** to validate the equivalences

   (a) $A \vee B \cong \sim (\sim A \wedge \sim B)$

 *(b) $A \rightarrow B \cong \sim (A \wedge \sim B)$

   (c) $A \leftrightarrow B \cong \sim (\sim (A \wedge B) \wedge \sim (\sim A \wedge \sim B))$

by which **ND** schemas can be reduced to $\mathbf{ND^-}$ schemas.

*(2) Paraphrase the introduction and elimination rules for $\rightarrow$ and $\leftrightarrow$ in terms of $\wedge$ and $\sim$ alone, and derive these paraphrases in the system $\mathbf{ND^-}$ (solutions given for $\rightarrow$ rules).

(3) Show that the soundness of $\mathbf{ND^-}$ is equivalent to the proposition that any satisfiable set of schemas is consistent.

3.5 COMPUTATIONAL COMPLEXITY OF PROOF SYSTEMS

The **computational complexity** of an algorithm is a measure of how efficient it is. This is expressed by giving a function relating the number of computation steps required to the size of the input. In the case of a proof system for **Prop**, the problem of determining its complexity may be stated as follows:

For proof system PS, find a function f such that the validity or invalidity of any inference containing n symbols can be determined using PS in at most $f(n)$ computation steps.

We have already noted that for the truth-table method, f is an exponential function. To be more exact, note that an inference such as

$$P_1, \ldots, P_{n-1}/P_n,$$

where $P_1, \ldots, P_n$ are distinct schematic letters, contains n symbols (each schematic letter counts as one symbol, and we do not count the $/$), and has a truth table with 2^n lines, so a 'brute force' application of the truth table method has at least exponential complexity. Moreover, no inference containing n symbols can contain *more* than n schematic letters, hence its truth table will have at most 2^n lines. So the complexity of this method is also at most exponential. Hence it is exactly exponential.

It can be shown that the other proof systems for **Prop** also all have exponential complexity. This is the worst case complexity: many inferences may be dealt with in less than exponential time, but there will always be some (the 'worst cases') which do require that long.

The problem of testing inferences in **Prop** for validity is an example of an important class of problems known as **NP-complete** problems. There are many different kinds of NP-complete problems, but for our purpose it suffices to restrict ourselves to *NP-complete decision problems*. A **decision problem** is a problem which, whatever the input, requires as output either 'yes' or 'no'. Thus, to give a numerical example, the problem

> Determine whether or not a given positive integer is prime

is a decision problem (since the input integer either is or is not prime, and that is all we are asking for), whereas the related problem

> Find the prime factors of a given positive integer

is not a decision problem (since the required output is a list of the factors of the input integer). Our problem,

> Determine whether or not a given inference on **Prop** is valid

is a decision problem, since all we require for an answer is 'yes' (i.e. it is valid) or 'no' (i.e. it is not valid), whichever is appropriate for the input argument. This decision problem is closely related to another important decision problem, the **satisfiability problem**:

> Determine whether or not a given schema in **Prop** is satisfiable.

The relationship between the two problems is that given an inference $P_1, \ldots, P_n / C$ (which is an appropriate input to the first problem), there is a schema $P_1 \wedge \cdots \wedge P_n \wedge \sim C$ which is unsatisfiable if and only if the inference is valid. So the second problem requires answer 'no' when given this schema if and only if the first problem requires answer 'yes' when given the inference. It is the second problem, the satisfiability problem, which we shall be concerned with below. From now on we shall abbreviate 'satisfiability problem' to SAT.

A decision problem is **NP** if it has a rather special property which we shall describe below. Let us call those inputs to the problem for which the correct output is 'yes' the **positive** inputs to the problem; and the ones for which the correct answer is 'no', the **negative** inputs. Now suppose that each positive input can be given a **certificate** which as it were vouches for the fact that the input is positive; a certificate must be some object or expression which can be tested in such a way that if it passes the test then the input with which it is associated must be positive. Anything will do for a certificate; for example, for the problem of testing whether a set of schemas is satisfiable, an appropriate certificate for any given positive input is an interpretation which satisfies it. For the problem

> Determine whether a given input integer is composite

an appropriate certificate would be a factor distinct from 1 or the input integer itself.

(A good certificate for the complementary problem of determining whether an integer is prime is harder to specify, and involves a fair bit of number theory to state.) It is most important that *negative* inputs do not have certificates: clearly nothing can truthfully certify that a negative input yields the answer 'yes', since it doesn't.

Of course, trivially, any positive input can be regarded as its own certificate, the certificate-checking process then being identical with the procedure for solving the original problem. But for many problems, there exist certificates for which it is comparatively easy, once a potential certificate has been given, to check whether or not it really is a certificate for a particular input. For these problems, this is usually much easier than actually solving the problem from scratch. For example, suppose you want to determine whether 5551 223 is composite. This is going to involve you in quite a lot of work. But if someone suggests to you that perhaps 1999 is a factor, it is quite easy for you to check whether or not it is. And as soon as you have determined that 1999 really *is* a factor of 5551 223, you have an answer to your problem: the latter number is composite.

Similarly, given a schema in **Prop**, together with an interpretation, it is easy enough to check whether or not that interpretation satisfies the schema. This will, in general, be much easier than finding an appropriate interpretation from scratch.

A decision problem Q is said to be NP (for 'non-deterministic polynomial') if it has the following special properties. First, each positive input to the problem has a certificate. Second, there is an algorithm A_Q which meets the following specification: A_Q takes as inputs any possible input I to Q together with any potential certificate C for that input, and delivers as output 'yes' if C *is* a certificate for I and 'no' if it is not. And third, the algorithm A_Q runs in polynomial time, that is, there is an integer k and a constant c such that the number of computation steps performed by A_Q in determining whether or not C is a certificate of I is at most cn^k, where n is some suitable measure of the size of I (e.g. the number of symbols it contains).

It is not hard to see that SAT is NP. A schema containing n symbols contains at most $n-1$ connectives (usually much less—the worst case is if the schema consists of $n-1$ negation signs followed by a schematic letter). Any interpretation is a potential certificate for the schema: to check whether or not it *is* a certificate, all we have to do is to compute the line of the truth table corresponding to that interpretation. But to compute *one line* of the truth table takes no more than $n-1$ steps (one for each connective occurring in the schema). So the function f giving the computational complexity of the certificate-checking algorithm for this problem is $f(n) = n-1 < 2n$.

But SAT is not just NP. It has another very special property, as follows. If Q is *any other* NP problem, there is a systematic procedure for converting any input I for Q into an input I' for SAT, in such a way that positive inputs for Q always get converted into positive inputs for SAT, and likewise negative to negative. Moreover, this conversion procedure itself runs in polynomial time. What this means is that *no NP problem is more than polynomially harder than SAT*. This is because, in order to solve an NP problem Q, all we have to do is, first, convert it to SAT, and then solve *that*; and the first step only takes polynomial time.

An **NP-complete** problem, then, is an NP problem such that any other NP problem can be converted into it (in the sense described above) in polynomial time. Now NP problems crop up all over the place; it would be of enormous practical use if we could find efficient ways of solving them. Of course, some NP problems are known to be solvable in polynomial time: this follows from the fact that any polynomial decision problem can be regarded as NP, by letting each positive input be its own certificate. Of more interest to us is the large class of NP problems for which no polynomial-time algorithm is known. This includes all the NP-complete problems.

The real importance of NP-completeness resides in the fact that *if any one NP-complete problem can be solved in polynomial time, then every NP problem can be solved in polynomial time*. To see this, remember that no NP problem is more than polynomially harder than any given NP-complete problem. This means that if the NP-complete problem is solvable in polynomial time, then the NP problem must be too, since 'polynomially harder than polynomial' is itself just a polynomial amount (i.e. the sum of two polynomial functions is again polynomial). So if ever anyone were to discover a polynomial solution to some NP-complete problem, we should immediately have polynomial solutions to a vast range of other problems, all the NP problems for which polynomial solutions had up to then not been known. And indeed, were this to be discovered, we should have no more use for the concept of NP problems, since the class of NP problems would have been shown to coincide with the class of polynomial problems. The question of whether or not these two classes coincide (the 'P = NP' problem) is one of the great unsolved problems in the theory of computing.

SAT has played an important part in the historical development of all these ideas, because not only is it an NP-complete problem, it was the very first NP-complete problem to be discovered—or more exactly, it was the very first problem demonstrated to be NP-complete. This was done by S. Cook in 1971, and hence the statement that SAT is NP-complete is known as **Cook's theorem**. Its proof lies outside the scope of this book; an accessible reference is Wilf (1988, Chapter 5).

Since SAT is NP- complete, the related problem of testing inferences for invalidity must also be NP-complete. This is because the latter problem (let us call it INVAL) can be converted into SAT by replacing each inference by the conjunction of its premises with its conclusion. Now if INVAL is NP-complete, the complementary problem VAL, of testing inferences for validity, must also be NP-complete, because if we had an algorithm for solving the latter, we could convert it into an algorithm for the former by simply swapping round 'yes' and 'no' answers. (Note that we are here only considering algorithms which are complete in the sense that they deliver an answer for every input, both positive and negative.) Thus the main topic of this chapter, proof systems for **Prop**, is closely related to matters of central concern to the theory of computation.

3.6 RESOLUTION

Our third and final proof method, **resolution**, differs from the other two we have looked at in that it is limited in its application to schemas of a special form. Recall first that a **literal** is either a schematic letter (in which case it is called a **positive** literal) or the negation of a schematic letter (a **negative** literal). Then we define a

clause to be a *disjunction of literals*. Thus typical clauses are P ∨ Q, ∼ P ∨ ∼ Q ∨ R, P ∨ Q ∨ ∼ R ∨ S. Since disjunction is commutative, so that, for example, P ∨ Q ∨ ∼ R is equivalent to P ∨ ∼ R ∨ Q, ∼ R ∨ P ∨ Q, etc., it is often convenient to think of a clause simply as a *set* of literals, in which the order they are written is immaterial. The clauses mentioned above can therefore be written as {P, Q}, { ∼ P, ∼ Q, R}, and {P, Q, ∼ R, S}. The point of this is that, for example, {P, Q} and {Q, P} are one and the same set (because sets do not come with their elements ordered in any way), whereas P ∨ Q and Q ∨ P are actually different, though equivalent, schemas. So in what follows, a clause is understood to be a set of literals, representing any of the equivalent schemas that can be obtained by disjoining the literals in some particular order.

A **unit clause** is a set consisting of just one literal, such as {P} or { ∼ Q}. For brevity, we write unit clauses without braces, i.e. instead of {P} and { ∼ Q} we write P and ∼ Q. In strict mathematical terms, it is quite incorrect to identify a literal with the set which contains that literal as its only member, but we allow ourselves this looseness of notation because experience shows that *in this context* no confusion is likely to arise. The **empty clause**, which we denote **F**, corresponds to a contradiction—it is false in every interpretation (this is because, when we add disjuncts to a clause, we increase the number of interpretations that will satisfy it—so if we remove all the disjuncts to produce the empty clause, no interpretation will satisfy it). In addition, any clause which contains a schematic letter and its negation (e.g. the clause {P, Q, ∼ P}) will be written as **T**; this clause corresponds to a tautology—it is true in every interpretation.

Any set of propositional calculus schemas can be converted to an equivalent set of clauses. All we have to do is first to convert to CNF (cf. the algorithm for converting to DNF in Section 2.9) and then separate the conjuncts: each conjunct is a clause.

Resolution is based on the validity of inferences of the following type:

$$\{P \vee Q, \sim Q \vee R\} \vDash P \vee R.$$

This is easily proved: for if the conclusion P ∨ R is false then P and R are both false, so that (a) if Q is false so is P ∨ Q, and (b) if Q is true, ∼ Q ∨ R is false. But Q must be either true or false, so in any case one of the premises is false. Hence if both premises are true, so is the conclusion.

We call the clause {P, R} the **resolvent** of the pair of clauses {P, Q} and { ∼ Q, R}, and the process of obtaining it **resolution**. We extend the definition so that whenever a clause C' contains the negation of a schematic letter A appearing unnegated in another clause C, we may resolve C and C' to produce a new clause which contains all the literals of C apart from A and all the literals of C' apart from ∼ A, i.e. if C contains A and C' contains ∼ A then the resolvent of C and C' is

$$\text{res}\,(C, C') = (C - \{A\}) \cup (C' - \{\sim A\}).$$

An alternative way of writing this is to put $C = K \cup \{A\}$ and $C' = K' \cup \{\sim A\}$, where A is not in K and ∼ A is not in K'; then we have

$$\text{res}\,(K \cup \{A\}, K' \cup \{\sim A\}) = K \cup K'.$$

Here are some examples:

$$\text{res}(\{P, \sim Q, R\}, \{Q, R, \sim S\}) = \{P, R, \sim S\}$$
$$\text{res}(Q, \{\sim P, \sim Q, R\}) = \{\sim P, R\}$$
$$\text{res}(P, \sim P) = \mathbf{F}$$
$$\text{res}(\{P, \sim Q\}, Q) = P$$
$$\text{res}(\{P, \sim Q, R\}, \{Q, \sim R, S\}) = \{P, R, \sim R, S\} = \{P, \sim Q, Q, S\} = \mathbf{T}.$$

Note that in the last case, there are two ways of resolving the two clauses; whenever this happens, both the resolvents reduce to **T**.

Resolution has been carefully defined so that if two clauses have a resolvent, then they jointly entail that resolvent, i.e.

$$\{C, C'\} \vDash \text{res}(C, C').$$

We have already shown this for the special case in which C and C' each contain two literals. More generally, to prove that

$$\{K \cup \{A\}, K' \cup \{\sim A\}\} \vDash K \cup K'$$

we suppose that the conclusion $K \cup K'$ is false, and observe that A must be either true or false; if A is true, then $K' \cup \{\sim A\}$ is false, whereas if A is false then $K \cup \{A\}$ is false, so in either case one of the premises is false. Hence if the premises are both true, so is the conclusion, and the entailment is proved.

How does all this help us? First note that, in general, if a set of schemas **S** entails a schema C, then the set **S** is satisfiable if and only if $\mathbf{S} \cup \{C\}$ is. This is because, first, if $\mathbf{S} \cup \{C\}$ is satisfiable then its subset **S** must be satisfiable too (since any interpretation that satisfies $\mathbf{S} \cup \{C\}$ thereby satisfies **S**), and second, if **S** is satisfiable, then C, being a logical consequence of **S**, is true in any interpretation which satisfies **S**, so any such interpretation must satisfy $\mathbf{S} \cup \{C\}$. Applying this general principle to sets of clauses, we may say that since $\{C, C'\}$ entails $\text{res}(C, C')$, $\{C, C'\}$ is satisfiable if and only if $\{C, C', \text{res}(C, C')\}$ is. More generally, if **S** is any set of clauses, and R is the resolvent of any two of them, then **S** is satisfiable if and only if $\mathbf{S} \cup \{R\}$ is. And so we have

---

The resolution principle

Let **S** be a set of clauses, and let **R(S)** be the result of adding to **S** all possible resolvents of pairs of members of **S**. Then **S** is satisfiable if and only if **R(S)** is satisfiable.

Example 3.6.1 We shall illustrate the use of the resolution principle by showing that the set

$$\mathbf{S} = \{\{\sim P, Q\}, \{\sim Q, R\}, P, \sim R\},$$

is unsatisfiable (remember that this set of clauses represents a set of schemas such as $\{\sim P \vee Q, \sim Q \vee R, P, \sim R\}$). First, we find all resolvable pairs of clauses in **S** and form their resolvents. There are three such pairs, as follows:

$$\text{res}(\{\sim P, Q\}, \{\sim Q, R\}) = \{\sim P, R\}$$
$$\text{res}(\{\sim P, Q\}, P) = Q$$
$$\text{res}(\{\sim Q, R\}, \sim R) = \sim Q$$

Hence we have

$$R(S) = \{\{\sim P, Q\}, \{\sim Q, R\}, P, \sim R, \{\sim P, R\}, Q, \sim Q\}.$$

We can see at a glance that **R(S)** is unsatisfiable, since it contains both Q and $\sim$ Q; and by the resolution principle this means that **S** itself is unsatisfiable. To finish the job off properly, though, we can apply the principle again to form the set:

$$R(R(S)) = \{\{\sim P, Q\}, \{\sim Q, R\}, P, \sim R, \{\sim P, R\}, Q, \sim Q, \sim P, R, F\}$$

where the last member, **F**, was obtained as the resolvent of Q and $\sim$ Q. The presence of **F** in **R(R(S))** means that this set is unsatisfiable, which means that **S** is.

Another way of stating the resolution principle, then, is as follows:

Let **S** be a set of clauses, and define the operation **R** as above. Then if for some integer *n* we have

$$F \in R^n(S)$$

then **S** is unsatisfiable.

In Example 3.6.1, many of the resolvents were computed unnecessarily: for example, the resolvent $\{\sim P, R\}$ did not contribute to the eventual production of the empty clause. In fact the only *relevant* resolutions are:

$$\text{res}(\{\sim P, Q\}, P) = Q$$
$$\text{res}(\{\sim Q, R\}, \sim R) = \sim Q$$
$$\text{res}(Q, \sim Q) = F.$$

These can be portrayed by means of a **resolution tree** as follows:

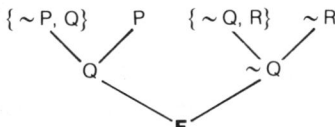

The precise definition of a resolution tree for a set of clauses is

---

A **resolution tree** over a set of clauses **S** is a binary tree in which each leaf is labelled by a member of **S** and each node other than a leaf is labelled by the resolvent of the clauses labelling its immediate descendents.

If in addition the root of the tree is labelled with **F** then we shall call the tree a **resolution refutation tree** or simply a **refutation tree** for **S**. If a set of clauses has a refutation tree, then it must be unsatisfiable: since, in addition, any set of propositional calculus schemas is equivalent to some set of clauses, resolution is a sound proof procedure for the propositional calculus.

Here is another example, in which instead of listing the sets **S**, **R(S)**, etc, we simply give the initial set **S** and a refutation tree:

$$\mathbf{S} = \{\{P, Q\}, \{\sim P, Q\}, \{\sim Q, R\}, \{\sim Q, \sim R, S\}, \{\sim Q, \sim R, \sim S\}\}$$

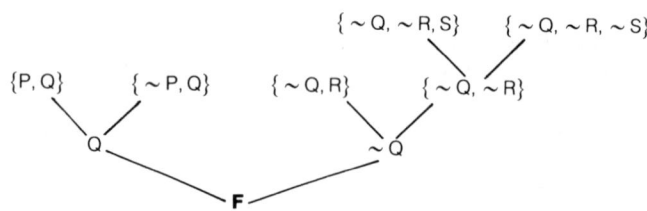

The existence of a refutation tree for **S** shows that **S** is unsatisfiable.

We have seen that resolution is a sound proof system. Is it also complete? Well, suppose there does *not* exist a refutation tree for some set **S**. Does this mean that **S** is *satisfiable*? To answer this, we must show that every unsatisfiable set of clauses has a refutation tree. We give below an algorithm for constructing a refutation tree for an arbitrary unsatisfiable set of clauses. After that we illustrate how the algorithm works by means of an example, and then we prove that the algorithm is correct.

Algorithm

Input: any set of clauses, **S**.
Output: if **S** is unsatisfiable, a refutation tree for **S**;
otherwise, a failure message.

(1) If for some schematic letter A, both A and $\sim$ A are members of **S**, then output the tree

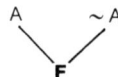

and halt. Otherwise go to (2).

(2) If every clause in **S** contains a negative literal, then output 'failure' and halt. Otherwise go to (3).

(3) Let $C = \{A_1, \ldots, A_n\}$ be a clause in **S** in which every literal is positive. Let $i = 1$.

(4) If $\sim A_i$ is in **S**, let T_i be the tree with the single node $\sim A_i$, and go to (8). Otherwise go to (5).

(5) For each clause K in $\mathbf{S} - \{C\}$, let $K' = K - \{\sim A_i\}$, and call K' the **counterpart** of K (note: if $\sim A_i$ does not occur in K, then $K' = K$). Let $\mathbf{S}_i$ be the set of counterparts thus constructed, i.e.

$$\mathbf{S}_i = \{K' \mid K \in \mathbf{S} - \{C\}\}$$

(6) Recursively call this algorithm with input $\mathbf{S}_i$. If the output to the recursive call is 'failure', then output 'failure' and halt. Otherwise let T_i' be the output refutation tree, and go to (7).

(7) Let T_i be the tree obtained from T_i' by replacing each leaf in the latter by the clause in **S** of which it is the counterpart, and adjusting the other nodes so that the resulting

tree is a resolution tree. If the root of T_i is **F**, go to (10).
(8)   If $i < n$, increase i by 1 and go to (4). Otherwise go to (9).
(9)   Output the tree

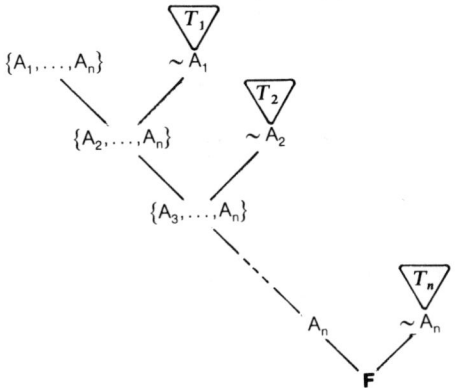

and halt.
(10)   Recursively call this algorithm for the input $\mathbf{S} - \{C\}$.

Example 3.6.2   We shall illustrate the algorithm using as input the set of clauses:

$$\mathbf{S} = \{\{P, Q, R\}, \{\sim P, S, T\}, \{\sim S, U\}, \sim T, \{\sim P, \sim U\}, \{\sim Q, W\}, \{\sim Q, \sim W\}, \sim R\}.$$

We follow the algorithm step by step:

(1) $\mathbf{S}$ has no contradictory pair of literals, so:
(2)   Not every clause in $\mathbf{S}$ has a negative literal, so:
(3)   Let $C = \{P, Q, R\}$, and let $i = 1$.
(4) $\sim P$ is not in $\mathbf{S}$, so:
(5) $\mathbf{S}_1 = \{\{S, T\}, \{\sim S, U\}, \sim T, \sim U, \{\sim Q, W\}, \{\sim Q, \sim W\}, \sim R\}$.
(6)   Recursively calling the algorithm for $\mathbf{S}_1$, and, to avoid confusion, labelling the steps (1'), (2'), etc, and the variables C', i', etc, we have:

    (1') $\mathbf{S}_1$ has no contradictory pair of literals, so:
    (2')   Not every clause in $\mathbf{S}_1$ has a negative literal, so:
    (3')   Let $C' = \{S, T\}$, and let $i' = 1$.
    (4') $\sim S$ is not in $\mathbf{S}_1$, so:
    (5') $\mathbf{S}_{11} = \{U, \sim T, \sim U, \{\sim Q, W\}, \{\sim Q, \sim W\}, \sim R\}$.
    (6')   Recursively calling the algorithm for $\mathbf{S}_{11}$:
        (1'') $\mathbf{S}_{11}$ contains both U and $\sim U$, so output the tree

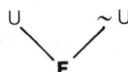

    (7')   Let T_{11} be the tree

$$\{\sim S, U\} \qquad \sim U$$
$$\searrow \qquad \swarrow$$
$$\sim S$$

    (8') $i' < 2$, so let $i' = 2$.

(4′) $\sim$ T is in $\mathbf{S}_1$, so let T_{12} be the tree $\sim$ T.

(8′) $i' = 2$, so:

(9′)   Output the tree:

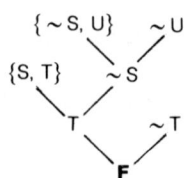

(7)   Let T_1 be the tree

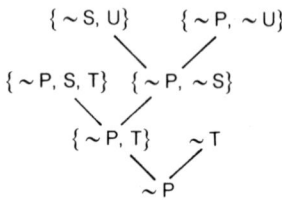

(8) $i < 3$ so let $i = 2$.

(4) $\sim$ Q is not in $\mathbf{S}$, so:

(5)   Let $\mathbf{S}_2 = \{\{\sim P, S, T\}, \{\sim S, U\}, \sim T, \{\sim P, \sim U\}, W, \sim W, \sim R\}$.

(6)   Recursively call the algorithm with input $\mathbf{S}_2$:

    (1′) $\mathbf{S}_2$ contains both W and $\sim$ W, so output the tree

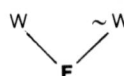

(7)   Let T_2 be the tree

(8) $i < 3$, so let $i = 3$.

(4) $\sim$ R is in $\mathbf{S}$, so let T_3 be the tree $\sim$ R.

(8) $i = 3$, so:

(9)   Output the tree:

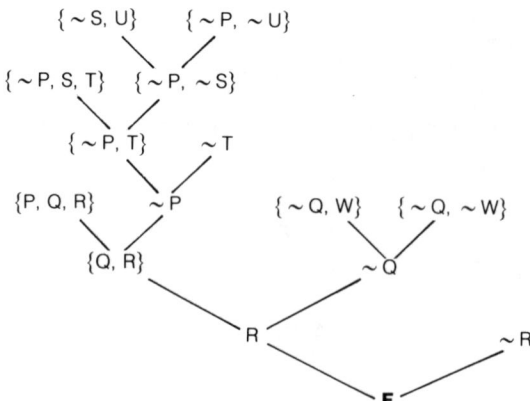

and halt.

To prove that the algorithm is correct (and hence that resolution is a complete proof system for **Prop**), we require a number of lemmas. First, to justify halting at step (2) if every clause contains a negative literal, we have:

Lemma 1. If every member of a set of clauses contains a negative literal, then that set of clauses is satisfiable.

Proof. To show this, consider the interpretation I which assigns the truth-value 'false' to every schematic letter appearing (either negated or unnegated) in any member of **S**. Now let C be any member of **S**; by hypothesis, C contains a negative literal, $\sim A$, say. Since A is false in the interpretation I, $\sim A$ is true under that interpretation, so I satisfies C. This goes for any clause in **S**, and hence I satisfies **S**—so **S** is satisfiable.

(To illustrate this, let $S = \{\{\sim P, Q\}, \{\sim Q, R, S\}, \{\sim R, S, T\}, \{\sim S, T\}\}$. Let I be the interpretation which makes P, Q, R, S, and T all false. Then it is easy to see that I satisfies each clause in **S**.)

The next lemma justifies the construction at step (7) of the algorithm.

Lemma 2. Let **S** be a set of clauses, and let L be any literal such that

(a)   L is a logical consequence of **S**, and
(b)   L is not a member of **S** (though it may, of course, be a member of one of the clauses in **S**).

Then the set $S' = \{C - \{L\} | C \in S\}$, obtained from **S** by removing L from each clause of **S** in which it occurs, is unsatisfiable.

Proof. Suppose that **S'** is *satisfiable*, and let I' be an interpretation satisfying **S'**. If L is false in I', set $I = I'$; otherwise let I be the interpretation which differs from I' only in that L is false in I but not I'. Now the only way that a clause can be true in I' but false in I is by containing the literal L. Therefore, since no clause of **S'** contains L, and I' satisfies **S'**, I must satisfy **S'** too. Now the clauses of **S** are just the clauses of **S'** with or without an extra disjunct L. Since adding a disjunct cannot make a true clause false, and I' satisfies S', I must satisfy **S** as well. So there is an interpretation satisfying **S** in which L is false: which means that L is not a logical consequence of **S**. Hence, if L *is* a logical consequence of **S**, then **S'** must be unsatisfiable, as required.

This lemma applies to our algorithm as follows. Since **S** is unsatisfiable, $S - \{C\}$ entails $\sim A_i$ for $i = 1, \ldots, n$ (where $C = \{A_1, \ldots, A_n\}$). Hence, by the lemma, the set S_i, obtained from $S - \{C\}$ by removing $\sim A_i$ from each clause in which it occurs, is unsatisfiable. This means we can recursively call the algorithm to find a refutation tree for S_i (here we are in effect assuming as an induction hypothesis that the algorithm is correct for any input containing fewer clauses than **S**). We then 'doctor' this refutation tree by putting back the occurrences of $\sim A_i$ which were removed when S_i was derived from $S - \{C\}$, and adjusting all the resolvents accordingly. The result, T_i, is a resolution tree over $S - \{C\}$. If any T_i still has root **F**, this means that $S - \{C\}$ is unsatisfiable, so we can abandon the clause C altogether and restart the algorithm with input $S - \{C\}$ (this is step (10) of our algorithm). If on the other hand none of the trees T_i has root **F**, then each T_i has root $\sim A_i$ (since in getting T_i from T'_i the only change made was to add $\sim A_i$ to some of the leaves and nodes of the tree), so we can construct the refutation tree for **S** as in step (9).

We have now done everything that is required to prove the correctness of the algorithm; all that remains to be done is to fill in a few details and present the induction in a mathematically impeccable form. This is left as an exercise for the reader.

The correctness of our algorithm means that *every* unsatisfiable set of clauses has a resolution refutation tree: we know this because the algorithm tells us how to construct one (note that the tree produced by the algorithm will not necessarily be the smallest possible tree for the set in question—what is important is that we can construct a tree at all). We now know that resolution is a complete proof system for the propositional calculus.

Resolution sequences

Some resolution trees have a particularly simple structure. Consider again our very first example, involving the set of clauses $S = \{P, \{\sim P, Q\}, \{\sim Q, R\}, \sim R\}$. The refutation tree we gave for this set is not by any means unique; another possibility is

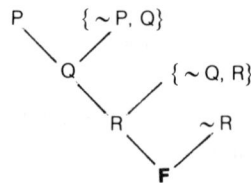

The nice thing about this tree is that *every resolution involves at least one element from* **S**. In other words, we never resolve together two clauses each of which is itself the result of an earlier resolution. A resolution tree having this property can be written out in a linear format as a *resolution sequence*:

$$P \xrightarrow{\{\sim P, Q\}} Q \xrightarrow{\{\sim Q, R\}} R \xrightarrow{\sim R} \mathbf{F}$$

A precise definition of a resolution sequence is

Let **S** be a set of clauses. A **resolution sequence** for **S** is a finite sequence of clauses $C_0, C_1, \ldots, C_n$ such that

(a) C_0 is a member of **S**, and

(b)   For $i = 1, \ldots, n, C_i$ is the resolvent of C_{i-1} with a member of **S**.

If $C_n = \mathbf{F}$ then $C_0, \ldots, C_n$ is called a **refutation sequence**.

Since a refutation sequence *is* a refutation tree, of a special kind, the existence of a refutation sequence for a set **S** is sufficient to prove that **S** is unsatisfiable. However, the converse of this does not hold: not every unsatisfiable set of clauses has a refutation sequence. A simple example is the set

$$\{\{P,Q\},\{\sim P,Q\},\{\sim Q,R\},\{\sim Q,\sim R\}\}$$

which has the refutation tree

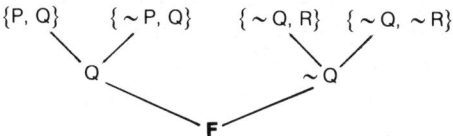

and hence is unsatisfiable, but has no refutation sequence. This is a consequence of the following lemma:

Lemma 3. If **S** has a refutation sequence, then **S** contains a unit clause.

Proof. Any refutation sequence must end in a resolution of the form res $(A, \sim A) = \mathbf{F}$, where A is a schematic letter. By the definition of a refutation sequence, one of A and $\sim$ A must be in **S**, so **S** does indeed contain a unit clause.

Resolution sequences are so much easier, and more efficient, to handle than resolution trees, that it is desirable to restrict our attention to a type of clause for which the existence or non-existence of a resolution sequence is an infallible criterion for determining whether or not a set is unsatisfiable.

A **Horn clause** is a clause containing at most one positive literal.

Examples of Horn clauses are

$$\mathbf{T}$$
$$\mathbf{F}$$
$$P$$
$$\sim P$$
$$\{\sim P,Q\}$$
$$\{\sim P,Q,\sim R\}$$
$$\{P,\sim Q,\sim R,\sim S\}$$

while some examples of non-Horn clauses are

$$\{P,Q\}$$
$$\{P,Q,\sim R\}$$
$$\{\sim P,Q,R\}$$
$$\{\sim P,Q,R,\sim S\}$$
$$\{P,Q,R,S,T\}$$

Our first result concerning Horn clauses is that the resolvent of two Horn clauses is itself a Horn clause. For suppose the number of positive literals in two clauses C

and C' are n and n' respectively. If res(C, C') exists, it is obtained by deleting a positive literal from one of C and C' and a negative literal from the other, and putting together the remaining literals. The number of positive literals in res(C, C') is therefore $n + n' - 1$. Now if C and C' are both Horn clauses, $n \leqslant 1$ and $n' \leqslant 1$, and hence $n + n' - 1 \leqslant 1$, i.e. res(C, C') is also a Horn clause.

We also have, as a straightforward corollary of Lemma 1, that if a set of Horn clauses is unsatisfiable, then it must either contain **F** or a positive unit clause (i.e. a schematic letter). This is because, by Lemma 1, it must contain a clause with no negative literal; but the only kind of Horn clauses of this kind are **F** and a schematic letter.

Our algorithm for constructing a refutation tree for any unsatisfiable set of clauses naturally applies to sets of Horn clauses as well as any other; and the result in this case is always a refutation *sequence*. This is because the clause C selected at step (3) has to be a positive unit clause, so the output at step (9) has the form

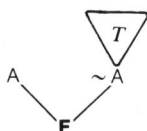

So long as the component T is a resolution sequence, then the whole thing will be a refutation sequence, so we can argue inductively that since the smallest possible output from the algorithm,

$$A \searrow \quad \sim A \swarrow$$
$$\mathbf{F}$$

is a refutation sequence, all possible outputs will be too. Thus if we restrict our use of resolution to the construction of resolution sequences, the method remains complete for Horn clauses, although as we have seen it is no longer complete for general clauses.

Example 3.6.3   We shall use the algorithm to construct a refutation sequence for the set $\mathbf{S} = \{P, \{\sim P, Q\}, \{\sim Q, \sim R\}, \{\sim P, R\}\}$.

We pass straight over steps (1) and (2); at (3) we select $C = P$. Since $\mathbf{S}$ does not contain $\sim P$, we pass over (4) to (5), at which we construct the set

$$\mathbf{S}_1 = \{Q, \{\sim Q, \sim R\}, R\},$$

and, at (6) recursively call the algorithm for $\mathbf{S}_1$.

In the recursive call, we pass over (1′) and (2′) as before, selecting $C' = Q$ at (3′). Since $\mathbf{S}_1$ does not contain $\sim Q$, we pass over (4′), constructing the set

$$\mathbf{S}_{11} = \{\sim R, R\}$$

at (5′). At (6′), we call the algorithm for this set.

At this level, we note that $\mathbf{S}_{11}$ contains both R and $\sim R$, and hence emerge from the secondary recursion with output tree T_{11} (here written as a refutation sequence):

$$\sim R \xrightarrow{R} \mathbf{F}$$

At (7') we insert into this sequence the occurrences of $\sim Q$ removed in going from $\mathbf{S}_1 - \{Q\}$ to $\mathbf{S}_{11}$, to obtain the sequence

$$\{\sim Q, \sim R\} \xrightarrow{\ R\ } \sim Q$$

Since $i = n = 1$ we pass over (8') to (9'), at which we output the sequence T_1:

$$\{\sim Q, \sim R\} \xrightarrow{\ R\ } \sim Q \xrightarrow{\ Q\ } \mathbf{F}$$

At this stage we re-emerge into the top level at step (7), at which we insert into this sequence the 'missing' occurrences of $\sim P$, giving us

$$\{\sim Q, \sim R\} \xrightarrow{\{\sim P, R\}} \{\sim P, \sim Q\} \xrightarrow{\{\sim P, Q\}} \sim P$$

Passing over (8), we finally reach (9), at which the output from the whole process is

$$\{\sim Q, \sim R\} \xrightarrow{\{\sim P, R\}} \{\sim P, \sim Q\} \xrightarrow{\{\sim P, Q\}} \sim P \xrightarrow{\ P\ } \mathbf{F}$$

This is the desired refutation sequence for **S**.

Of course, this refutation sequence is by no means unique. In fact, for this set **S** we can find a refutation sequence beginning with whichever of its members we choose. Some examples are as follows:

$$\{\sim P, R\} \xrightarrow{\{\sim Q, \sim R\}} \{\sim P, \sim Q\} \xrightarrow{\{\sim P, Q\}} \sim P \xrightarrow{\ P\ } \mathbf{F}$$

$$\{\sim P, Q\} \xrightarrow{\{\sim Q, \sim R\}} \{\sim P, \sim R\} \xrightarrow{\{\sim P, R\}} \sim P \xrightarrow{\ P\ } \mathbf{F}$$

$$P \xrightarrow{\{\sim P, R\}} R \xrightarrow{\{\sim Q, \sim R\}} \sim Q \xrightarrow{\{\sim P, Q\}} \sim P \xrightarrow{\ P\ } \mathbf{F}$$

Nor do these examples exhaust the possibilities. Altogether there are 12 distinct refutation sequences for this set: two beginning with P, four with $\{\sim Q, \sim R\}$, and three with each of the other two clauses. They can all be collected together in a single diagram, the **resolution network** for the set, as follows:

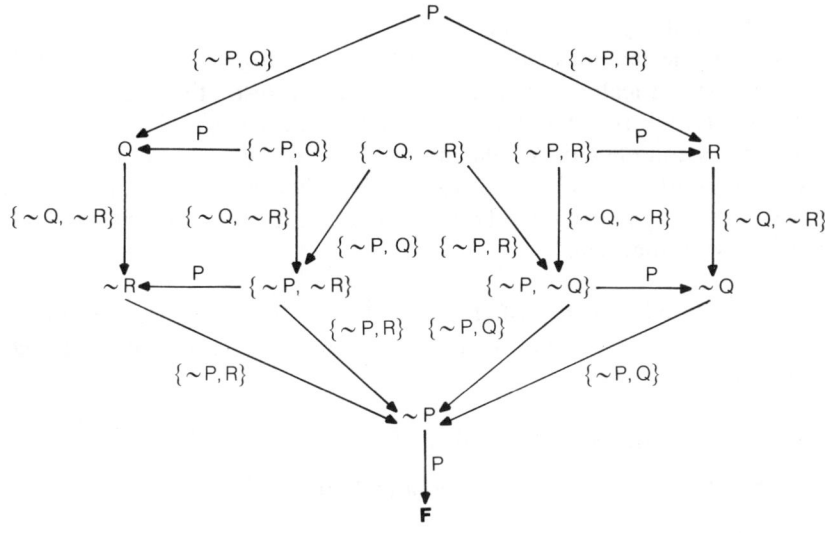

This network has a node for every clause occurring in any resolution sequence for S, and an arc for every resolution. So any possible resolution sequence can be traced out in the network; the presence of **F** in the network means that **S** is unsatisfiable.

If a set of Horn clauses is satisfiable, then we can still draw the resolution network for that set, but now it won't contain **F**. For example, the set

$$\{\{\sim P, Q\}, \{\sim P, S\}, \{\sim Q, R\}, \{\sim S, R\}, P\}$$

has the following resolution network:

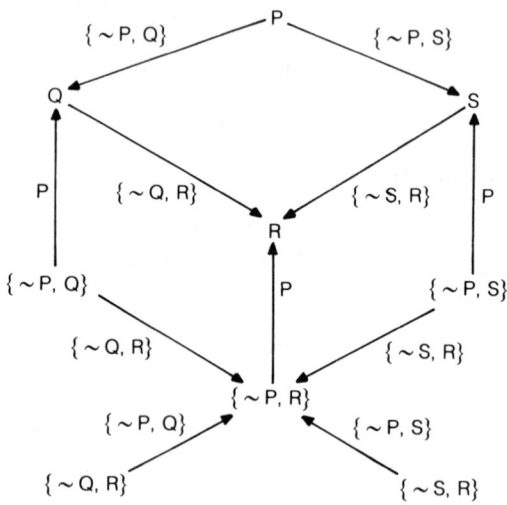

The eight possible resolution sequences for this set are all contained in the network; since **F** does not occur in the network, none of them is a refutation sequence, so the set is satisfiable (putting P, Q, R, and S all true satisfies it).

We thus have the makings of a procedure for testing a set of Horn clauses for satisfiability: systematically construct the complete resolution network, stopping when either the term **F** is encountered (in which case the original set is unsatisfiable), or the network is complete (in which case, if **F** is not there, the set is satisfiable).

For even a modest-sized set, this procedure is rather inefficient; but luckily it turns out not to be necessary to explore all possible resolution sequences. This is vouchsafed us by the following theorem.

Theorem. Let **S** be a finite unsatisfiable set of Horn clauses, and let *C* be any member of **S**. Then *either* **S** − {*C*} is unsatisfiable, *or* there is a refutation sequence for **S** beginning with *C*.

Proof. We use induction on the number of clauses in **S**:

Base case: **S** = {A, ∼A} for some schematic letter A. In this case the refutation sequences

$$A \xrightarrow{\sim A} \mathbf{F}$$

and

$$\sim A \xrightarrow{A} \mathbf{F}$$

give us what is required.

Induction step: Assume the result holds for all sets containing fewer than n clauses, and let $|\mathbf{S}| = n$.

Now whenever C occurs in a refutation sequence for $\mathbf{S}$, some literal in C corresponds to a contradictory literal in the preceding term of the sequence—but this contradictory literal must ultimately have come from some other member C' of $\mathbf{S}$, which means that C and C' could be resolved together directly. This means that if $\mathbf{S} - \{C\}$ contains no clause that can be resolved with C, then C does not figure in any refutation sequence for $\mathbf{S}$. But in this case, any refutation sequence for $\mathbf{S}$ is also a refutation sequence for $\mathbf{S} - \{C\}$, so $\mathbf{S} - \{C\}$ is unsatisfiable.

Suppose, on the other hand, that $\mathbf{S} - \{C\}$ is satisfiable. Then we know from the above argument that $\mathbf{S}$ must contain a clause C' such that C and C' have a resolvent, say R_1. Let $\mathbf{S}'$ be the set obtained by replacing C and C' in $\mathbf{S}$ with R_1, i.e.

$$\mathbf{S}' = (\mathbf{S} - \{C, C'\}) \cup \{R_1\}.$$

Now $\mathbf{S}'$ has $n - 1$ members, and $\mathbf{S}' - \{R_1\}$ is satisfiable (since it is a subset of $\mathbf{S} - \{C\}$), so by the induction hypothesis there is a refutation sequence for $\mathbf{S}'$ beginning with R_1:

$$R_1 \xrightarrow{C_1} R_1 \xrightarrow{C_2} \cdots \xrightarrow{C_{n-1}} R_n = \mathbf{F}$$

We can now immediately construct a refutation sequence for $\mathbf{S}$ beginning with C:

$$C \xrightarrow{C'} R_1 \xrightarrow{C_1} R_2 \xrightarrow{C_2} \cdots \xrightarrow{C_{n-1}} R_n = \mathbf{F}$$

and the proof is complete.

This theorem will only help us if we can be sure of picking a clause C in $\mathbf{S}$ for which $\mathbf{S} - \{C\}$ is unsatisfiable. Under certain circumstances this will indeed be the case. In particular, suppose we are given a set $\mathbf{P}$ of clauses known to be satisfiable, and wish to determine whether some schematic letter A not in $\mathbf{P}$ logically follows from $\mathbf{P}$. This will be the case just if $\mathbf{P}' = \mathbf{P} \cup \{A\}$ is unsatisfiable. Since $\mathbf{P}' - \{A\} = \mathbf{P}$ is satisfiable, the theorem tells us that $\mathbf{P}'$ will be unsatisfiable if and only if it has a refutation sequence beginning with $\sim A$. So all we need do is search systematically for refutation sequences of this kind.

Example 3.6.4 Given that $\mathbf{P} = \{\{Q, \sim S\}, P, \{\sim P, Q\}, \{\sim P, \sim Q, R\}\}$ is satisfiable, determine whether $\mathbf{P} \vDash R$.

We seek a refutation sequence for $\mathbf{P} \cup \{\sim R\}$, beginning with $\sim R$. The portion of the resolution network accessible from $\sim R$ is

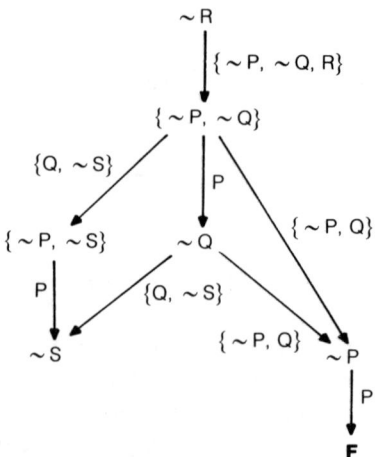

So a suitable refutation sequence is

$$\sim R \xrightarrow{\{\sim P,\,\sim Q,R\}} \{\sim P,\,\sim Q\} \xrightarrow{\{\sim P,Q\}} \sim P \xrightarrow{P} F$$

Hence $\mathbf{P} \cup \{\sim R\}$ is unsatisfiable, so $P \vDash R$.

Example 3.6.5 Given that $\mathbf{P} = \{P, \{\sim P, \sim R, S\},\ \{\sim P, \sim Q, T\},\ \{\sim Q, R\},\ \{\sim S, U\},\ \{\sim T, U\}\}$ is satisfiable, determine whether or not $\mathbf{P}$ entails U. The resolution network starting from $\sim U$ is

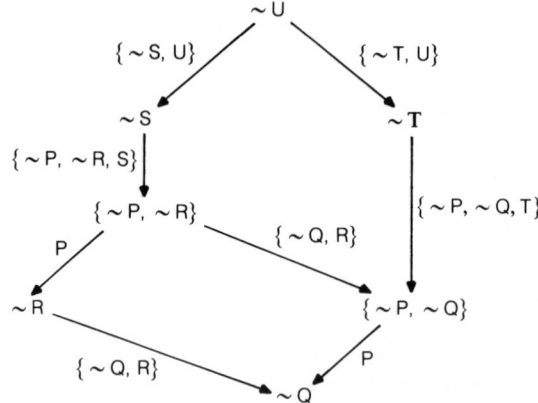

No resolution sequence beginning with $\sim U$ terminates in **F** so $\mathbf{P} \cup \{\sim U\}$ is satisfiable, i.e. U is *not* a logical consequence of $\mathbf{P}$.

So far we have simply exhibited the completed resolution networks (or relevant portions thereof) without indicating how they were obtained. We really need a method of generating these networks systematically. It turns out to be easier to control our strategy if we think of the network as expanded into a *tree*. We do this by replacing every converging pair of arcs, of the form

by a parallel pair of the form

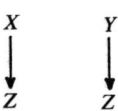

In this way the network acquires more nodes; but it does become a tree, which allows us to exploit existing tree-searching algorithms.

Our two networks above, when rewritten as trees, come out looking like this:

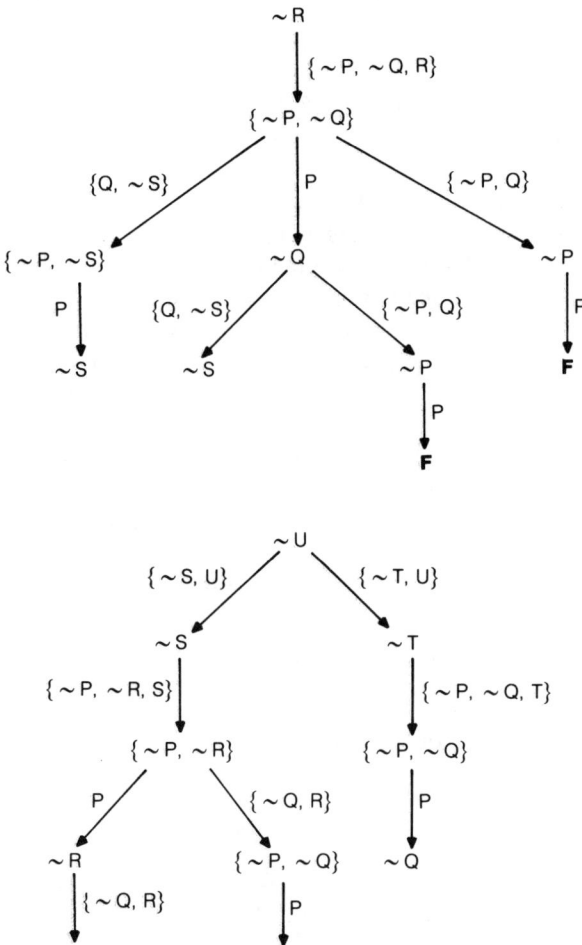

We have adopted the following convention for drawing the trees: where two or more branches fan out from one node, we arrange them from left to right in the order that their respective arc labels are presented in the original set. Thus, for example, in the last tree above, the initial fork at $\sim$U is drawn with the arc labelled $\{\sim S, U\}$ to the left of the arc labelled $\{\sim T, U\}$ because $\{\sim S, U\}$ precedes $\{\sim T, U\}$ in the original presentation of the set.

A natural way of searching through a tree is to work downward through the branches, always taking (say) the leftmost branch at any fork; on reaching a leaf, **backtrack** to the last fork encountered and try the leftmost branch that has not yet been explored. To illustrate, the nodes of the tree

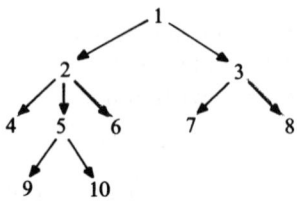

would be traversed in the order $1, 2, 4, 5, 9, 10, 6, 3, 7, 8$. Equally, if we are *constructing* the tree in this systematic fashion, then this is the order in which the nodes will be created.

The strategy we have presented here is a *left-to-right depth-first search with backtracking* or, more simply, a **pre-order traversal** of the tree. Many other strategies are possible, of course, but we shall confine our attention to this one.

An important point to notice is that relative to our chosen strategy, the order in which the clauses are given to use can be quite significant. In Example 3.6.4, with the clauses ordered as they are, the pre-order traversal does not reach an **F** until over half the tree has already been searched. If the clauses had been given in the order $\{\{\sim P, \sim Q, R\}, \{\sim P, Q\}, \{Q, \sim S\}, P\}$, then the traversal would find **F** at the end of the very first branch explored, thus enabling the search to be terminated. A different strategy might favour the original ordering of the clauses over this one, but any particular strategy is going to perform efficiently with some clause orderings and not with others.

Sometimes, it is not just efficiency that is at stake. Suppose we want to determine whether the set

$$\mathbf{P} = \{\{\sim P, Q\}, \{\sim Q, P\}, P\}$$

entails Q. Using our 'left-to-right' strategy, we would spend all our time exploring the infinite left-most branch of the resolution tree:

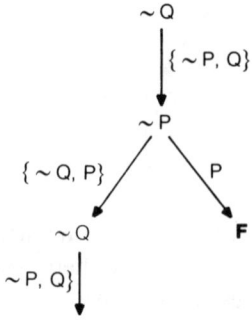

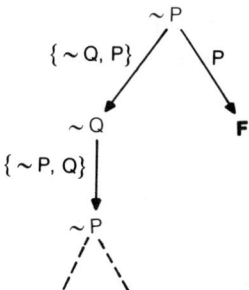

We would never succeed in finding any of the infinitely many occurrences of **F** in the tree!

The fact that the tree is infinite reflects the fact that the resolution network for $\mathbf{P} \cup \{\sim Q\}$, beginning with $\sim Q$, contains a loop:

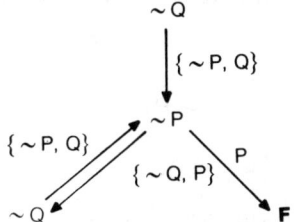

However, if the original set of clauses is presented in the order

$$\{P, \{\sim P, Q\}, \{\sim Q, P\}\}$$

then the tree becomes

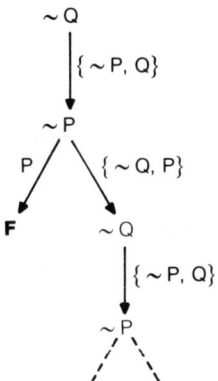

and our strategy manages to find **F** before entering the loop in the network.

Logic programming

All these ideas are highly relevant to the discipline known as **logic programming**. In logic programming, Horn clauses are interpreted as facts, queries, or rules. A positive unit clause such as P or Q is interpreted as a **fact**. A clause with a positive literal and one or more negative literals is interpreted as a **rule**; this is possible since the clause

$$\{\sim A_1, \ldots, \sim A_n, A\}$$

represents a schema such as

$$\sim A_1 \vee \cdots \vee \sim A_n \vee A$$

and this in turn is equivalent to

$$A_1 \wedge \cdots \wedge A_n \rightarrow A.$$

So our clause can be thought of as a *rule* to the effect that 'A is true if $A_1, \ldots, A_n$ are all true'. In logic programming it is customary to write this as

$$A \leftarrow A_1, \ldots, A_n.$$

The operation of resolution then takes the form: the resolvent of

$$A \leftarrow A_1, \ldots, A_n$$

and

$$A_m \leftarrow B_1, \ldots, B_k$$

is

$$A \leftarrow A_1, \ldots, A_{m-1}, B_1, \ldots, B_k, A_{m+1}, \ldots, A_n.$$

To illustrate, consider the following set of rules:

(1) If it rains, the picnic will be cancelled.
(2) If the picnic is cancelled, we shall go to the museum.
(3) If we go to the museum and it rains, we shall go by bus.
(4) If we go by bus, we shall need money.

and the fact

(5) It will rain.

Can we infer from (1)–(5) that we shall need money?
    Using the key

$$P = \text{'It will rain'}$$
$$Q = \text{'The picnic will be cancelled'}$$
$$R = \text{'We will go to the museum'}$$
$$S = \text{'We will go by bus'}$$
$$T = \text{'We will need money'}$$

we can rewrite (1)–(5) in logic programming notation as

(1) $Q \leftarrow P.$
(2) $R \leftarrow Q.$
(3) $S \leftarrow Q, R.$
(4) $T \leftarrow S.$
(5) $P.$

In clausal form, we are concerned with the set

$$\{\{\sim P, Q\}, \{\sim Q, R\}, \{\sim Q, \sim R, S\}, \{\sim S, T\}, P\}.$$

The query 'Will we need money?' is posed in the negative form

$$\sim T.$$

By resolving $\sim T$ with $\{\sim S, T\}$ we obtain $\sim S$; so our query has been replaced by a new query 'Will we go by bus?'. This resolves in turn with $\{\sim Q, \sim R, S\}$ to give $\{\sim Q, \sim R\}$, so that our query now consists of *two* questions, namely 'Will the picnic be cancelled?' and 'Will we go to the museum?'. Next, we resolve with $\{\sim Q, R\}$ to reduce the query to $\sim Q$ ('Will the picnic be cancelled?'), which resolves with $\{\sim P, Q\}$ to give $\sim P$ ('Will it rain?'). This finally resolves with P to give the empty clause **F**. Curiously enough, this corresponds to the answer 'yes' rather than to 'no'! This is because all our queries are put in *negative* form, so we can think of **F** as $\sim$**T** and hence as representing the query 'Is **T** true?'—to which the answer is 'yes', since **T** is defined to be *always* true.

What we have just done, of course, is to construct the refutation sequence

$$\sim T \xrightarrow{\{\sim S, T\}} \sim S \xrightarrow{\{\sim Q, \sim R, S\}} \{\sim Q, \sim R\} \xrightarrow{\{\sim Q, R\}} \sim Q \xrightarrow{\{\sim P, Q\}} \sim P \xrightarrow{P} \mathbf{F}.$$

The restriction to Horn clauses means that we are not easily able to handle very natural kinds of rules such as

> If we go to the museum and it is not raining, we shall walk.

Using U for 'We will walk', this rule is

$$(6a) \quad R \wedge \sim P \rightarrow U.$$

This is equivalent to the *non-Horn* clause

$$(6b) \quad R \rightarrow P \vee U,$$

i.e. to $\{\sim R, P, U\}$; and we know that introducing non-Horn clauses means that we can no longer be sure that every unsatisfiable set has a refutation sequence, so that the procedure we have adopted would become incomplete.

Logic programming, in practice, makes use of an extra rule of inference called **negation by failure** in order to overcome this problem. This, and the further development of logic programming, lie outside the scope of this book. For details, see Hogger (1984) and Lloyd (1984).

Exercise 3.6

(1) Resolve the following pairs of clauses:

    *(a) $\{P, Q\}, \{\sim Q, R, S\}$
    *(b) $\{\sim P, \sim Q, R\}, \{P, R\}$
    (c) $\{P, Q, R\}, \{S, \sim Q, T\}$
    *(d) $\{P, Q\}, \sim Q$
    *(e) $\{P, Q, \sim R, S\}, \{T, \sim R, \sim P\}$
    *(f) $\{\sim P, \sim Q, R\}, \{P, Q, S\}$
    (g) $P, \{\sim P, \sim Q\}$
    (h) $\{P, Q\}, \{\sim P, \sim Q\}$
    (i) $\{P, Q, \sim R \sim S, T\}, \{U, P, \sim S, R, W\}$

*(2) Which of the following are Horn clauses?

(a) $\{P, \sim Q, \sim R\}$
(b) $\{\sim P, Q, \sim R, S\}$
(c) Q
(d) $\{P, R\}$
(e) $\{\sim P, \sim Q, \sim R\}$
(f) $\{P, \sim Q, \sim R, S, \sim T\}$
(g) $\sim R$
(h) $\{\sim P, \sim Q, R, \sim S, \sim T\}$

(3) Use resolution to determine which of the following sets of clauses are unsatisfiable:

*(a) $\{P \lor Q, \sim R \lor S, \sim P \lor S, \sim Q \lor R, \sim S\}$
*(b) $\{P \lor Q \lor R, \sim S, \sim Q \lor S, \sim P \lor S\}$
(c) $\{\sim P, P \lor Q, P \lor \sim U, \sim Q \lor \sim R \lor S, \sim S, R \lor T, \sim T \lor U\}$
(d) $\{P \lor \sim Q, \sim P \lor Q, P, \sim Q\}$
(e) $\{\sim P \lor Q, \sim P \lor R, \sim Q \lor \sim R \lor S, \sim Q \lor \sim R \lor T, \sim S \lor \sim T \lor U, \sim U\}$
(f) $\{P, \sim P \lor Q, \sim P \lor \sim Q \lor R, \sim R \lor \sim S \lor T, S \lor T, \sim P \lor T, \sim R \lor T\}$
(g) $\{P \lor Q, \sim P \lor R \lor S, \sim R \lor T, \sim R \lor U, \sim U \lor S \lor \sim T\}$

*(4) Find a resolution sequence which refutes the same set of clauses as the following resolution tree:

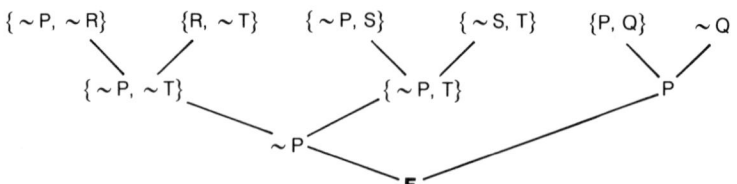

Can you do the same thing for the tree generated in Example 3.6.2?

*(5) Convert the following inference into clausal form and use resolution to determine whether or not it is valid:

$$P \lor Q$$
$$P \rightarrow R \lor S$$
$$R \rightarrow T \land U$$
$$\underline{U \land \sim S \rightarrow \sim T}$$
$$\sim S \rightarrow Q$$

(6) Use the method of Question (5) to validate the inferences in Exercise 3.2, Question (1).

4 The Predicate Calculus

4.1 OBJECTS, PROPERTIES, AND RELATIONS

Consider the following inference:

No integer is greater than its own square
$\frac{1}{2}$ is greater than its own square

$\frac{1}{2}$ is not an integer

The logical form of this inference is given by the schema:

No A is B
c is B

c is not an A

Intuitively, this is plainly valid, since if the conclusion were false but the second premiss were true, then c would be both an A and a B, which would make the first premiss false: so if both premisses are true, the conclusion must be true too.

What makes this inference valid? The word 'no' in the first premiss plays an important role here. For example, if 'no' were replaced by 'some' or 'every', then the inference would become invalid. But there is no way of representing the statement-schema 'No A is B' in the formalism of the propositional calculus. It looks like a negation, and indeed it is; but consider what it is the negation *of*. What statement is *denied* by an assertion that no A is B? The best answer is 'Some A is B'. Thus to assert that *no* integer is greater than its own square amounts to nothing more than to deny that *some* integer is greater than its own square (i.e. to deny that there is an integer greater than its own square). This merely pushes the problem of representation back a step: for we still need to represent 'Some A is B', and this is something that cannot be done in the propositional calculus.

Both 'integer' and 'greater than its own square' denote **properties** that an object may or may not have. If it seems odd to think of the word 'integer' as denoting a property, you can think of it as denoting the property of *being* an integer. The first premiss of our inference says that no object has both these properties together, that being an integer and being greater than its own square do not occur together as properties of one and the same object. Or, to put it slightly differently, that the class of integers has no elements in common with the class of objects which are greater than their own squares. The schematic letters A and B in our inference schema are thus to be interpreted as referring to properties.

The third schematic letter, c, is different, which is why we used a lower-case letter in this case. The expression '$\frac{1}{2}$' does not denote a property, it denotes an **object** (albeit an abstract, mathematical object). Objects are not the same as properties: objects *have* properties. Thus $\frac{1}{2}$ has the property of being greater than its own square (since $\frac{1}{2} > \frac{1}{2} \times \frac{1}{2} = \frac{1}{4}$), and this is precisely what is stated by the second premiss of our inference; while the conclusion of the inference tells us that $\frac{1}{2}$ does *not* have the property of being an integer.

We could interpret our inference schema differently by choosing different readings for A, B, and c. For example, for A we could have the property of being a man, for B the property of being an island, and for c the object known as the Isle of Man. We then have the valid, if somewhat pointless, inference:

> No man is an island
> The Isle of Man is an island
> ───────────────────────────
> The Isle of Man is not a man

Another interpretation gives us

> No fish breathes air
> My pet lungfish breathes air
> ───────────────────────────
> My pet lungfish is not a fish

This one is still valid, but it is not sound, since the first premiss is false.

Now consider the following inference:

> John is a poet
> Mary is a computer scientist
> John is married to Mary
> ──
> Some poet is married to some computer scientist

Again, this is intuitively valid; what schema does it represent? If we use a schematic letter R in place of the word 'married', then we get the schema:

> a is a P
> b is a Q
> a is R to b
> ────────────────────
> Some P is R to some Q

In this schema there are letters for objects (a, b) and for properties (P, Q) as before; but we also have the letter R which behaves rather differently. The expression 'is married to' does not refer to a property of single objects but to a **relation** between *pairs* of objects.

Other relations are expressed by the expressions 'is taller than', 'is the father of', 'is half of', 'is greater than', 'is a subset of', 'is north of', 'loves', 'hates', and so on. We shall allow all these relations as interpretations of the schematic relation 'is R to', even though, in English, not all of them begin with 'is' or end with 'to'. Thus two more instantiations of the schema above are

> 8 is a cube
> 16 is a square
> 8 is half of 16
> _____
>
> Some cube is half of some square

and

> Alnwick is an English town
> Dumfries is a Scottish town
> Alnwick is north of Dumfries
> _____
>
> Some English town is north of some Scottish town

If instead of the first two premisses we had said 'Alnwick is in England' and 'Dumfries is in Scotland', we would have to be careful how we phrased the conclusion. We could not say 'Some in England is north of some in Scotland', since this is not good English; nonetheless, it is clear enough what is meant, and a suitable paraphrase, in acceptable English, would be 'Somewhere in England is north of somewhere in Scotland'.

Relations can also involve more than two objects. For example, in the statements 'Okehampton is between Exeter and Plymouth', 'Charles is the child of Phillip and Elizabeth', and 'The greatest common divisor of 15 and 20 is 5', we have **three-place** relations, which could be schematized as 'a is R to b and c'. In principle there is no reason why we should not have relations involving as many terms as we want, though in practice two- and three-place relations are the most frequently encountered.

In this section we have identified several kinds of schematic letter which are needed for displaying the logical form of inferences which go beyond the resources of the propositional calculus. We need to talk about

- objects
- properties of objects
- relations between objects.

These form the subject matter of the branch of logic known as the predicate calculus.

Exercise 4.1 Each of the following inferences is an interpretation of one of the schemas shown below. Identify which schema corresponds to each inference, indicating clearly what objects, properties or relations are represented by the schematic letters. Determine intuitively which of the inferences are valid.

Schemas

(a) Some A is a B
No C is a B
_____
No A is a C

(b) a is R to b
b is R to c
_____
a is R to c

(c) Either nothing is A or everything is A
If a is A then b is not A
_____
a is not A

    (d)   Every A is either a B or a C
           No B is a D
           No C is a D

           No A is a D

Inferences

  *(1)   Every rectangle is either a square or an oblong
         No square has unequal diagonals
         No oblong has unequal diagonals

         No rectangle has unequal diagonals

  *(2)   Julia is taller than David
         David is taller than Rosemary

         Julia is taller than Rosemary

   (3)   Some Yugoslavs speak Albanian
         No Greeks speak Albanian

         No Yugoslavs are Greeks

   (4)   Every science examination lasts either 2 hours or 3 hours
         No two-hour examination can take place in the Library
         No three-hour examination can take place in the Library

         No science examination can take place in the Library

  *(5)   Either nobody goes or everyone goes
         If John goes, Mary will not go

         John will not go

   (6)   Some of the lecturers know the answer
         None of the students knows the answer

         None of the lecturers is a student

   (7)   Josie lives next door to Maggie
         Maggie lives next door to Sue

         Josie lives next door to Sue

   (8)   You shall have none of them or all of them
         If you have the red one then you shall not have the blue one

         You shall not have the red one

4.2 NAMES AND PREDICATES

We have used the schematic letter c in place of 'John', '$\frac{1}{2}$', 'The Isle of Man', and 'my pet lungfish'. All these expressions are intended to pick out a unique individual object, a particular person, number, island, fish or whatever other kind of object we want to refer to. I say 'intended' because the best of intentions sometimes fail, and actually I do not own a pet lungfish, so when *I* say 'my pet lungfish' I fail to refer to anything at all. Such failures aside, though, there is little problem, in practical terms, in identifying an expression which is supposed to pick out a unique individual. Such expressions are called **referring expressions**.

In ordinary language, a referring expression may be simple or complex, depending on *how* it picks out the individual that it refers to. A *simple* referring expression is a **proper name** such as 'Hamlet' or 'William Shakespeare'. It refers *directly* to the individual it names. A *complex* referring expression is a **definite description** such as 'Shakespeare's greatest play' or 'the author of *Hamlet*'. A definite description picks out an individual by *describing* it, i.e. by mentioning a property that it alone possesses. Shakespeare is not just *an* author of *Hamlet*, he is *the one and only* author of that play.

In the predicate calculus, we shall use lower-case letters like a, b, and c to play the part of proper names. These letters are called **individual constants**: 'individual' because they are supposed to refer to individuals of some kind or other, and 'constant' because each of them is supposed to refer to the *same* individual at all of its occurrences under a given interpretation (cf. the infelicitous use of the name 'Bach' to denote two different individuals in the example given in Section 1.2).

Although individual constants will typically be replaced as proper names when we interpret predicate calculus schemas by means of English sentences, there is no reason why we should not also replace them by definite descriptions, as we did when we interpreted the schematic letter c as referring to my pet lungfish in one of the examples in Section 4.1. If we do this, though, we are denying ourselves the possibility of representing the logical structure of the definite description itself. For example, the definite description 'my pet lungfish' characterizes the object it refers to as a lungfish and as belonging to me as a pet: no proper name can do that. We shall discuss later what options are available to us if we wish to represent the internal logical complexity of definite descriptions (see below, Section 4.4). Sometimes what looks like a definite description can really only be regarded as a proper name. The title of a book, story, or play, for example *The Hound of the Baskervilles*, must always be treated as a proper name, whatever its linguistic form.

Turning now to properties, we note that in English there are many different ways of ascribing a property to an object. Here are some examples:

'John is a poet' ascribes to John the property of being a poet.
'The president is bald' ascribes to the president the property of baldness.
'Jane can swim' ascribes to Jane the property of being able to swim.
'Jonathan was born in 1984' ascribes to Jonathan the property of having been born in 1984.
'Rosemary likes eggs' ascribes to Rosemary the property of liking eggs.
'127 is prime' ascribes to the number 127 the property of being a prime number.

These ones are easy enough to analyse because in each case the object to which a property is ascribed is the subject of the sentence. But a sentence can be understood as ascribing a property to an object even though that object is named by a part of the sentence other than the subject:

'Everybody admires David' ascribes to David the property of being admired by everybody.

Note that it would be a mistake to analyse this example as ascribing the property of admiring David to an individual called 'Everybody'. There is no such individual; the role of the word 'everybody' in this statement is quite different, as we shall see.

A statement involving a relation and two names can be analysed in two different ways as ascribing a property to an object:

'John loves Mary' ascribes to John he property of loving Mary; but it also ascribes to Mary the property of being loved by John.

Of course, these two analyses amount to the same thing: if they did not, the sentence 'John loves Mary' would be ambiguous. *It is extremely important for logic that the same statement can be analysed in more than one way.* Here is another example:

'John's brother was best man at Charles's wedding' ascribes to John the property of having a brother who was best man at Charles's wedding; to John's brother it ascribes the property of having *been* best man at Charles's wedding; and to Charles it ascribes the property of having had John's brother as best man at his wedding.

Any statement which ascribes a property to an object can be analysed as consisting of a **predicate** and a **referring expression** (either a name or a definite description). The latter refers to the object, the former denotes the property. The predicate consists of everything that is left when the referring expression is removed from the statement, including the gap which indicates where it was removed from. Thus

'John is a poet' can be broken down into the name 'John' and the predicate '... is a poet'.

'Everyone admires David' can be broken down into the name 'David' and the predicate 'Everyone admires ...'.

'John loves Mary' can be analysed in two ways, according as it is broken down into

(i)   the name 'John' and the predicate '... loves Mary', or
(ii) the name 'Mary' and the predicate 'John loves ...'.

Warning: This use of the term 'predicate' in logic is rather different from the use of the same term in linguistics; in English grammar, for example, it is customary to analyse the sentence 'John loves Mary' as breaking down uniquely into the *subject* 'John' and the *predicate* 'loves Mary'. It would not be correct, in a grammatical (as opposed to logical) analysis, to split it up into 'John loves' and 'Mary'. Moreover, the grammatical subject need not count as a referring expression in the logical analysis, e.g. the subject of 'Everybody admires David' is 'everybody', but this is not a referring expression, because it does not pick out a unique individual object (though '... admires David' *is* a predicate)—for more on such expressions, see Section 4.3.

The predicate calculus avoids all this verbal complexity by using a simple uniform notation for ascribing a property to an object. A property is denoted by a **predicate letter** such as F, G, or P, together with an **argument place** corresponding to the gap '...' which we wrote in the predicates given above: thus we have F(____), G(____), and P(____), where '____' represents the argument place.

To ascribe the property denoted by F(___) to the object denoted a, we write F(a). We say that a is the **argument** of the predicate F(___) in this case (note that some authors omit the parentheses, simply writing Fa). Thus the schema F(a) could be presented as the logical analysis of

'John is a poet', with F(___) for '...is a poet', and a for John;

'Everyone admires David', with F(___) for 'Everyone admires...', and a for David.

'$\frac{1}{2}$ is greater than its own square', with F(___) for '...is greater than its own square', and a for $\frac{1}{2}$.

Relations can be handled in a similar way to properties. A statement like 'John is married to Mary' can be broken down into the two names 'John' and 'Mary' and a **two-place** (or **binary**) predicate '___ is married to...', which is what remains when the names are removed. Note that the argument places are distinguished from each other by being written differently ('___' and '...'): this is because they will, in general, be filled by different names. If we had written '___ is married to ___', we would have implied that the two argument places have to be filled by the *same* name, giving us nonsensical statements such as 'John is married to John' (i.e., 'John is married to himself').

Our earlier example, 'John's brother was best man at Charles's wedding', can similarly be analysed as consisting of the two names 'John' and 'Charles' and the two-place predicate

'___'s brother was best man at ...'s wedding',

which denotes the relation that holds between any two people such that the brother of the first was best man at the wedding of the second. Alternatively, the same statement can be analysed as consisting of the referring expressions 'John's brother' and 'Charles', and the two-place predicate

'___ was best man at...'s wedding'.

Note that the order of the two names in a relational statement is, in general, significant. Thus 'John loves Mary' says something different from 'Mary loves John'. The exceptions involve *symmetric* relations, which hold between a pair of objects irrespective of the order in which they are named. An example is '___ is married to...': clearly 'John is married to Mary' is exactly equivalent to 'Mary is married to John'.

In the predicate calculus, a two-place relation is denoted by a predicate letter with *two* argument places (e.g., G(...,___)), a three-place relation by a predicate letter with three argument places (e.g., Q(___,...,___)), and so on. To ascribe the two-place relation denoted by G(...,___) to the objects denoted by a and b (taken in that order), we write G(a,b). This schema can be given as the logical analysis of

'John loves Mary', with G(...,___) for '... loves ___', a for 'John', and b for 'Mary';

'24 is a multiple of 3', with G(...,___) for '...is a multiple of ___', a for '24', and b for '3'.

A subtlety: we could also interpret G(a, b) as 'John loves Mary' by reading G(...,___) as '_____ loves...', a as 'Mary', and b as 'John'. Study this example carefully to see how it differs from the previous analysis of the same sentence. This example shows us that it is of the utmost importance to pay attention to the roles played by the different argument places in a relational predicate.

Note that any predicate letter, such as F, can be used to represent a one-place predicate, a two-place predicate, or indeed a predicate with any number of argument places. In setting up a notation for a particular application of the predicate calculus, it has to be decided at the outset how many argument places each predicate letter is to have, and for as long as one uses that particular notation, one must stick with the decision. (In mathematically rigorous treatments it is usual to distinguish the number of argument places of a predicate by means of a numerical superscript, so that F^1 would be a letter for a one place predicate, F^2 for a two-place, and so on. Thus $F^1, F^2, F^3, \ldots$ could all occur in the same system without risk of ambiguity.)

Three series of adjectives are available for referring to the number of argument places a predicate has, derived from English, Latin, and Greek. For one, two, three, and n argument places, they are

English	Latin	Greek
one-place	unary	monadic
two-place	binary	dyadic
three-place	ternary	triadic
n-place	n-ary	n-adic

Exercise 4.2

(1) Analyse each of the following statements into referring expressions and predicates is as many ways as possible (e.g., 'John loves Mary' can be analysed as (a) '...loves Mary' with argument 'John', (b) 'John loves...' with argument 'Mary', and (c) '...loves _____' with arguments 'John' and 'Mary'.)

   *(a) Antony met Carol in Leeds.
   *(b) Charles lives in the biggest city in England.
    (c) The King of Spain is not bald.
    (d) Ben's mother is married to Frances's brother.
    (e) Three squared is the successor of two cubed.
    (f) Somebody has stolen Jeremy's bicycle.
   *(g) The author of 'The Merchant of Venice' was an Englishman.
    (h) The greatest common divisor of 28 and 42 is prime.

(2) Using the following readings of the predicate letters:

$$E(\underline{\quad}) \qquad \text{'}\underline{\quad}\text{ is even'}$$
$$P(\underline{\quad}) \qquad \text{'}\underline{\quad}\text{ is prime'}$$
$$M(\underline{\quad}, \ldots) \qquad \text{'}\underline{\quad}\text{ is a multiple of...'}$$
$$G(\underline{\quad}, \ldots) \qquad \text{'}\underline{\quad}\text{ is greater than...'}$$

and taking a, b, c, d,... to stand for the numbers 1, 2, 3, 4,..., respectively, write English versions of the statements represented by each of the following schemas, and decide which ones are true and which false under this interpretation (e.g., P(d) says '4 is prime', which is false):

*(a)	E(b)	(b)	P(c)
(c)	M(d, b)	*(d)	G(a, e)
(e)	~ M(b, c)	(f)	E(b) ∧ P(b)
(g)	M(c, c)	*(h)	M(f, b) ∨ P(f)
(i)	G(d, d)	*(j)	E(i) ↔ P(i)

(3) Using the translations given in question (2), write predicate calculus schemas to represent the following statements (e.g., to represent the statement '5 is not a multiple of 2', we write ~ M(e, b)):

- (a)   4 is even.
- *(b)   7 is prime.
- *(c)   6 is a multiple of 3.
- (d)   7 is not even.
- *(e)   Either 3 is prime or 2 is greater than 4.
- (f)   If 5 is a multiple of 2 then 7 is a multiple of 3.
- (g)   If 6 is even then 6 is a multiple of 2.
- (h)   4 is prime if and only if 5 is even.
- *(i)   If 8 is a multiple of 2 then either 8 is prime or 8 is greater than 2.
- (j)   If 7 is greater than 6, and 6 is greater than 5, then 7 is greater than 5.

4.3 QUANTIFIERS

Consider again the inference

> John is a poet
> Mary is a computer scientist
> John loves Mary
> _____
> Some poet loves some computer scientist

We are now in a position to write logical schemas to represent each of the three premisses, for example:

> P(a)
> Q(b)
> R(a, b)

where P(____), Q(____), and R(____,...) are interpreted to mean '____ is a poet', '____ is a computer scientist', and '____ loves...', respectively, and a names John and b names Mary. What about the conclusion?

The word 'some' plays an important part here, so we obviously need a way of representing this. The way we do so is by **regimenting** the English sentence into the following form:

> For some x, x is a poet and for some y, y is a computer scientist and x loves y.

The letters x and y here are important because they show how the various bits of the sentence are linked together. Thus since x belongs with the predicate '____ is a poet' and occupies the first argument place of '____ loves ...', we know that it is a poet who loves; while y, which occupies the second argument place of '____ loves ...', belongs with the predicate '...is a computer scientist', and hence it is a computer scientist who is loved. If we swapped round the terms of the relation, to get

> For some x, x is a poet and for some y, y is a computer scientist and y loves x,

then we would have a different statement, namely 'Some computer scientist loves some poet'. The symbols x and y here are called **bound variables**.

In symbols, we write $\exists$ to represent the phrase 'for some'. We can thus represent the conclusion of our inference by the predicate calculus schema

$$\exists x(P(x) \land \exists y(Q(y) \land R(x,y))).$$

The symbol $\exists$ is known as the **existential quantifier**: it may be read as 'for some', 'there is' or 'there exists'. We could have regimented our conclusion into either of the forms

There is an x such that x is a poet and
> there is a y such that y is a computer scientist and x loves y;

There exists an x such that x is a poet and
> there exists a y such that y is a computer scientist and x loves y;

and either of these would also represent the schema just given.

Here are some more statements whose predicate calculus representations use an existential quantifier. In each case we give a version in ordinary English, a regimented version, and the corresponding predicate calculus schema:

(1) There exists a poet.
     For some x, x is a poet.
     $\exists x P(x)$.
(2) Some poet is a computer scientist.
     For some x, x is a poet and x is a computer scientist.
     $\exists x(P(x) \land Q(x))$.
(3) Some computer scientist is a poet.
     For some x, x is a computer scientist and x is a poet.
     $\exists x(Q(x) \land P(x))$.

Examples (2) and (3) differ only in the order of the conjuncts. Since we know that conjunction is a commutative truth-function, it follows that the schemas in (2) and (3) are equivalent. And indeed, the English sentences 'Some poet is a computer scientist' and 'Some computer scientist is a poet' do say the same thing: only the emphasis is different. In terms of set theory, both sentences say that the class of poets and the class of computer scientists have non-empty intersection.

(4) Someone loves Mary.
     For some x, x loves Mary.
     $\exists x R(x, b)$.
(5) Mary loves somebody.
     For some x, Mary loves x.
     $\exists x R(b, x)$.

(6)   Mary loves a poet.
      For some x, x is a poet and Mary loves x.
      $\exists x(P(x) \wedge R(b,x))$.
(7)   Mary does not love a computer scientist.
      It is not the case that for some x, x is a computer scientist and Mary loves x.
      $\sim \exists x(Q(x) \wedge R(b,x))$.

Examples (6) and (7) show that the existential quantifier can sometimes be used to represent phrases like 'a poet', which contain the indefinite article ('a' or 'an'). As a rule of thumb, if the indefinite article is preceded by part of the verb 'to be' (e.g. 'am', 'is', or 'are'), then all we need is a predicate, but if it is preceded by another verb, or if it introduces the subject of the sentence, then we need a quantifier.

Thus we have

'Mary is a poet'	$P(b)$
'Mary loves a poet'	$\exists x(P(x) \wedge R(b,x))$
'A poet loves Mary'	$\exists x(P(x) \wedge R(x,b))$

(8)   No computer scientist loves Mary.
      It is not the case that for some x, x is a computer scientist and x loves Mary.
      $\sim \exists x(Q(x) \wedge R(x,b))$

Example (8) illustrates how the existential quantifier can be used with negation to represent expressions like 'no computer scientist' or 'no poet'. The negation of the statement 'Some computer scientist loves Mary' could be read as

> It is not the case that some computer scientist loves Mary;

but this says precisely the same as 'No computer scientist loves Mary'. Hence the correct representation of *this* sentence is as given in (8).

We can now give a predicate calculus representation of the whole of the inference we began this chapter with. The inference was:

> No integer is greater than its own square
> $\frac{1}{2}$ is greater than its own square
> _____
> $\frac{1}{2}$ is not an integer

Regimenting the first premiss, we have

> It is not the case that for some x, x is an integer and x is greater than its own square.

Then, putting $G(\underline{})$ for '____ is greater than its own square', $I(\underline{})$ for '____ is an integer', and a for '$\frac{1}{2}$', the inference comes out as

> $\sim \exists x(I(x) \wedge G(x))$
> $G(a)$
> _____
> $\sim I(a)$

As a further refinement, we could note that another way of saying, for example, '$\frac{1}{2}$ is greater than its own square' is '$\frac{1}{2}$ is greater than the square of $\frac{1}{2}$'. So the predicate

'____ is greater than its own square' is just a special case of the two-place predicate '____ is greater than the square of ...', in which the two argument places are filled by two copies of the same name. Putting Q(____,...) to mean '____ is greater than the square of ...', we then have

$$\sim \exists x(I(x) \wedge Q(x,x))$$
$$Q(a,a)$$
$$\overline{\sim I(a).}$$

In English, 'some' can go with either a singular or a plural noun. We can say either 'Some poet is a computer scientist' or 'Some poets are computer scientists'. You may be wondering how we can distinguish these statements in predicate calculus notation. The answer is that we cannot! The existential quantifier is used to assert the mere *existence* of something with a certain property. It does not distinguish between existence of just one such thing and the existence of more than one. Thus *both* the statements above would be represented by the schema $\exists x(P(x) \wedge Q(x))$. And if you had to give an English version of this schema (with the letters P and Q interpreted as above), you could give any of:

Some poet is a computer scientist.
Some poets are computer scientists.
There is a poet who is a computer scientist.
There are poets who are computer scientists.
There exists a poet who is a computer scientist.
There exist poets who are computer scientists.

(Indeed, in view of the commutativity of conjunction, we could swap 'poet' with 'computer scientist' throughout to obtain another six correct readings of the schema.). From the point of view of the predicate calculus, all these statements say the same thing, that the class of poets and the class of computer scientists have at least one member in common. Any differences in meaning you may detect among the English sentences are matters of nuance and emphasis, which go beyond the strict concerns of logic (but see below, pp. 212–213).

Binding and scope

An existential quantifier, together with its bound variable, e.g., $\exists x$, can be thought of as a **higher-level predicate** which takes as its argument not a referring expression but an ordinary **first-level predicate** such as P(____). Thus the schema $\exists x P(x)$ can be thought of as the result of applying the higher-level predicate $\exists x$ to the first-level predicate P(____).

A general first-level predicate may be thought of as the result of removing one or more occurrences of a referring expression from a complete schema; for example, P(____) can be obtained by removing a from P(a). Similarly, if we remove both occurrences of b from $Q(b) \wedge R(a,b)$, we obtain the **complex** first-level predicate $Q(\underline{\quad}) \wedge R(a,\underline{\quad})$, which denotes (under our interpretation) the property of being a computer scientist whom John loves; and this complex predicate can be the argument of a quantifier $\exists y$, giving the schema

$$\exists y(Q(y) \wedge R(a, y)).$$

Note how the bound variable of the quantifier fills in the argument places of the predicate. The function of the bound variable is precisely to **bind** the quantifier to those argument places. An alternative notation could be

$$\exists (Q(\underline{\quad}) \wedge R(a, \underline{\quad}))$$

in which the bindings are drawn explicitly. This shows us that *which* letters we use as the bound variable is immaterial, so long as the bindings are clear. Thus the three schemas

$$\exists x(Q(x) \wedge R(a, x))$$
$$\exists y(Q(y) \wedge R(a, y))$$
$$\exists z(Q(z) \wedge R(a, z))$$

are equivalent, being alternative ways of representing the binding structure shown explicitly above.

This schema can occur as a conjunct within a larger schema such as

$$P(a) \wedge \exists y(Q(y) \wedge R(a, y)),$$

which under our interpretation says that John is a poet and loves some computer scientist. Removing both occurrences of a from this larger schema, we obtain a new complex (first-level) predicate

$$P(\underline{\quad}) \wedge \exists y(Q(y) \wedge R(\underline{\quad}, y))$$

which denotes the property of being a poet who loves a computer scientist. This predicate can in turn be the argument of a new existential quantifier, giving us a schema with the following binding structure:

$$\exists (P(\underline{\quad}) \wedge \exists y(Q(y) \wedge R(\underline{\quad}, y))).$$

In predicate calculus notation, the bindings must be indicated by means of a bound variable. As before, which variable we use is immaterial, *so long as the bindings are clear*. If we used y, the bindings would *not* be clear, since it would be impossible to tell which argument places are bound to which of the two quantifiers. So we may use any variable apart from y, for example:

$$\exists x(P(x) \wedge \exists y(Q(y) \wedge R(x, y))),$$
$$\exists z(P(z) \wedge \exists y(Q(y) \wedge R(z, y))).$$

This way of thinking of the structure of a quantified schema, i.e. as the result of applying a quantifier to a first-level predicate, makes it clear what the correct notion of *scope* should be in the case of quantifiers. The scope of a quantifier is simply the first-level predicate to which it is applied. Thus in

$$\exists x(P(x) \wedge \exists y(Q(y) \wedge R(x, y))),$$

the scope of $\exists x$ is $P(\underline{\quad}) \land \exists y(Q(y) \land R(\underline{\quad}, y))$, while the scope of $\exists y$ is $Q(\underline{\quad}) \land R(\ldots, \underline{\quad})$.

The universal quantifier

In addition to the existential quantifier, the predicate calculus has a symbol $\forall$, called the **universal quantifier**. This is used for constructing schemas which can represent statements such as

> Every integer has a square.
> All men are mortal.
> John has read every Shakespeare play.
> Each house has a roof.

The English words 'all', 'every', and 'each' are fairly reliable indicators that a universal quantifier should be used in the predicate calculus representation of a sentence. As with the existential quantifier, it is helpful to regiment the sentences first into an intermediate form. In the regimented forms, we use 'for every x', 'for every y', etc. Thus our examples above become

> For every x, if x is an integer, then x has a square.
> For every x, if x is a man then x is mortal.
> For every x, if x is a Shakespeare play, then John has read x.
> For every x, if x is a house, then x has a roof.

All of these can be represented by the predicate calculus schema

$$\forall x(P(x) \rightarrow Q(x)),$$

given suitable interpretations of $P(\underline{\quad})$ and $Q(\underline{\quad})$. Note that while the existential quantifier characteristically governs a conjunction, the universal quantifier governs a conditional. Thus while 'Some man is mortal' has the form $\exists x(P(x) \land Q(x))$, 'Every man is mortal' has the form $\forall x(P(x) \rightarrow Q(x))$.

Note what happens if you get these the wrong way round: the schema $\exists x(P(x) \rightarrow Q(x))$ says that something either lacks the property denoted by P or has the property denoted by Q—thus the existence of something with property P is enough to make the schema true, as is the existence of something lacking the property Q, so this represents a much weaker statement than $\exists x(P(x) \land Q(x))$. On the other hand, the schema $\forall x(P(x) \land Q(x))$ is much too *strong*: it says that everything has both the property P and the property Q. Sometimes we do want to say something of this form, but for the statements presented in English and in regimented form above, what is wanted is the schema with the conditional rather than the conjunction.

As with the existential quantifier, the universal quantifier makes no distinction between singular and plural. In English we can say 'All men are mortal' or 'Every man is mortal', and the difference is one of nuance which the predicate calculus ignores. We choose 'every' for our regimentation because 'all' does not always indicate the presence of a universal quantifier in logical form. A good test is whether the sentence can be rephrased using 'every' instead of 'all'. We cannot rephrase the sentence

All the children play together

as

Every child plays together

(which doesn't even make sense!), hence its logical representation will not be of the form $\forall x(P(x) \rightarrow Q(x))$.

Even the 'every' test is not infallible: it would be a mistake to analyse the sentence

Every house has a different colour

as having the form $\forall x(P(x) \rightarrow Q(x))$. The trouble is that '... has a different colour' does not denote a property which each individual house either does or does not possess. Any house has a colour, to be sure; but it can only have a *different* colour in the context of a comparison with something else.

The relationship between $\forall$ and $\exists$

We represented the sentence 'No computer scientist loves Mary' by the schema

$$\sim \exists x(Q(x) \wedge R(x, b))$$

because we noted that to assert that no computer scientist has a certain property is precisely to deny that *some* computer scientist has the property. Another way of approaching this statement is to note that if no computer scientist has a certain property, then every computer scientist *lacks* that property. In the case that the property is that of loving Mary, this suggests that we could regiment 'No computer scientist loves Mary' into the form

For every x, if x is a computer scientist, then x does not love Mary.

This suggests

$$\forall x(Q(x) \rightarrow \sim R(x, b))$$

as the predicate calculus representation of our sentence.

In the predicate calculus, the combination $\sim \exists x$ is regarded as equivalent to the combination $\forall x \sim$. This means that each quantifier can be expressed in terms of the other, since $\exists x$ is equivalent to $\sim \sim \exists x$ which is equivalent to $\sim \forall x \sim$, and likewise $\forall x$ is equivalent to $\sim \exists x \sim$. Sometimes, indeed, one quantifier is *defined* in terms of the other, i.e. either $\forall x$ is defined to be an abbreviation for $\sim \exists x \sim$, or $\exists x$ is defined to be an abbreviation for $\sim \forall x \sim$.

Whichever treatment we choose, we obtain the equivalences

$$\sim \exists x(Q(x) \wedge R(x, b)) \cong \forall x \sim (Q(x) \wedge R(x, b))$$
$$\cong \forall x(Q(x) \rightarrow \sim R(x, b)),$$

thus vindicating our earlier analyses (for the second equivalence here, cf. Exercise 2.8, Question (1j).

A note on 'some' and 'any'

The combination of quantifiers with negation is complicated in English by the existence of the word 'any'. In English, to negate a sentence such as 'John ate some apples', we do not usually say 'John did not eat some apples'; instead we either say 'John ate no apples' or, more usually, 'John did not eat any apples'. So when it occurs within the scope of negation, 'some' typically becomes 'any'. But there is another way of reading 'any' (with a particular stress and intonation) which gives it a completely different meaning: 'John didn't eat *any* apples, he only ate Granny Smiths'. Colloquially, we can distinguish these two readings of 'any' as follows:

> John didn't eat any apples at all.
> John didn't eat any old apples.

When 'any' occurs in an *un*negated sentence, it usually functions as a universal quantifier rather than an existential one. Thus 'Anyone who steals will be punished' says something logically the same as 'Everyone who steals will be punished'. As so often, there is a difference in nuance here, but this does not get expressed in the predicate calculus representation, which for both sentences has the form $\forall x(P(x) \to Q(x))$.

A further point to beware of is that expressions like 'someone' or 'a person', which suggest an existential quantifier, can sometimes also be used where logic demands a universal quantifier. Our most recent example could also be written as 'A person who steals will be punished', which is, perhaps, ambiguous between an existential reading and a universal one. Again, 'A multiple of 2 is an even number' means that *every* multiple of 2 is even, not merely that *some* multiple of 2 is even. So this sentence has the form $\forall x(P(x) \to Q(x))$.

Be especially careful with the combinations 'if some', 'if a', and 'if there is'. At first sight, the statement

> If a number is prime and greater than 2 then it is odd

appears to have the form

$$\exists x(P(x) \wedge G(x)) \to \cdots$$

(where $P(\underline{\quad})$ stands for '$\underline{\quad}$ is prime' and $G(\underline{\quad})$ for '$\underline{\quad}$ is greater than 2'), but this cannot be right, since there is no schema that can fill the gap '...': we want a translation of 'it is odd', but we cannot give one because we do not know what 'it' refers to. In fact 'it' does not refer to anything in this sentence; its role here is analogous to that of a bound variable in the predicate calculus, binding the argument place of '$\underline{\quad}$ is odd' to the quantifier represented by 'a number'. This shows us that, despite appearances, the connective 'if...then $\underline{\quad}$' must fall *within* the scope of the quantifier; and since this statement is a generalization about *all* numbers (i.e., it says that every prime greater than 2 is odd), the quantifier must be universal. We thus arrive at the analysis

$$\forall x(P(x) \wedge G(x) \to O(x)).$$

The moral to be drawn from all this is that representing English sentences in predicate calculus notation cannot be done mechanically. You must always consider the *meaning* of the English sentence and try to construct a schema which has precisely the correct logical form for that meaning. In this respect, translation into predicate calculus does not differ from translation into any other language. In both cases, our aim is not really to represent an English sentence; it is to represent its meaning.

Combination of existential and universal quantifiers

Quantifiers of both kinds occur in statements such as

Every boy loves some girl.
Some boy loves every girl.
Every girl is loved by some boy.
Some girl is loved by every boy.

The first of these may be straightforwardly regimented as

(1) For every x, if x is a boy then for some y, y is a girl and x loves y

and hence may be represented by the schema

$$\forall x(B(x) \to \exists y(G(y) \wedge L(x,y)))$$

(where the predicate letters have the obvious interpretations).
   The second sentence is less straightforward. Does it mean

(2) For some x, x is a boy and for every y, if y is a girl then x loves y?

Or does it mean

(3) For every y, if y is a girl then for some x, x is a boy and x loves y?

You may find the difference between these two meanings hard to spot at first, but once you have spotted it you should have less difficulty with other similar distinctions.
   Version (2) says that there is some boy, let us call him 'Tom', who loves every girl, i.e. Tom loves Anna, Tom loves Barbara, Tom loves Caroline, etc. This reading corresponds to the schema

$$\exists x(B(x) \wedge \forall y(G(y) \to L(x,y))).$$

   Version (3), on the other hand, says that whichever girl you choose, some boy loves her: thus, for example, Robert loves Anna, Simon loves Barbara, Tom loves Caroline, and so on, with no girl remaining unloved. This reading corresponds to the schema

$$\forall y(G(y) \to \exists x(B(x) \wedge L(x,y))).$$

A less ambiguous way of expressing version (3) in English would be 'Every girl is loved by some boy' (i.e. the third of our sentences above).
   The fourth sentence, 'Some girl is loved by every boy', is also ambiguous between

(4) For some y, y is a girl and for every x, if x is a boy then x loves y,

which corresponds to the schema

$$\exists y(G(y) \wedge \forall x(B(x) \to L(x,y))),$$

and the regimentation (1) already given.
   The four English sentences thus correspond to the four predicate calculus schemas

$$\forall x(B(x) \to \exists y(G(y) \wedge L(x,y)))$$
$$\exists x(B(x) \wedge \forall y(G(y) \to L(x,y)))$$

$$\forall y(G(y) \to \exists x(B(x) \land L(x,y)))$$
$$\exists y(G(y) \land \forall x(B(x) \to L(x,y)))$$

but the correspondence is not straightforwardly one-to-one because some of the English sentences are ambiguous. In general, any English sentence containing two quantifying expressions of different kinds, one within the scope of the other, is liable to be ambiguous. Anyone who spends much time constructing predicate calculus representations of such sentences will soon acquire a heightened sensitivity to such ambiguities—it may truly be said that if you study logic then your native language will never seem quite the same again! Indeed, you may find yourself becoming unbearably pedantic. Even so, this heightened sensitivity is, on the whole, something to be valued rather than despised.

Here are some more examples involving universal and existential quantifiers together.

Example 4.3.1 In this example, we shall use I(____) for '____ is an integer' and G(____, ...) for '____ is greater than...'. Then

$$\forall x(I(x) \to \exists y(I(y) \land G(y,x)))$$

says that for every integer there is a greater one—so that in effect it says there is no upper limit to the size of integers.

If we rearrange the schema slightly, bringing the existential quantifier to the front together with the first conjunct in its scope, we obtain

$$\exists y(I(y) \land \forall x(I(x) \to G(y,x))),$$

which says something completely different, viz. that some particular integer is greater than every integer. This statement is of course false.

Example 4.3.2 Likewise, the statement 'Everyone has a mother' must be analysed as having the form

$$\forall x \exists y M(y,x)$$

(where M(____, ...) is read as '____ is the mother of...'), and *not* as

$$\exists x \forall y M(x,y)$$

which would mean that there is a universal Mother who is mother to everyone.

Example 4.3.3 Now consider the statement 'Every set of positive integers has a least member'. To express this in the predicate calculus, we shall use the predicates S(____), M(____, ...), and L(____, ...) to stand for '____ is a set of positive integers', '____ is a member of...', and '____ is less than or equal to...', respectively. A preliminary regimentation gives us

> For every x,
>> if x is a set of positive integers
>>> then for some y,
>>>> y is the least member of x.

We need to analyse this further by regimenting '___ is the least member of...' to

> ___ is a member of...,
>> and
>> for every z,
>>> if z is an member of...
>>> then ___ is less than or equal to z.

(the *least* member of a set is the one that is less than or equal to each member of the set). Our complete regimentation is thus

> For every x,
>> if x is a set of positive integers,
>>> then for some y,
>>>> y is a member of x
>>>> and
>>>> for every z,
>>>>> if z is a member of x
>>>>> then y is less than or equal to z.

Conversion to predicate calculus notation is now easy:

$$\forall x (S(x) \to \exists y (M(y, x) \wedge \forall z (M(z, x) \to L(y, z)))).$$

This schema, with three levels of quantifier embedding, is typical of the complexity of statements in mathematics (some are even more complex); everyday language, by contrast, usually gets by with no more than two levels.

Exercise 4.3

*(1) Translate into predicate calculus notation the inferences given in Exercise 4.1.

(2) In the two lists of schemas shown below, each schema on the left is equivalent to exactly one of the schemas on the right. Find all the equivalent pairs:

$\exists x P(x) \wedge \exists x Q(x)$	$\exists y (P(y) \wedge \forall x R(x, y))$
$\exists x P(x) \wedge \forall y Q(y)$	$\forall z (P(z) \to \exists u R(u, z))$
$\exists x (P(x) \wedge \forall y R(x, y))$	$\exists x P(x) \wedge \exists y Q(y)$
$\forall x (P(x) \to \exists y R(x, y))$	$\forall y \exists z R(y, z) \to \exists w P(w)$
$\forall x (P(x) \to \exists y R(y, x))$	$\exists u (P(u) \wedge \forall w R(u, w))$
$\forall y (\exists z R(y, z) \to P(y))$	$\exists z P(z) \wedge \forall z Q(z)$
$\exists x (P(x) \wedge \forall y R(y, x))$	$\forall x (\exists y R(x, y) \to P(x))$
$\forall x \exists y R(x, y) \to \exists x P(x)$	$\forall x (P(x) \to \exists z R(x, z))$

(3) Use the equivalence between $\sim \forall x$ and $\exists x \sim$ to rewrite the following schemas, (i) using existential quantifiers only, and (ii) using universal quantifiers only:

*(a) $\forall x (P(x) \to \exists y (Q(y) \wedge R(x, y)))$
*(b) $\sim \forall x \exists y (F(x, y) \vee G(x, y))$
(c) $\exists x \forall y F(x, y) \vee \forall x G(x, x)$
(d) $\forall x \forall y (R(x, y) \to R(y, x))$
(e) $\forall x \exists y (R(x, y) \vee \forall z S(x, y, z))$

(4) Write predicate calculus schemas to represent the following statements:

*(a) Mary admires every professor.
(b) Some professor admires Mary.

*(c)   Mary admires herself.

  (d)   No student attended every lecture.

*(e)   No lecture was attended by every student.

*(f)   No lecture was attended by any student.

  (g)   Every telephone number in Exeter has five or six figures.

*(h)   No language has more words than English.

*(i)   Daniel admires anyone who admires him.

  (j)   Daniel admires anyone who admires himself.

  (k)   John admires anyone who admires both David and Daniel.

*(l)   No even number greater than two is prime.

  (m)   If nobody makes a cake, everyone will be hungry.

*(n)   If anyone makes a cake, no-one will be hungry.

*(o)   If anyone makes a cake, John will eat it.

  (p)   If anyone makes a noise, the teacher will punish him.

  (q)   All boys, and some girls, want to become rich.

  (r)   John does not know anyone who admires him.

*(s)   No child has seen everything that some adults have seen.

4.4   FUNCTIONS AND FUNCTION SYMBOLS

In Section 4.2 we distinguished between simple referring expressions (proper names) and complex ones (definite descriptions), and introduced individual constants as the predicate calculus analogue of the former category. We also noted that we could use an individual constant to refer to an object picked out in English by means of a definite description, but that in doing this we sacrificed the internal logical structure of the definite description itself. In this section we will look at one way in which the predicate calculus can handle definite descriptions so as to respect their intrinsic logical complexity.

Suppose we wish to represent the statement

Mary's father is a poet

in predicate calculus notation. One way of analysing this sentence is to treat the expression 'Mary's father' as a proper name. On this analysis, the logical structure of the statement is given by the schema $P(a)$, where $P(\_\_\_)$ stands for '$\_\_\_$ is a poet' and a denotes Mary's father. This is fine so far as it goes: for example, it enables us to exhibit the schema under which the inference

Mary's father is a poet
Every poet admires Shakespeare
_____
Mary's father admires Shakespeare

is valid, namely $P(a), \forall x(P(x) \rightarrow Q(x, b))/Q(a, b)$.

However, for some purposes this analysis does not go far enough. Consider, for example, the inference

$$\frac{\text{Mary's father is a poet}}{\text{Someone's father is a poet}}$$

The conclusion here consists of the existential quantifier applied to the complex predicate '...'s father is a poet'. This predicate can be obtained from the sentence 'Mary's father is a poet' by removing the name 'Mary'. If we represent this sentence, as we did above, by P(a), where a stands for 'Mary's father', then we will be quite unable to do anything to this representation that corresponds to *removing the name* '*Mary*', for there is no component in P(a) that denotes Mary and hence plays the part of the name 'Mary'.

What we really need is a way of splitting the sentence 'Mary's father is a poet' into *three* components, one for 'Mary', one for 'father', and one for 'poet'. A ploy we might try is to regiment the sentence into the form

For some *x*, *x* is Mary's father and *x* is a poet,

which gives us the predicate calculus schema

$$\exists x (F(x, a) \wedge P(x)),$$

where F(_____, ...) represents '_____ is ...'s father'.

The trouble with this is that it does not actually tell us that Mary has only one father: we could equally well read this schema as 'Some of Mary's fathers are poets'! We have to appeal to our knowledge of the world—of the nature of fatherhood—to derive the *uniqueness* of the individual who fathered Mary, but this knowledge is not built into the representation of '_____ is ...'s father' as an ordinary two-place relation.

It is because Mary has exactly one father that the definite description 'Mary's father' can be used. Both definite descriptions and names share the property of picking out unique individuals, and both can be arguments of predicates: that is why they are grouped together as the category of referring expressions.

Many definite descriptions either are, or can be paraphrased as, expressions having the general form 'The _____ of ...', for example 'Mary's father' (i.e. 'The father of Mary'), 'The King of Spain', 'Shakespeare's greatest play' (i.e. 'The greatest play of Shakespeare'). and 'The author of *Hamlet*'. If we remove the proper name from an expression of this form, we obtain what is known as a **function expression**. Thus 'The father of ...', The King of ...', 'The greatest play of ...', and 'The author of ...' are all function expressions.

Just as a (unary) predicate has an argument place which can be filled by a referring expression to yield a statement, so a function expression has an argument place which can be filled by a referring expression to yield another referring expression. Put 'Mary' as argument to the function expression 'the father of ...', and we get the referring expression 'the father of Mary'. Put *this* referring expression as argument to the function expression 'The greatest play of ...', and we get the referring expression 'The greatest play of the father of Mary', i.e. 'Mary's father's greatest play'.

A function expression may have more than one argument place: consider the definite description 'the sum of 2 and 3'. This can be analysed as the result of filling the two argument places in the binary function expression 'the sum of ____ and ...' with the two proper names '2' and '3'.

In the predicate calculus, the role of referring expressions is played by a category of expressions known as **terms**. Like a referring expression, a term may be simple or complex. A simple term is just an individual constant. In order to construct complex terms, we introduce a new class of schematic letters known as **function symbols**. These play the part of function expressions. A function symbol has one or more argument places, and the result of filling up the argument places with individual constants (or previously constructed terms) is a complex term.

Thus the function symbol $f(\text{____})$ combines with the individual constant a to yield the term $f(a)$. If $f(\text{____})$ is read as 'the father of ____', and a as 'Mary', then the term $f(a)$ can be read as 'The father of Mary', i.e. 'Mary's father'. We can now write

$$\begin{array}{ll} \text{Mary's father is a poet} & P(f(a)) \\ \text{Someone is a poet} & \exists x P(x) \\ \text{Someone's father is a poet} & \exists x P(f(x)) \end{array}$$

The third of these schemas is obtained by applying the existential quantifier to the complex predicate $P(f(\text{____}))$ obtained by removing a from $P(f(a))$.

We can use a function symbol to give a more complete analysis of our earlier statement '$\frac{1}{2}$ is greater than its own square'. The predicate '... is greater than its own square' can be broken up into a relational predicate '... is greater than ____' and a function expression 'the square of ...'. Writing these as $G(..., \text{____})$ and $q(...)$ respectively, and putting a for '$\frac{1}{2}$', we get $G(a, q(a))$ for '$\frac{1}{2}$ is greater than its own square' and $\sim \exists x(I(x) \land G(x, q(x)))$ for 'No integer is greater than its own square'.

Some more examples to illustrate the use of function symbols follow.

Example 4.4.1   Let $f(\text{____})$ and $m(\text{____})$ stand for 'the father of ____' and 'the mother of ____' respectively, and let a denote Mary. Then $f(f(a))$ and $f(m(a))$ denote Mary's paternal and maternal grandfathers respectively, while $m(f(a))$ and $m(m(a))$ denote her two grandmothers.

Example 4.4.2   If we use $s(\text{____})$ for 'the successor of ____', and a for zero, then the natural numbers

$$0, 1, 2, 3, 4, \ldots$$

can be represented as

$$a, s(a), s(s(a)), s(s(s(a))), s(s(s(s(a)))), \ldots$$

(The successor of an integer is one more than it, e.g. the successor of 17 is 18.)

Example 4.4.3   Suppose we want to represent the (true) statement

Every perfect square is either a multiple of 4 or
the successor of a multiple of 4.

A perfect square is the square of an integer; so to begin with, we can regiment the statement into the form

For every integer x,
the square of x is either a multiple of 4
or the successor of a multiple of 4.

How do we say that a number is the successor of a multiple of 4? One way of doing this is to say that the *predecessor* of the number is a multiple of four. This gives us

> For every integer x,
> either the square of x is a multiple of 4
> or the predecessor of the square of x is a multiple of 4.

Putting

I(____)	for	'____ is an integer',
q(____)	for	'the square of ____',
p(____)	for	'the predecessor of ____',
F(____)	for	'____ is a multiple of 4',

we finally obtain

$$\forall x(I(x) \rightarrow (F(q(x)) \vee F(p(q(x))))).$$

So far all the function expressions we have used have been *unary*, i.e. they have required only a single argument to become completed into a complex name. Two-place function expressions are very common in arithmetic, witness for example

'the sum of ____ and ...'	(i.e. '____ + ...')
'the product of ____ and ...'	(i.e. '____ × ...')
'the difference of ____ and ...'	(i.e. '____ − ...')
'____ to the power ...'	(i.e. '____ ...')

When using the predicate calculus to represent mathematical statements, it is customary to deviate from the strict syntax of predicate calculus schemas by way of compromise with the standard mathematical notation; thus we could write

$$\forall x \forall y \forall z(x > y \rightarrow x + z > y + z)$$

to represent the statement that adding the same number to both sides of an inequality preserves the inequality. Here we have a two-place predicate '____ > ...' and a two-place function symbol '____ + ...', written in **infix** notation, i.e. with the symbol itself coming *between* its arguments rather than before them as in the standard **prefix** notation. Moreover, these symbols are to be interpreted so that they have their usual mathematical meaning. *There are many pitfalls in making compromises of this kind,* and we avoid it in this book, except in Section 6.3, where we shall discuss the logic of elementary arithmetic. The pitfalls arise, roughly speaking, because such compromises make it hard to prevent one's *logical* understanding of the schemas from being 'contaminated' by one's mathematical habits and intuitions. This need not matter so long as one is doing *mathematics*—so mathematicians make free use of logical symbols such as the quantifiers and connectives. But when doing logic, it is important to aim for a clear understanding of the purely logical aspects of one's schemas, and for this it helps to use as logically pure a notation as possible.

Exercise 4.4

(1) Turn back to the first question of Exercise 4.2 and identify every occurrence of a definite description in the sentences given there. Analyse each definite description into a function symbol and its argument(s). Convert all the sentences in the question into predicate calculus notation.

(2) To translate statements about sets into predicate calculus notation, we may use the following key:

a	the empty set
u(____,...)	the union of ____ and ...
i(____,...)	the intersection of ____ and ...
M(____,...)	____ is a member of ...
S(____,...)	____ is a subset of ...
E(____,...)	____ is equal to ...

What facts about sets are expressed by the following schemas?

*(a)   ~M(a, a)
  (b)   E(a, u(a, a))
*(c)   ∀xE(x, u(a, x))
  (d)   ∃x∀yE(y, u(x, y))
*(e)   ∀x∀yE(u(x, y), u(y, x))
  (f)   ∀x∀yS(i(x, y), y)

How could you use the function symbol i to represent the intersection of *three* sets A, B, and C? Is there more than one way of doing this?

(3) Use the key in Question 2 to represent the following statements as predicate calculus schemas:

  (a)   The empty set is a subset of every set.
*(b)   Nothing is a member of the empty set.
  (c)   One set is a subset of another if and only if every member of the first is a member of the second.
*(d)   If something is a member of both of two sets then it is a member of the intersection of those sets.
  (e)   Any set is a subset of its union with any other set.

4.5 FORMAL SYNTAX OF THE PREDICATE CALCULUS

We are now in a position to formalize the language of the predicate calculus, just as we did for the propositional calculus in Section 2.10. We define a **first-order predicate calculus** to be a formal system with the following ingredients:

(1) *Lexicon*

  (a)   a set, possibly empty, of **individual constants** (usually lower-case letters from the beginning of the alphabet, e.g. a, b, c);
  (b)   a set, possibly empty, of **function symbols** (usually lower-case letters from the middle of the alphabet, e.g. f, g, h, p, q);
  (c)   a non-empty set of **predicate letters** (usually upper-case letters, e.g. B, F, P, Q, R);
  (d)   a non-empty, potentially infinite, set of **variables** (usually lower-case letters from the end of the alphabet, e.g. u, v, w, x, y, z);
  (e)   the two **quantifiers** ∀ and ∃;
  (f)   the five **connectives** ~, ∧, ∨, →, and ↔;
  (g)   three **punctuation marks** (,), and ,.

Each function symbol and predicate has associated with it a positive integer, called its **arity**, which determines the number of argument places it has. (Some authors use the words **valency** or **polyadicity** instead of 'arity'.)

(2)   *Rules of formation*

(a)   A **term** is either an individual constant (a **simple term**) or $f(t_1, \ldots, t_n)$, where f is a function symbol of arity n and $t_1, \ldots, t_n$ are terms. Terms are the formal equivalent of what we earlier called referring expressions: the simple terms correspond to (simple) names, the others to definite descriptions. Note that the definition of a term is recursive: the set of terms is built up in stages, starting with the simple terms (stage 0), then terms constructed from function symbols and simple terms (stage 1), then terms constructed from function symbols and terms from stages 0 and/or 1, and so on.

(b)   If $t_1, \ldots, t_n$ are terms and P is a predicate letter of arity n, then $P(t_1, \ldots, t_n)$ is a **schema** (a **simple schema**)

(c)   If A and B are schemas then so are $\sim A$, $(A \wedge B)$, $(A \vee B)$, $(A \rightarrow B)$, and $(A \leftrightarrow B)$, exactly as in the propositional calculus.

(d)   A **unary predicate** is obtained by replacing one or more terms in a schema by the **argument place** '____'. Note that unary predicates are not expressions of the language, since the argument place '____' is not a member of the lexicon; we use unary predicates as stepping-stones towards the construction of quantified schemas.

If P is a predicate letter of arity 1 then $P(\_\_\_)$ is a **simple unary predicate** (since it can be derived from $P(a)$ by replacing a by '____'). This is the simplest case; a more typical example would be something like $Q(\_) \wedge R(a, \_)$.

We shall use upper-case Greek letters such as Φ and Ψ (phi and psi) when we want to refer to a unary predicate in general, without wanting to specify any particular example.

(e)   If $\Phi(\_\_\_)$ is a unary predicate and x is any variable not appearing in $\Phi(\_\_\_)$, then $\forall x \Phi(x)$ and $\exists x \Phi(x)$ are schemas.

The set of schemas constructed according to these rules of formation from the sets C, F, P, and V of individual constants, function symbols, predicates, and variables respectively, is called the **first-order language** determined by C, F, P, and V. We shall denote it $L(C, F, P, V)$. Note that if C is empty, then the rules given do not allow us to construct any schemas at all, since the necessary starting-point of schema construction is the simple schemas, and each of these contains at least one individual constant. In this case, in order to generate a language, we introduce, temporarily, one individual constant a, construct the language $L' = L(\{a\}, F, P, V)$, and then define $L(\varnothing, F, P, V)$ to consist of those elements of L' which do not contain a. (For example, if L contains a predicate letter P, then L' will contain the schema $P(a)$, and hence also $\forall x P(x)$, where x is in V. So $L(\varnothing, F, P, V)$ will now contain the latter schema, since this does not contain a, but not the former, which does.)

We now give an extended example to illustrate schema construction in a first-order language.

Example 4.5.1 Let $C = \{a\}$, $F = \{s, u\}$, $P = \{E, G\}$, and $V = \{v, w, x, y, z, \ldots\}$, let the arities of s and E be 1, and of u and G be 2. Consider the language $L(C, F, P, V)$.
The terms of the language include the series

$$a, s(a), s(s(a)), s(s(s(a))), s(s(s(s(a)))), \ldots;$$

the double series

$$u(a, a), u(a, s(a)), u(a, s(s(a))), \ldots,$$
$$u(s(a), a), u(s(a), s(a)), u(s(a), s(s(a))), \ldots,$$
$$u(s(s(a)), a), u(s(s(a)), s(a)), u(s(s(a)), s(s(a))), \ldots,$$

and a great many others besides, such as

$$s(s(u(s(s(u(a, s(a)))), s(u(s(a), s(s(a))))))).$$

Some simple schemas are

   (1) $E(a)$
   (2) $G(a, a)$
   (3) $E(s(s(a)))$
   (4) $G(s(a), a)$
   (5) $E(u(a, a))$
   (6) $G(u(a, a), a)$

From these we can use the connectives to give complex schemas such as

   (7) $E(a) \rightarrow E(s(s(a)))$ (from (1) and (3))
   (8) $G(a, a) \wedge G(a, a) \rightarrow G(a, a)$ (from three copies of (2))

Some unary predicates formed from the schemas constructed so far are:

   (2′) $G(a, \_)$ (from (2))
   (5′) $E(u(\_, \_))$ (from (5))
   (6′) $G(u(a, \_), \_)$ (from (6))
   (7′) $E(\_) \rightarrow E(s(s(\_)))$ (from (7))
   (8′) $G(a, a) \wedge G(a, \_) \rightarrow G(a, \_)$ (from (8))

Applying a quantifier to each of these unary predicates gives us some new schemas:

   (9) $\exists y G(a, y)$ (from (2′))
   (10) $\forall x E(u(x, x))$ (from (5′))
   (11) $\forall y G(u(a, y), y)$ (from (6′))
   (12) $\forall x (E(x) \rightarrow E(s(s(x))))$ (from (7′))
   (13) $\forall z (G(a, a) \wedge G(a, z) \rightarrow G(a, z))$ (from (8′))

We now have more material for combining together with the connectives, allowing us to construct, for example, the schema

   (14) $G(a, a) \rightarrow \forall y G(u(a, y), y)$ (from (2) and (11))

Some further unary predicates are

   (13′) $\forall z (G(a, \_) \wedge G(\_, z) \rightarrow G(a, z))$ (from (13))
   (14′) $G(\_, a) \rightarrow \forall y G(u(\_, y), y)$ (from (14))

which in turn give us the schemas

$\quad$ (15)$\quad$ $\forall y \forall z (G(a, y) \wedge G(y, z) \rightarrow G(a, z))$ $\qquad$ (from (13'))
$\quad$ (16)$\quad$ $\forall x (G(x, a) \rightarrow \forall y G(u(x, y), y))$ $\qquad$ (from (14'))

Finally, from these schemas we obtain the unary predicates

$\quad$ (15')$\quad$ $\forall y \forall z (G(\_, y) \wedge G(y, z) \rightarrow G(\_, z))$ $\qquad$ (from (15))
$\quad$ (16')$\quad$ $\forall x (G(x, \_) \rightarrow \forall y G(u(x, y), y))$ $\qquad$ (from (16))

from which in turn we construct the schemas

$\quad$ (17)$\quad$ $\forall x \forall y \forall z (G(x, y) \wedge G(y, z) \rightarrow G(x, z))$ $\qquad$ (from (15'))
$\quad$ (18)$\quad$ $\forall w \forall x (G(x, w) \rightarrow \forall y (G(u(x, y), y)))$ $\qquad$ (from (16'))

If we omitted the constant a from the lexicon to give a language with no individual constants, then of the 18 schemas above, only (10), (12), (17) and (18) would occur in the new language, since only these contain no occurrences of a.

In the rules of formation, we needed to define unary predicates in order to specify the construction of quantified schemas. There is nothing to stop us from generalizing to predicates of any arity. All that is necessary is to allow more than one kind of argument place to be substituted for constants in a schema. The arity of the resulting predicate is then just the number of distinct types of argument place that are used.

For example, if we substitute argument-places for the three occurrences of c in the schema

$$Q(c, b) \wedge R(a, c) \rightarrow S(a, b, c)$$

we may obtain a unary predicate, namely

$$Q(\_\_\_, b) \wedge R(a, \_\_\_) \rightarrow S(a, b, \_\_\_),$$

three different binary predicates:

$$Q(\ldots, b) \wedge R(a, \_\_\_) \rightarrow S(a, b, \_\_\_),$$
$$Q(\_\_\_, b) \wedge R(a, \ldots) \rightarrow S(a, b, \_\_\_),$$
$$Q(\_\_\_, b) \wedge R(a, \_\_\_) \rightarrow S(a, b, \ldots),$$

or a ternary predicate:

$$Q(\_\_\_, b) \wedge R(a, \ldots) \rightarrow S(a, b, \text{---}).$$

Another generalization is from function symbols to **functors**. A functor is obtained from a term by replacing one or more individual constants by argument places. The arity of the functor is the number of different types of argument place used.

For example, by removing a from the term $f(a, g(b, c))$ we obtain the unary functor $f(\_\_\_, g(b, c))$; by removing both a and b we can obtain either a unary functor $f(\_\_\_, g(\_\_\_, c))$ or a binary functor $f(\_\_\_, g(\ldots, c))$. And there are five possible functors that can be obtained by removing a, b, *and* c, as follows:

$\qquad$ Unary:$\quad$ $f(\_\_\_, g(\_\_\_, \_\_\_))$
$\qquad$ Binary:$\quad$ $f(\_\_\_, g(\_\_\_, \ldots))$
$\qquad\qquad\qquad$ $f(\_\_\_, g(\ldots, \_\_\_))$
$\qquad\qquad\qquad$ $f(\_\_\_, g(\ldots, \ldots))$
$\qquad$ Ternary:$\quad$ $f(\_\_\_, g(\ldots, \text{---}))$

We use lower-case Greek letters like ϕ and ψ, together with the appropriate number of argument places, to represent functors. If $\phi(\underline{\quad})$ stands for $f(\underline{\quad}, g(\underline{\quad}, c))$, then $\phi(a)$ is the term $f(a, g(a, c))$, $\phi(b)$ is $f(b, g(b, c))$, and so on. And if $\psi(\underline{\quad}, \ldots)$ stands for $f(\underline{\quad}, g(\ldots, c))$, then $\psi(a, a)$ is the same term as $\phi(a)$. We shall meet functors again in Section 6.2.

Exercise 4.5

(1) In the language of Example 4.5.1, which of the following expressions are correctly formed schemas? Explain why the incorrect ones are wrong.

    *(a) $\forall x E(x)$
    *(b) $\exists y (G(a, y) \wedge E(x))$
    *(c) $\forall x \exists y (G(x, a))$
    (d) $\forall x \exists y (G(x, a) \vee G(y, x))$
    (e) $G(s(a, a), a)$
    *(f) $E(u(a, s(a))) \rightarrow G(a, a)$
    *(g) $\exists x (u(a, x) \vee s(x))$
    (h) $\forall x (E(x) \vee E(s(x)))$
    (i) $\sim (E(x) \wedge E(s(x)))$
    *(j) $\forall s E(s(a))$
    (k) $G(a, E(a))$
    (l) $\forall x G(u(x, x), s(x)) \vee G(s(a), u(a, a))$

*(2) Write down all the stages involved in the construction of the schema

$$\exists x (P(x) \wedge \forall y Q(x, y)) \vee (P(a) \wedge \forall y Q(b, y)).$$

(3) Write down all the predicates obtainable from each of the following schemas:

    *(a) $P(a, b)$
    *(b) $P(a) \rightarrow R(a, b)$
    *(c) $P(a) \rightarrow \exists x R(a, x)$
    (d) $Q(a, b, f(a, b))$
    (e) $R(f(a), g(b))$
    (f) $P(a, b) \rightarrow \forall x (Q(a, x) \vee R(b, c, x))$

and state the arity of each predicate you obtain.

(4) Write down all the functors obtainable from each of the following terms:

    *(a) $f(a, b)$
    (b) $f(a, g(a, b))$
    (c) $h(a, b, c)$
    *(d) $g(f(a), h(f(b)))$
    (e) $f(g(a), f(b, g(c)))$

and state the arity of each functor you obtain.

4.6 FORMAL SEMANTICS OF THE PREDICATE CALCULUS

We have repeatedly used the idea of an *interpretation* of the symbols occurring in a first-order language. Now that we have established a formal definition of such languages, we must, as we did with the propositional calculus, give a formal definition of an interpretation as well.

Consider the first-order language L introduced at the end of the last section. There are many ways of interpreting the symbols a, s, u, E and G which occur in the lexicon of this language. Any interpretation must be consistent with the syntax, in the sense that, for example, individual constants must be interpreted as referring expressions (i.e. as referring to unique individuals), unary predicates must be interpreted so that they denote properties, and so on.

For example, we could interpret the language as talking about *people*—to be definite, let us say *all people who have lived up to now*. Then a must refer to one specific person, for example you, the reader of this book; s(_) could be read as 'the father of ____', u(____,...) as 'the latest shared male ancestor of ____ and ...', E(_) as '____ is now alive', and G(____,...) as '____ is older than ...'.

It should now be clear enough, intuitively, what the various simple schemas of L mean under this interpretation. For example, E(a) says that you, the reader, are now alive, E(s(s(a))) says that your paternal grandfather is alive, G(s(a), a) says that your father is older than you, and so on.

Likewise, you should not have too much trouble with complex schemas, so long as they are not *too* complex. For example,

$$\forall x (E(x) \rightarrow E(s(s(x))))$$

says that if someone is alive, then so is their paternal grandfather (a blatant falsehood, if ever there were one!), and

$$\forall x \forall y \forall z (G(x, y) \wedge G(y, z) \rightarrow G(x, z))$$

says that if one person is older than a second, who is older than a third, then the first person is older than the third (in short, it says that 'older than' is a *transitive* relation, which is indeed the case).

Exercise 4.6a   Determine the meanings under this interpretation of all the other schemas given in our example, and decide which are true and which are false.

What we require of a *formal* interpretation is a systematic way of computing the meanings of complex schemas from the meanings of their constituents. We do this by providing an algorithm which enables us to determine, for each schema of the language, whether or not it is to count as true under the interpretation. It is the manner in which this truth value is computed that corresponds to the meaning of the schema.

A **formal interpretation** for a first-order language L has two ingredients. First, we must specify a **domain** (or **universe**), D, which is simply the set of objects which we want L to be interpreted as talking about. In the example we have just looked at, the domain was the set of all people who have ever lived; but we could have chosen it to be *any set whatever*, including the empty set. We shall see below a very different interpretation of the same language, in which the domain consists not of people but of numbers.

Second, we must specify an **interpretation function**, I, which establishes a correspondence between the simple components of the language (i.e. the constants, function symbols, and predicate letters) and the domain. We shall discuss informally how it does this for each kind of simple constituent before giving the formal definitions.

Individual constants

The interpretation of individual constants is straightforward. They correspond to names in ordinary language, and the role of a name is to name something. So our interpretation function must assign to each individual constant some object from the domain. The object assigned to the constant a is $I(a)$; this is an element of D. We can say that a names $I(a)$ under the interpretation (D, I).

Function symbols

A one-place function symbol corresponds to an expression like 'the father of ____', which combines with a name to produce a definite description. We can regard a definite description as a *complex* name: like a name, it picks out a unique object, which is related in a certain way to the object picked out by the argument (e.g. 'the father of Mary' refers to a person who is related in a particular way to Mary). Under the interpretation I, then, we want a function symbol f to refer to a way of picking out an object related in a certain way to the object named by its argument. Mathematically, what we require is a *function* which maps each object in the domain into another (or possibly the same) object.

A function may be expressed either by giving a *rule* (as when we define the function 'the square of ____' by the rule: multiply ____ by itself), or by listing each possible argument alongside the result of applying the function (e.g. the function denoted by the expression 'the capital of ____' can be given by means of a list, of which the following:

$$
\begin{array}{l}
\text{France} \longrightarrow \text{Paris} \\
\text{Spain} \longrightarrow \text{Madrid} \\
\text{Italy} \longrightarrow \text{Rome} \\
\text{Egypt} \longrightarrow \text{Cairo} \\
\text{USA} \longrightarrow \text{Washington}
\end{array}
$$

is a sample). For our purposes, it does not matter which method we use: what is important is that we specify a function which assigns an element of D to each element of D; and it is a function of this kind that is assigned by the interpretation function I to a one-place function symbol such as f. In symbols, we write

$$I(f): D \to D$$

to mean that the interpretation of f is a function from D to D.

Function symbols of higher arity are treated similarly. A two-place function symbol such as u in our example corresponds to a function which maps a *pair* of domain elements onto a domain element (e.g. in our interpretation, it maps the pair consisting of Prince Charles and Prince Andrew onto Prince Philip, their latest common male ancestor). Thus for a two-place function symbol u we write

$$I(\mathsf{u}): D^2 \to D,$$

where D^2 is the set of all *ordered pairs* of elements from D (i.e. in symbols, $D^2 = \{(a,b) | a \in D, b \in D\}$).

Generalizing, an *n*-place function symbol g is interpreted as a function which maps each ordered *n*-tuple of domain elements onto a domain element:

$$I(\mathsf{g}): D^n \to D.$$

Terms

The interpretation function I is only defined on the simple constituents of the language; to interpret terms we have to *extend* I to a function I' which can be defined uniquely in terms of I. The function I' will supply an interpretation for the complex expressions in the language. In relation to the simple constituents, I' behaves identically to I. We now specify how it operates on terms.

A term corresponds to a referring expression in ordinary language, so we should expect I' to assign to it an object of the domain. The interpretation of function symbols has been carefully constructed to make this possible. Consider the term s(a). We know that $I(\mathsf{a})$ is an element of D, and that $I(\mathsf{s})$ is a function which maps elements of D onto elements of D. This means, in particular, that $I(\mathsf{s})$ maps $I(\mathsf{a})$ onto an element of D, which may be denoted $I(\mathsf{s})[I(\mathsf{a})]$, i.e. the result of applying the function $I(\mathsf{s})$ to the domain element $I(\mathsf{a})$. It is *this* object that is assigned, under the extended interpretation I', to the term s(a):

$$I'(\mathsf{s}(\mathsf{a})) = I(\mathsf{s})[I(\mathsf{a})].$$

In general, we may describe what happens as follows. A term consists of an *n*-place function symbol f together with *n* terms $\mathsf{t}_1, \ldots, \mathsf{t}_n$ as its arguments. The interpretation I assigns to f a function $I(\mathsf{f})$ which maps ordered *n*-tuples of domain objects to domain objects. Assume now that each of the terms t_i has already been assigned an object of the domain as its interpretation under I', and denote this object $I'(\mathsf{t}_i)$. Then the list $I'(\mathsf{t}_1), \ldots, I'(\mathsf{t}_n)$ constitutes an ordered *n*-tuple of domain objects. The function $I(\mathsf{f})$ maps this *n*-tuple onto a domain object $I(\mathsf{f})[I'(\mathsf{t}_1), \ldots, I'(\mathsf{t}_n)]$, and it is this domain object that is assigned to the term f($\mathsf{t}_1, \ldots, \mathsf{t}_n$) under the interpretation I':

$$I'(\mathsf{f}(\mathsf{t}_1, \ldots, \mathsf{t}_n)) = I(\mathsf{f})[I'(\mathsf{t}_1), \ldots, I'(\mathsf{t}_n)].$$

Predicates and simple schemas

We begin by considering just unary predicates. A unary predicate $\Phi(\_)$ is intended to refer to a *property* which each element of the domain either possesses or lacks. Like functions, properties can be specified in two different ways, either by means of a *rule* for deciding whether or not an object has the property (e.g. the rule for '... is even' is: if 2 divides ... without remainder then ... is even), or by simply *listing* those objects which have the property. In either case, the effect of our specification is to pick out a *subset* of the domain, whose members are exactly those domain elements which possess the property in question.

Formally, it is much easier to talk about subsets than about properties, so in our formal interpretation we ignore every aspect of a property apart from the subset of domain elements which it determines. The formal interpretation of a simple unary predicate P(_) is thus simply a subset of the domain:

$$I(P) \subseteq D.$$

Note that we write $I(P)$ rather than $I(P(\_))$ to aid readability—but this is yet another abuse of notation!

In our example, the predicate E(_) was intended to mean '____ is alive'. This corresponds to a formal interpretation in which $I(E)$ is stipulated to be the set of people who are alive. Note that $I(E)$ does not tell us the *meaning* of the word 'alive'—we have to know that already in order to determine who goes into the set $I(E)$! The formal interpretation merely enables us to put formally on record our intention that E(_) should mean '____ is alive', and to do this in way that, as we shall see, will enable us to characterize the logical properties of inferences containing this predicate.

We are now able to interpret some of the *simple schemas* of the language, namely those formed from a simple unary predicate and a term. As in Section 2.10, we shall speak of a schema A being *satisfied by*, or *true under*, the interpretation (D, I), or else *falsified by* or *false under* (D, I). We shall write

$$\vDash_{D,I} A$$

in the former case, and

$$\nvDash_{D,I} A$$

in the latter.

As with function symbols, the interpretation of predicates has been carefully designed so as to enable us to define what it means for a simple schema to be satisfied by an interpretation. Consider the schema P(a). We know that the interpretation of P is a subset $I(P)$ of D, and that the interpretation of a is an element $I(a)$ of D. Now either $I(a)$ is a member of $I(P)$ or it is not. We want P(a) to be interpreted as saying that the object named a has the property denoted by P. So clearly P(a) must be true under the interpretation (D, I) just so long as the object named a *does* have the property denoted by P. But the object named a is $I(a)$, and the set of objects with the property denoted by P is $I(P)$, so it follows that P(a) is true under (D, I) just so long as $I(a)$ is a member of $I(P)$. We thus have the rule

$$\vDash_{D,I} P(a) \text{ if and only if } I(a) \in I(P).$$

More generally, since the argument of P(_) can be any term, not just a constant, we can put

$$\vDash_{D,I} P(t) \text{ if and only if } I'(t) \in I(P).$$

Predicates with more than one argument place are interpreted similarly. A predicate of arity n expresses a relation among n objects. Such a relation can be specified as a set of ordered n-tuples, i.e. as a subset of D^n, so we have, for a predicate $Q(\underline{\quad},\ldots,\underline{\quad})$ of arity n,

$$I(Q) \subseteq D^n.$$

The schema $Q(t_1,\ldots,t_n)$ will be true under (D, I) so long as the n-tuple $(I'(t_1),\ldots,I'(t_n))$ formed from the interpretations of the t_i belongs to $I(Q)$, the set of n-tuples determined by the n-place relation denoted by Q. So our general rule for simple schemas is

$$\vDash_{D,I} Q(t_1,\ldots,t_n) \text{ if and only if } (I'(t_1),\ldots,I'(t_n)) \in I(Q).$$

Complex schemas

The rules of formation specify two ways of constructing complex schemas: one using connectives, the other using quantifiers. The interpretation of the former kind of complex schema is exactly as in the propositional calculus: simply use the rules (2)–(6) on p. 66 (Section 2.10). The construction of quantified schemas involves complex unary predicates, so first we must discuss the interpretation of these.

Recall that for an *simple* unary predicate $P(\underline{\quad})$, $I(P)$ is a set of domain elements, and (D, I) satisfies $P(t)$ if and only if $I'(t)$ is a member of $I(P)$. Now suppose we augment the lexicon of our language L by adding a **dummy constant** $*$. Then the language itself will be enlarged to a language L^*. Consider all the possible ways in which we could extend the interpretation I of L to the language L^*. For each domain element d, there is an interpretation I_d of L^* which behaves identically to I in relation to the simple expressions of L, and in addition assigns to $*$ the domain element d, i.e. we have

$$I_d(*) = d.$$

Then for a simple unary predicate $P(\_)$, we have

$$\vDash_{D,I_d} P(*) \text{ if and only if } d \in I(P).$$

Since d can be *any* domain element here, it follows that the membership of $I(P)$ is uniquely determined if we know the truth value of $P(*)$ under each of these extended interpretations I_d, since we have

$$I(P) = \{d \in D \mid \vDash_{D,I_d} P(*)\}$$

(i.e. $I(P)$ is the set of domain elements d for which $P(*)$ comes out true if $*$ is interpreted as d).

Of course, for a simple predicate $P(\_)$ this is a pointless move: we cannot tell whether $P(*)$ is true under I_d unless we *already* know what $I(P)$ (and hence $I_d(P)$) is. But for *complex* predicates this does provide us with exactly what we need in order to specify their interpretations.

Consider, for example, the complex unary predicate $Q(\_) \wedge R(a, \_)$. Using our dummy constant, we can construct a schema $Q(*) \wedge R(a, *)$ in the extended language L^*. Given an interpretation I of L, we can form extensions I_d which are interpretations of L^*. For each d in D, the schema $Q(*) \wedge R(a, *)$ comes out as either true or false under the interpretation I_d. The ones for which it comes out true are precisely those domain elements which have the property denoted by the complex predicate $Q(\_) \wedge R(a,\_)$. We therefore specify the interpretation of this complex predicate in I to *be* exactly the set of domain elements d for which $Q(*) \wedge R(a,*)$ is true under I_d, i.e.

$$I'(Q(\_) \wedge R(a,\_)) = \{d \in D \mid \vDash_{D,I_d} Q(*) \wedge R(a,*)\}.$$

And for a general unary predicate $\Phi(\_)$, we put

$$I'(\Phi) = \{d \in D \mid \vDash_{D,I_d} \Phi(*)\}.$$

Note that if for Φ here we put a simple predicate $P(\_)$, then this definition agrees exactly with what we have already found to be the case above.

Our second kind of complex schema is obtained by prefixing a unary predicate with a quantifier whose variable binds into the argument place(s) of the predicate. We want the schema $\forall x \Phi(x)$ to say that *every* domain element has the property denoted by $\Phi(\_)$, and we want $\exists x \Phi(x)$ to say that *some* domain element has that property. But the set of domain elements which have the property denoted by $\Phi(\_)$ is just $I'(\Phi)$. So for $\forall x \Phi(x)$ to true, we require that $I'(\Phi)$ should equal the whole domain; while for $\exists x \Phi(x)$ to be true we only require that $I'(\Phi)$ should be non-empty. Our semantic rules for the quantifiers are therefore as follows:

$$\vDash_{D,I} \forall x \Phi(x) \text{ if and only if } I'(\Phi) = D;$$
$$\vDash_{D,I} \exists x \Phi(x) \text{ if and only if } I'(\Phi) \neq \varnothing.$$

Note: We have to introduce the dummy constant $*$ and make use of the extended interpretations I_d in order to specify how complex predicates, and hence quantified schemas, are interpreted under I, but it is important to notice that the final results—the interpretations of a complex predicate or quantified schema—are interpretations in I, not in I_d. The method we have used here for specifying these interpretations is a little unorthodox, but ultimately equivalent to (and, I believe, easier to understand than) the conventional way which requires us to admit schemas with *unbound* (or *free*) variables and to complicate the formal semantics with *variable assignments*. For a concise account of the conventional method, see for example Lloyd, 1984.

Here we collect together all the rules for the formal interpretation of a first-order language $L = (C, F, P, V)$.

A **formal interpretation** of $L = (C, F, P, V)$ is a pair (D, I), where D is any set (the domain of the interpretation), and I is a function on $C \cup F \cup P$ such that

    (a)   For each $c \in C$, $I(c) \in D$,

    (b)   For each f of arity n in F, $I(f): D^n \to D$,

    (c)   For each P of arity n in P, $I(P) \subseteq D^n$.

The interpretation function I is extended to a function I' on terms by the rule:

$$I'(f(t_1, \ldots, t_n)) = I(f)[I'(t_1), \ldots, I'(t_n)]$$

Satisfaction is defined for a simple schema by the rule

$$\vDash_{D,I} P(t_1, \ldots, t_n) \text{ if and only if } (I'(t_1), \ldots, I'(t_n)) \in I(P)$$

and for complex schema whose main functor is a connective by

$$\vDash_{D,I} \sim A \text{ if and only if } \nvDash_{D,I} A$$
$$\vDash_{D,I} A \wedge B \text{ if and only if both } \vDash_{D,I} A \text{ and } \vDash_{D,I} B$$
$$\vDash_{D,I} A \vee B \text{ unless both } \nvDash_{D,I} A \text{ and } \nvDash_{D,I} B$$
$$\vDash_{D,I} A \rightarrow B \text{ unless both } \vDash_{D,I} A \text{ and } \nvDash_{D,I} B$$
$$\vDash_{D,I} A \leftrightarrow B \text{ if and only if both } \vDash_{D,I} A \rightarrow B \text{ and } \vDash_{D,I} B \rightarrow A$$

For a complex unary predicate,

$$I'(\Phi) = \{ d \in D \mid \vDash_{D, I_d} \Phi(*) \};$$

For a complex schema whose main schema is a quantifier,

$$\vDash_{D,I} \forall x \Phi(x) \text{ if and only if } I'(\Phi) = D;$$
$$\vDash_{D,I} \exists x \Phi(x) \text{ if and only if } I'(\Phi) = \varnothing.$$

Example 4.6.1   We shall illustrate the formal semantics by referring back to the example language introduced in Section 4.5.1. We have already mentioned one possible interpretation of this language, in which the domain is the set of all people who have ever lived. We now consider another, perhaps more natural, interpretation of the same language.

In this interpretation, the domain D is the set of *natural numbers* (i.e. the non-negative integers $0, 1, 2, 3$, etc.). We want to read a as '0', s(_) as 'the successor of _____', u(_,...) as '_____ plus...', E(_) as '_____ is even', and G(_____,...) as '_____ is greater than...'. Formally, we have:

$$D = \mathbf{N}$$
$$I(a) = 0$$
$$I(s)[n] = n + 1$$
$$I(u)[m, n] = m + n$$
$$I(E) = \{ n \in \mathbf{N} \mid n \text{ is even} \} = \{ 0, 2, 4, 6, 8, 10, \ldots \}$$
$$I(G) = \{ (m, n) \in \mathbf{N}^2 \mid m > n \}$$

Under this interpretation, the sequence of terms a, s(a), s(s(a)),... are interpreted as follows:

$$I'(a) = I(a) = 0$$
$$I'(s(a)) = I(s)[I'(a)] = 0 + 1 = 1$$
$$I'(s(s(a))) = I(s)[I'(s(a))] = 1 + 1 = 2$$
$$I'(s(s(s(a)))) = I(s)[I'(s(s(a)))] = 2 + 1 = 3$$
$$\vdots$$

so this sequence of terms denotes the natural numbers $0, 1, 2, 3, \ldots$.

Other terms can be interpreted similarly; for example,

$$I'(u(s(a), u(a, s(a)))) = I(u)[I'(s(a)), I'(u(a, s(a)))]$$
$$= I'(s(a)) + I'(u(a, s(a)))$$
$$= 1 + I(u)[I'(a), I'(s(a))]$$
$$= 1 + (0 + 1)$$
$$= 2$$

We turn now to some of the schemas (1)–(18) discussed in Section 4.5.

(1)   E(a). Since $I(a)$ is a member of $I(E)$, we have $\vDash_{D,I} E(a)$, i.e. schema (1) is satisfied by this interpretation. Note that the *reason* why it is satisfied is that 0 is even; that is why we can say that under this interpretation $E(a)$ *says* that 0 is even.

(2)   G(a, a). In order for this schema to be true in (D, I), we should need that $(I(a), I(a))$ is a member of $I(G)$, in other words that $0 > 0$. For this reason, the meaning of G(a, a) under this interpretation is that 0 is greater than 0. Since this is not the case, we have $\nvDash_{D,I} G(a, a)$.

(3)   E(s(s(a))). We have already seen that $I'(s(s(a))) = 2$, so under this interpretation E(s(s(a))) says that 2 is even. Since 2 *is* even, $\vDash_{D,I} E(s(s(a)))$.

The reader should determine in the same way which of the schemas (4), (5), and (6) are satisfied by (D, I).

Since (1) and (3) are both satisfied by (D, I), so is

(7) $E(a) \rightarrow E(s(s(a)))$

using the rule for $\rightarrow$; and similarly, since (2) is falsified by (D, I),

(8) $G(a, a) \wedge G(a, a) \rightarrow G(a, a)$

is satisfied.

Turning now to the complex predicates, we note that

$$I'(G(a, \_)) = \{d \in D \mid \vDash_{D, I_d} G(a, *)\}.$$

Now $G(a, *)$ is satisfied only by those interpretations I_d for which $(0, d)$ is in $I(G)$, i.e. for which $0 > d$. But this is *false* for every natural number d, since no natural number is less than zero, and hence $I'(G(a, \_))$ is empty. Using the rule for the existential quantifier, we thus have that

(9) $\exists y G(a, y)$

is not satisfied in (D, I).

Once again, notice that the *reason* why (9) is not satisfied in (D, I) is that no natural number is less than zero. For that reason we can say that under the interpretation (D, I), schema (9) *denies* that no natural number is less than zero, i.e. it says that *some* natural number is less than zero. Observe how the *meaning* of this schema under (D, I) depends both on the interpretation function I which assigns 'meanings' (of a kind) to the simple constituents of the language, and on the semantic rules by which the derived interpretation I' is defined in terms of I. The former component (the *base component* of the semantics) determines that a refers to 0 and that G means 'is greater than'; the latter component (the *compositional component* of the semantics) is what determines that ∃ means 'there is' or 'for some'.

Continuing, the interpretation of the predicate

$$E(\_) \rightarrow E(s(s(\_)))$$

is the set of elements d in D for which $E(*) \rightarrow E(s(s(*)))$ is satisfied in (D, I_d). Now an arbitrary natural number d is either even or odd. If d is odd then $E(*)$ is not satisfied by (D, I_d), so by the rule for the conditional, $E(*) \rightarrow E(s(s(*)))$ *is* satisfied in this case. And if d is even, then so is $d + 2$, which means that both $E(*)$ and $E(s(s(*)))$ are satisfied by (D, I_d) (since $I'_d(s(s(*))) = d + 2$), and hence the conditional $E(*) \rightarrow E(s(s(*)))$ is satisfied as well. Thus this schema is satisfied by (D, I_d), whichever d in D we choose, which means that

$$I'(E(\_) \rightarrow E(s(s(\_)))) = D.$$

This is turn means that the schema

$$(12) \quad \forall x (E(x) \rightarrow E(s(s(x))))$$

is satisfied by (D, I), using the rule for the universal quantifier.

What makes (12) satisfied by (D, I) is precisely the fact that, for any natural number d, if d is even, then so is $d + 2$. That is why we can say that under this interpretation (12) says that for any natural number d, if d is even so is $d + 2$.

Speaking generally, we can characterize the meaning of a schema A under an interpretation (D, I) as follows. The schema A says that the condition C obtains, where C is a necessary and sufficient condition for the satisfaction of A by (D, I). Philosophers refer to this characterization of meaning as the *truth-conditional* account of meaning.

Note the power of the formal semantics: in the earlier parts of this chapter we relied on intuition to determine the meaning of a schema under a given interpretation of the language. The formal semantic rules go a long way towards *replacing* intuition. In doing so, the formal semantics becomes the ultimate arbiter of what the schemas in a first-order language mean. Any first-order schema is completely unambiguous under a given formal interpretation of the language. Of course, we use ordinary English sentences to convey the meanings of schemas informally; the fact that these sentences are liable to be ambiguous is neither here nor there. The ambiguity of English does not get transmitted back to the predicate calculus; which is why the predicate calculus can be such a useful means of expression in contexts where it is of the utmost importance to avoid all possible ambiguity. A first-order schema is not simply a translation of an English sentence into funny symbols; the predicate calculus is a language in its own right. Because it is totally artificial, its definitions and rules can be completely controlled. Therein lie both its power and its limitations.

Exercise 4.6   All the questions in this exercise refer to the language $L = L(C, F, P, V)$, where

$$C = \{a, b, c\}$$
$$F = \{f, g\}$$
$$P = \{P, Q\}$$
$$V = \{u, v, w, x, y, z, \ldots\}$$

and P and f have arity 1 and Q and g have arity 2.

(1)   Consider the interpretation (D, I) of L such that D is the set of non-empty strings of letters from the English alphabet (so, for example, typical members of D are 'aberr', 'qqqq', 'y', and 'intersubstitutability')

I(a) = 'a'
I(b) = 'the'
I(c) = 're'

I(f)   is defined as follows:
    —if d is of length 1 (e.g., d = 'y') then I(f)[d] = d
    —otherwise, I(f)[d] is the result of removing the first letter from d (so, for example, I(f)['spot'] = 'pot').

I(g)   maps the pair (d, d') to the result of concatenating d with d' (e.g., I(g)['cur', 'few'] = 'curfew').

I(P)   is the set of all English words (so e.g., 'apt', 'pat', and 'tap' are in I(P), but 'atp', 'pta', and 'tpa' are not).

I(Q)   is the set of pairs (d, d') such that d is a substring of d' (e.g., ('rat', 'crate'), ('art', 'part'), ('ear', 'ear'), but not ('ear', 'ar'), ('rat', 'raft') or ('ear', 'wtggf')).

In this interpretation, determine the truth values of the following schemas (i.e., 'true' for a satisfied schema, 'false' for a falsified one), and express in English what they say:

  *(a) P(a)
  (b) P(b)
  (c) P(c)
  (d) P(f(a))
  *(e) P(f(b))
  (f) P(f(c))
  *(g) P(g(a, b))
  (h) P(g(a, c))
  (i) P(g(b, c))
  (j) P(g(f(b), c))
  *(k) P(g(f(c), f(b)))
  *(l) Q(c, b)
  (m) Q(f(c), f(b))
  (n) Q(f(f(c)), f(f(b)))
  *(o) $\forall x Q(x, x)$
  (p) $\forall x Q(f(x), x)$
  *(q) $\exists x Q(x, f(x))$
  (r) $\forall x(P(x) \rightarrow P(f(x)))$
  (s) $\exists x(P(x) \wedge P(f(x)))$
  (t) $\exists x \exists y(P(x) \wedge P(y) \wedge P(g(x, y)))$
  *(u) $\exists x \exists y(P(x) \wedge P(y) \wedge P(g(x, y)) \wedge P(g(y, x)))$
  (v) $\exists x \exists y(\sim P(x) \wedge \sim P(y) \wedge P(g(x, y)) \wedge P(g(y, x)))$

(2)   Consider the interpretation (N, J) of L in which
  N is the set of natural numbers

$J(a) = 0$
$J(b) = 1$
$J(c) = 3$
$J(f): n \rightarrow n^2$
$J(g): m, n \rightarrow m + n$

$J(\mathrm{P}) = \{n \in N \mid n \text{ is even}\}$
$J(\mathrm{Q}) = \{(m, n) \in N \times N \mid m \text{ divides } n\}$

In this interpretation, determine the truth values of each of the schemas (a)–(v) in Question (1), and express in English what they say.

(3)   Write down schemas in the language L which, when interpreted in accordance with the interpretation (N, J) given in Question (2) express the following facts (in which 'number' should be understood to mean 'natural number' throughout):

   (a)   Every odd number divides its own square.
*(b)   Not every square divides the number it is the square of.
   (c)   If X divides Y, then X^2 divides Y^2.
*(d)   The square of an odd number is odd.
   (e) $(X + Y)^2 = X^2 + 2XY + Y^2$.
*(f)   If the sum of two numbers is even, so is the sum of their squares.
   (g)   There is an odd number X such that $X + 3$ divides $X + 9$.

4.7 SATISFACTION, ENTAILMENT, AND EQUIVALENCE IN THE PREDICATE CALCULUS

In this section we show how the formal semantics for the predicate calculus enables us to characterize the general logical notions we introduced in Chapter 1. Fundamental to this is the notion, defined in the last section, of a schema A being *satisfied* by an interpretation (D, I), written $\vDash_{D,I} A$. We also say in this case that (D, I) is a **model** *for* A.

Recall from the first chapter that a set of statements is said to be satisfiable so long as it is possible for all the statements to be true at the same time. In the predicate calculus, 'at the same time' means 'under one interpretation'. So for the predicate calculus, our definition of satisfiability is as follows:

> A set **S** of first-order schemas is *satisfiable* if and only if there is a formal interpretation (D, I) of the language of **S** such that $\vDash_{D,I} A$ for every A in **S**.

By 'the language of **S**' we mean a first-order language whose lexicon includes all the constants, functions, predicates, and variables occurring in **S**. If (D, I) satisfies each member of **S**, we say that (D, I) satisfies, or is a model for, **S**. So a terser statement of the definition of satisfiability is:

> A set of first-order schemas is satisfiable if and only if it has a model.

Example 4.7.1   Consider the set

$$\mathbf{S}_1 = \{\forall x(\mathrm{P}(x) \to \mathrm{Q}(x)), \exists x\mathrm{P}(x), \exists x \sim \mathrm{Q}(x)\}.$$

It is easy to find a model for this set. Put $D = \{0, 1\}$, $I(P) = \{0\}$, and $I(Q) = \{0\}$. We can tabulate the truth values of the schemas $P(*)$, $\sim Q(*)$, and $P(*) \rightarrow Q(*)$ in the extended interpretations (D, I_0) and (D, I_1) as follows:

	I_0	I_1
$P(*)$	T	F
$\sim Q(*)$	F	T
$P(*) \rightarrow Q(*)$	T	T

From this we see that

$$I'(P(\_)) = \{0\} \neq \varnothing$$
$$I'(\sim Q(\_)) = \{1\} \neq \varnothing$$
$$I'(P(\_) \rightarrow Q(\_)) = \{0, 1\} = D.$$

It follows immediately that each element of $\mathbf{S}_1$ is true in (D, I), and hence that $\mathbf{S}_1$ is satisfiable.

Note that satisfiability is a property that a set of schemas has *absolutely*, i.e. not relative to an interpretation; but it has this property by virtue of there being an interpretation which satisfies it. In this respect, satisfiability and unsatisfiability are quite different from truth (i.e. satisfaction) and falsehood (i.e. falsification): a schema is not true or false absolutely, but only relative to an interpretation.

Example 4.7.2   Consider the set of schemas

$$\mathbf{S}_2 = \{\forall x (P(x) \rightarrow Q(x)), \exists x P(x), \sim \exists x Q(x)\}.$$

Let (D, I) be any interpretation which satisfies the first two schemas in $\mathbf{S}_1$. Since (D, I) satisfies $\exists x P(x)$, we know that $I(P)$ must be non-empty. This means that we may postulate a domain element d belonging to $I(P)$. Further, since (D, I) satisfies $\forall x (P(x) \rightarrow Q(x))$, we have

$$I'(P(\_) \rightarrow Q(\_)) = D,$$

so in particular

$$d \in I'(P(\_) \rightarrow Q(\_)).$$

But this just means that

$$\vDash_{D, I_d} P(*) \rightarrow Q(*).$$

Now since d is in $I(P)$, we also have

$$\vDash_{D, I_d} P(*)$$

so by the semantic rule for $\rightarrow$ we must also have

$$\vDash_{D, I_d} Q(*).$$

This means that d is in $I'(Q(\_))$, which is therefore non-empty. Hence (D, I) satisfies $\exists x Q(x)$. But this shows that (D, I) does *not* satisfy $\sim \exists x Q(x)$. We have shown that any interpretation which satifies the first two schemas in $\mathbf{S}_2$ fails to satisfy the third schema in that set. Hence no interpretation satisfies all three members of $\mathbf{S}_2$, and so $\mathbf{S}_2$ is unsatisfiable.

As we saw in Chapter 1, entailment can be defined straightforwardly in terms of satisfiability, the rule being that a set of schemas **S** entails a schema A just so long as $\mathbf{S} \cup \{\sim A\}$ is unsatisfiable. In terms of our model theory, this means that **S** entails A so long as $\mathbf{S} \cup \{\sim A\}$ has no models. Equivalently, **S** entails A so long as every model for **S** satisfies A (otherwise some model for **S** satisfies $\sim A$ and hence is a model for $\mathbf{S} \cup \{\sim A\}$).

From the unsatisfiability of $\mathbf{S}_2$ above we know that the set consisting of the first two schemas in $\mathbf{S}_2$ entails the negation of the third schema of $\mathbf{S}_2$. We have thus shown that the inference

$$\frac{\begin{array}{l}\forall x(P(x) \rightarrow Q(x)) \\ \exists x P(x)\end{array}}{\exists x Q(x)}$$

is valid.

Applying the rule of $\rightarrow$-introduction to this inference, we see that the inference

$$\frac{\forall x(P(x) \rightarrow Q(x))}{\exists x P(x) \rightarrow \exists x Q(x)}$$

is valid, i.e. that

$$\forall x(P(x) \rightarrow Q(x)) \vDash \exists x P(x) \rightarrow \exists x Q(x).$$

We cannot reverse the entailment here; that is, the second schema does *not* entail the first. To show this, we need only exhibit an interpretation which satisfies the former but not the latter. Put $D = \{0, 1\}$, $I(P) = \{0\}$, and $I(Q) = \{1\}$. Then

$$\vDash_{D,I} \exists x Q(x)$$

so

$$\vDash_{D,I} \exists x P(x) \rightarrow \exists x Q(x).$$

But $I'(P(\_) \rightarrow Q(\_)) = \{1\} \neq D$, so (D, I) does not satisfy $\forall x(P(x) \rightarrow Q(x))$.

In general, since A logically implies B so long as every model for A is a model for B, and B logically implies A so long as every model for B is a model for A, then A is *equivalent* to B just so long as A and B have exactly the same models, i.e. are satisfied by the same interpretations. Writing **M**(A) to denote the set of all models for A (i.e. the set of all interpretations which satisfy A), we thus have

---

For any two first-order schemas A and B,

$A \vDash B$ if and only if $\mathbf{M}(A) \subseteq \mathbf{M}(B)$

and

$A \cong B$ if and only if $\mathbf{M}(A) = \mathbf{M}(B)$

In Section 4.3 we noted that the combination $\forall x \sim$ is equivalent to $\sim \exists x$. We are now in a position to demonstrate this formally. Specifically, we can show that for any unary predicate $\Phi(\_\_)$,

$$\forall x \sim \Phi(x) \cong \sim \exists x \Phi(x).$$

For suppose

$$\vDash_{D,I} \forall x \sim \Phi(x).$$

This means that $I'(\sim \Phi(\_)) = D$. Let d be any element of D. By the semantic rule for negation, (D, I_d) satisfies $\sim \Phi(*)$ if and only if (D, I_d) falsifies $\Phi(*)$. Since $I'(\sim \Phi(\_)) = D$, we know that (D, I_d) satisfies $\sim \Phi(*)$ for every d in D. Hence it satisfies $\Phi(*)$ for *no* d in D, which means that $I'(\Phi(\_))$ is empty. Hence (D, I) does not satisfy $\exists x \Phi(x)$, i.e.

$$\vDash_{D,I} \sim \exists x \Phi(x).$$

Each step in this reasoning can be reversed, so (D, I) satisfies $\forall x \Phi(x)$ if and only if it satisfies $\sim \exists x \Phi(x)$, so these two schemas have the same models and hence are equivalent, as required.

Consideration of which interpretations satisfy a given schema leads us naturally on to model-theoretic definitions of logical truth and logical falsehood. Remember that a schema is logically true if it logically *cannot* be false. In terms of our model theory, this means that it is satisfied by every interpretation. Similarly, a schema is logically false if it is falsified by every interpretation.

An example of a logically true schema is

$$\forall x (F(x) \rightarrow \exists y F(y)).$$

To show that this is logically true, let (D, I) be an arbitrary interpretation of the language of this schema. (D, I) satisfies $\exists y F(y)$ so long as $I'(F(\_))$ is non-empty, i.e. so long as $I(F)$ is non-empty. Let e be any member of D. If e in $I(F)$ then, from what we have just said, (D, I) satisfies $\exists x F(y)$, so (D, I_e) satisfies $F(*) \rightarrow \exists y F(y)$. If, on the other hand, e is not in $I(F)$, then (D, I_e) satisfies $\sim F(*)$, so again (D, I_e) satisfies $F(*) \rightarrow \exists y F(y)$. Since e is either in $I(F)$ or not in $I(F)$, it follows that

$$I(F(\_) \rightarrow \exists y F(y)) = D$$

and hence that

$$\vDash_{D,I} \forall x (F(x) \rightarrow \exists y F(y)).$$

Since (D, I) was chosen arbitrarily, it follows that the schema is true in every interpretation, and hence is logically true.

As with the propositional calculus, we write $\vDash A$ to mean that a schema A is logically true; but note that we do not use the term 'tautology' here unless the schema in question is a substitution-instance of a tautology of the propositional calculus (so that, for example,

$$\forall x F(x) \lor \sim \forall x F(x)$$

is a tautology, but

$$\forall x (F(x) \lor \sim F(x))$$

is not, although it, too, is logically true). We thus have

A first-order schema A is **logically true** if and only if it is satisfied by every interpretation, i.e.

$$\vDash A \text{ if and only if for every interpretation } (D, I),\ \vDash_{D,I} A$$

A first-order schema A is **logically false** if and only if it is falsified by every interpretation, i.e.

$$\nvDash A \text{ if and only if for every interpretation } (D, I),\ \nvDash_{D,I} A$$

Exercise 4.7

(1) Determine which of the following sets of schemas are satisfiable. For each of those which are, give an interpretation which satisfies it.

*(a) $\{\forall x \exists y P(x, y), \forall x \sim P(x, x)\}$

*(b) $\{\exists x \forall y P(x, y), \forall x \sim P(x, x)\}$

(c) $\{\forall x(P(x) \vee Q(x)), \sim \exists x P(x), \sim Q(a)\}$

(d) $\{\exists x P(x), \forall x(P(x) \to Q(x)), \forall x \sim Q(x)\}$

(e) $\{\forall x(P(x) \to \exists y Q(x, y)), \forall x \forall y(Q(x, y) \to R(y)), P(a) \wedge \sim R(a)\}$

(2) Classify the following schemas as logically true, logically false, or neither.

*(a) $\forall x(P(x) \vee \sim P(x))$

*(b) $\forall x P(x) \vee \forall x \sim P(x)$

*(c) $\exists x(P(x) \wedge \sim P(x))$

(d) $\exists x P(x) \wedge \exists x \sim P(x)$

(e) $\forall x P(x) \to \exists x P(x)$

(f) $\forall x P(x) \wedge \exists x \sim P(x)$

(g) $\forall x \exists y(P(x, y) \vee \forall z \sim P(x, z))$

(h) $\exists x \forall y \exists z(Q(x, z) \wedge \sim Q(x, y))$

(3) Express in terms of **M**(A) and **M**(B) the following sets of interpretations:

*(a) $\mathbf{M}(A \wedge B)$

(b) $\mathbf{M}(A \vee B)$

*(c) $\mathbf{M}(A \wedge \sim B)$

(d) $\mathbf{M}(A \to B)$

5 Proof Systems for the Predicate Calculus

In this chapter, we consider in turn each of the proof systems introduced for the propositional calculus in Chapter 3, and show how it can be extended so as to handle the predicate calculus. In each case, what we have to do is to *add* to the existing rules: since the propositional connectives are all also present in the predicate calculus, with the same semantic rules, all the proof rules relating to these connectives can be included without change in a proof system for the predicate calculus. So in what follows, we shall be concentrating on establishing suitable proof rules for the quantifiers.

5.1 NATURAL DEDUCTION

Just as in Section 3.2 we gave an introduction rule and an elimination rule for each of the connectives, so here we must specify introduction and elimination rules for the two quantifiers. Of the four rules that are required, two are entirely straightforward; the other two require more discussion. The straightforward rules are ∀-elimination and ∃-introduction.

∃-introduction

Suppose that in the course of a derivation we assert a schema of the form $\Phi(t)$, where $\Phi(\_)$ is a unary predicate and t is any term. For this to be true in an interpretation, we require that $I'(t) \in I'(\Phi)$, and hence that $I'(\Phi)$ is non-empty. But if *that* is the case, then $\exists x \Phi(x)$ must also be true in that interpretation, and we have the rule

$$(\exists\text{-intro}) \qquad \frac{\Phi(t)}{\exists x \Phi(x)}$$

This rule corresponds to an inference such as

$$\frac{\text{Buzz Aldrin has been to the Moon}}{\text{Someone has been to the Moon}}$$

Its validity is unproblematic.

∀-*elimination*

Suppose now that in the course of a derivation we have asserted a schema of the form $\forall(x)\Phi(x)$, and let t be any term. In order for $\forall x\Phi(x)$ to be true in an interpretation, we need $I'(\Phi(\_)) = D$, and hence in particular $I'(t) \in I'(\Phi(\_))$. This means that $\Phi(^*)$ is true if * is interpreted to mean $I'(t)$, and hence that $\Phi(t)$ is true (since t *is* interpreted to mean $I'(t)$). So $\Phi(t)$ is a logical consequence of $\forall x\Phi(x)$, and this justifies our rule

$$\forall\text{-elim} \quad \left|\begin{array}{c} \forall x(\Phi(x)) \\ \hline \Phi(t) \end{array}\right.$$

This rule is exemplified by an inference such as

$$\frac{\text{Everyone needs sleep}}{\text{Buzz Aldrin needs sleep}}$$

Example 5.1.1   Validate the inference

$$\frac{\forall x R(x, a)}{\exists x R(a, x)}$$

(e.g. from 'Everyone admires David' it follows that David admires somebody, namely himself). The formal derivation is as follows:

> (1) $\forall x R(x, a)$ (premiss)
> (2) $R(a, a)$ (1, ∀-elim: a/x)
> (3) $\exists x R(a, x)$ (2, ∃-intro: x/a)

Notice that in the annotation for ∀-elimination, we write 'a/x' to indicate that the variable x is replaced by the term a; and likewise, with ∃-introduction, we write 'x/a' to indicate that the term a is replaced by the variable x.

Example 5.1.2   Validate the inference

$$\begin{array}{c} \forall x P(x) \lor \forall x \sim P(x) \\ \sim P(a) \\ \hline \sim P(b) \end{array}$$

An informal interpretation of this inference is

> $\dfrac{\begin{array}{l}\text{Either everyone will come or no-one will come}\\ \text{John will not come}\end{array}}{\text{Mary will not come}}$

The derivation is as follows:

> (1) $\forall x P(x) \lor \forall x \sim P(x)$ (premiss)
> (2) **SUBDERIVATION**
>   (2.1) $\forall x P(x)$ (assumption)
>   (2.2) **SUBDERIVATION**
>   (2.2.1) $P(b)$ (assumption)
>   (2.2.2) $P(a)$ (2.1, ∀-elim: a/x)

$$(2.2.3) \quad \sim P(a) \qquad \text{(premiss)}$$

(2.3) $\quad \sim P(b) \qquad$ (2.2, $\sim$-intro)

(3) **SUBDERIVATION**

(3.1) $\quad \forall x \sim P(x) \qquad$ (assumption)

(3.2) $\quad \sim P(b) \qquad$ (3.1, $\forall$-elim: b/x)

(4) $\quad \sim P(b) \qquad$ (1, 2, 3, $\vee$-elim)

Since the first premiss is a disjunction, the overall structure of this derivation is that of $\vee$-elimination. This determines the assumptions of the two subderivations: they are the two disjuncts of the first premiss. The desired conclusion of the whole derivation, and hence of each subderivation, is $\sim P(b)$. In the second subderivation, this follows immediately from the assumption; but in the first subderivation it has to be derived indirectly using $\sim$-introduction.

Where two or more applications of $\forall$-elimination occur in successive steps of a derivation, they can be condensed into a single step. Thus instead of

(1) $\quad \forall x \forall y \forall z P(x, y, z)$

(2) $\quad \forall y \forall z P(a, y, z) \qquad$ (1, $\forall$-elim: a/x)

(3) $\quad \forall z P(a, b, z) \qquad$ (2, $\forall$-elim: b/y)

(4) $\quad P(a, b, c) \qquad$ (3, $\forall$-elim: c/z)

we can write

(1) $\quad \forall x \forall y \forall z P(x, y, z)$

(2) $\quad P(a, b, c) \qquad$ (1, $\forall$-elim: a/x, b/y, c/z)

The same thing can be done with $\exists$-introduction, allowing us to write, for example,

(1) $\quad P(a, b, c)$

(2) $\quad \exists x \exists y \exists z P(x, y, z) \qquad$ (1, $\exists$-intro: x/a, y/b, z/c)

We now turn to the other two quantifier rules, $\exists$-elimination and $\forall$-introduction.

$\forall$-introduction

To motivate this rule, we need to consider under what circumstances we should be entitled to infer a schema $\forall x \Phi(x)$ from previously asserted schemas.

Example 5.1.3   Consider the following inference:

> Every square is a rectangle
> Every rectangle has equal diagonals
> ———————————————————
> Every square has equal diagonals

Informally, one might argue for the validity of this as follows:

(1) Suppose PQRS is a square.

(2) If PQRS is a square then it is a rectangle.

(3) So PQRS is a rectangle.

(4) If PQRS is a rectangle, then it has equal diagonals.

(5) So PQRS has equal diagonals.

(6) So if PQRS is a square, then it has equal diagonals.
(7) Therefore every square has equal diagonals.

This reasoning is valid because 'PQRS' is not really being used to refer to any particular square (even if there really is a square of that name). 'PQRS' is being used as a **dummy name** that can be understood to be referring to any square whatever. And because it is being used in this way, we are justified in making the generalization from line 6 to line 7, in which the implicit generality of the dummy name 'PQRS' is replaced by the explicit generality of a quantifier.

In formal terms, suppose that in some derivation one has just asserted a schema $\Phi(a)$. So long as it is possible to regard a here as playing the role of a dummy name, then we can legitimately go on to assert $\forall x\Phi(x)$. Under what circumstances *is* it legitimate to treat a as a dummy name? Essentially what is required is that a is not already involved in the derivation in a substantial way, so that in effect it *does not matter* which individual is referred to by a; only then will what is asserted using the constant a be completely general, justifying the introduction of a universal quantifier. So what we require is that a does *not* occur in

(i)   the predicate $\Phi(\_)$;
(ii)   any of the premisses;
(iii) the assumption of any current subderivation.

By a **current subderivation** is meant a subderivation which has been begun but not finished, i.e. a subderivation whose assumption has not yet been discharged. (For example, in Example 5.1.2, at line (2.2.2) the subderivations (2) and (2.2) are both current; by the time we reach (2.3), the former is still current, but not the latter; and neither is current by the time we reach line (3).)

This all leads us on to the formulation of the rule:

$$(\forall\text{-intro}) \quad \dfrac{\Phi(a)}{\forall x\Phi(x)}$$

(so long as a does not occur in
(i) $\Phi(\_)$
(ii)   a premiss
(iii) the assumption of a current
       subderivation)

Using this rule, we can formalize the argument given in Example 5.1.3 as follows:

(1) $\forall x(P(x) \rightarrow Q(x))$ (premiss)
(2) $\forall x(Q(x) \rightarrow R(x))$ (premiss)
(3)   SUBDERIVATION
       (3.1) $P(a)$ (assumption)
       (3.2) $P(a) \rightarrow Q(a)$ (1, $\forall$-elim: a/x)
       (3.3) $Q(a)$ (3.1, 3.2, $\rightarrow$-elim)
       (3.4) $Q(a) \rightarrow R(a)$ (2, $\forall$-elim: a/x)
       (3.5) $R(a)$ (3.3, 3.4, $\rightarrow$-elim)
(4) $P(a) \rightarrow R(a)$ (3, $\rightarrow$-intro)
(5) $\forall x(P(x) \rightarrow R(x))$ (4, $\forall$-intro: x/a)

The move from line 4 to line 5 is an application of $\forall$-introduction, in which $\Phi(\_) = P(\_) \rightarrow Q(\_)$. The constant a that is replaced by the bound variable does not occur in $\Phi(\_)$, in either of the two premisses, or in the assumption of a current sub-derivation (it *does* occur in the assumption of the subderivation at 3, but this is no longer current by the time we get to 5). Hence the conditions for applying the rule are all met.

The conditions for applying $\forall$-introduction are carefully designed to ensure that the constant which gets replaced by the variable is playing the role of a dummy name in the schema to which the rule is applied. To see more clearly why they are necessary, it is instructive to examine what happens if we violate the conditions.

If we violated condition (i), we could 'prove' that 'Everyone admires himself' implies 'Everyone admires David', as follows:

(1)	$\forall x R(x, x)$	(premiss)
(2)	$R(a, a)$	(1, $\forall$-elim: a/x)
(3)	$\forall x R(x, a)$	(2, $\forall$-intro: x/a—incorrectly applied)

If we violated condition (ii), we could 'prove' that 'Mary admires David' implies 'David admires Mary':

(1)	$R(a, b)$	(premiss)
(2)	$\forall y R(a, y)$	(1, $\forall$-intro: y/b—incorrectly applied)
(3)	$\forall x \forall y R(x, y)$	(2, $\forall$-intro: x/a—incorrectly applied)
(4)	$\forall y R(b, y)$	(3, $\forall$-elim: x/b)
(5)	$R(b, a)$	(4, $\forall$-elim: a/y)

and condition (iii) prevents us from using this fraudulent reasoning within a sub-derivation, which would enable us to 'prove' that *all* admiration is mutual, as in

(1)	SUBDERIVATION	
	(1.1) $R(a, b)$	(assumption)
	$\vdots$	
	(1.5) $R(b, a)$	(as above)
(2)	$R(a, b) \rightarrow R(b, a)$	(1, $\rightarrow$-intro)
(3)	$\forall x \forall y (R(x, y) \rightarrow R(y, x))$	(2, $\forall$-intro: x/a, y/b)

Note that the application of $\forall$-introduction at line 3 is correct: neither a nor b occurs in line 3 itself, and there are no premisses or current subderivations. The error in this argument lies solely within the subderivation.

$\exists$-elimination

For this rule, we need to consider under what circumstances we can infer something from a schema of the form $\exists x \Phi(x)$. An inference to illustrate this is:

Example 5.1.4

> Anyone who admires Jane admires David
> Someone admires Jane
> ———————————————
> Someone admires David

An informal way of arguing this is as follows:

(1) Someone admires Jane.
(2) Suppose it were Mary who admired Jane.
(3) Then since anyone who admires Jane admires David,
     Mary would admire David.
(4) So in that case someone would admire David.
(5) So in any case someone admires David.

Here, as before, 'Mary' functions as a dummy name. What is happening is that we are told that (1) someone admires Jane; we assume 'for the sake of the argument' that (2.1) the 'someone' who admires Jane is Mary, and we go on to derive from this assumption the conclusion (2.3) that someone admires David. Since the truth of this conclusion does not depend on its being Mary, rather than anyone else, who admires Jane, we are then entitled to derive the same conclusion (3) from the original premiss (1).

Formally, suppose in a derivation one has just asserted a schema $\exists x\Phi(x)$. Let a be any individual constant satisfying the conditions (i)–(iii) governing the application of $\forall$-introduction, and let A be any schema, *not containing* a, which can be derived from $\Phi(a)$. Then we are entitled to infer A from $\exists x\Phi(x)$ also. For in just these circumstances, we can use a as a dummy name to instantiate the bound variable in $\exists x\Phi(x)$. We are told that *something* has the property denoted by $\Phi(\_)$, and we are just *calling* it a. No error is introduced here, because the restrictions ensure that the name a is not already playing a part in the derivation at this point.

We can thus state the rule of $\exists$-elimination as follows:

$$(\exists\text{-elim}) \quad \begin{array}{|l} \exists x\Phi(x) \\ \quad \left|\begin{array}{l} \Phi(a) \\ \hline A \end{array}\right. \\ \hline A \end{array}$$

(so long as a does not occur in
  (i) $\Phi(\_)$
  (ii) a premiss
  (iii) the assumption of a current
        subderivation
  (iv) A)

Using this rule, we can formalize the argument given in Example 5.1.4 as follows:

(1) $\forall x(P(x) \rightarrow Q(x))$ (premiss)
(2) $\exists x P(x)$ (premiss)
(3) **SUBDERIVATION**
     (3.1) $P(a)$ (assumption)
     (3.2) $P(a) \rightarrow Q(a)$ (1, $\forall$-elim: a/x)
     (3.3) $Q(a)$ (3.1, 3.2, $\rightarrow$-elim)
     (3.4) $\exists x Q(x)$ (3.3, $\exists$-intro: x/a)
(4) $\exists x Q(x)$ (2, 3, $\exists$-elim: x/a)

To check that the condition for $\exists$-elimination are met, note that

(i)   a does not occur in $\Phi(\_) = P(\_)$;
(ii) a does not occur in either premiss;

(iii) a does not occur in the assumption of a subderivation
current at line 4 (there are none);

(iv) a does not occur in A = ∃xQ(x).

We conclude this section with some more examples of derivations by natural deduction in the predicate calculus. Work through these derivations step by step, being especially careful to check that the conditions for ∀-introduction or ∃-elimination are met wherever these rules are applied.

Example 5.1.5   Prove that the schema

$$\forall x(F(x) \to \exists y F(y))$$

is logically true.

We must derive this schema without using any premisses. A suitable derivation is:

       (1)   SUBDERIVATION
           (1.1)   F(a)   (assumption)
           (1.2)   ∃yF(y)   (1.1, ∃-intro: y/a)
       (2)   F(a) → ∃yF(y)   (1, →-intro)
       (3)   ∀x(F(x) → ∃yF(y))   (2, ∀-intro: x/a)

Example 5.1.6   Show that ∼∀xF(x) logically implies ∃x ∼ F(x).

       (1)   SUBDERIVATION
           (1.1)   ∼∃x ∼ F(x)   (assumption)
           (1.2)   SUBDERIVATION
                (1.2.1)   ∼ F(a)   (assumption)
                (1.2.2)   ∃x ∼ F(x)   (1.2.1, ∃-intro: x/a)
                (1.2.3)   ∼∃x ∼ F(x)   (1.1, rep)
           (1.3)   F(a)   (1.2, ∼-elim2)
           (1.4)   ∀xF(x)   (1.3, ∀-intro: x/a)
           (1.5)   ∼ ∀xF(x)   (premiss)
       (2)   ∼ ∼∃x ∼ F(x)   (1, ∼-intro)
       (3)   ∃x ∼ F(x)   (2, ∼-elim)

The converse of this result also holds, as we know from Section 4.3, and indeed it is possible to use our natural deduction rules to justify the following four derived rules of inference (see Exercise 5.1, Question (1)):

     (NA) $\dfrac{\sim \forall x \Phi(x)}{\exists x \sim \Phi(x)}$ (EN) $\dfrac{\exists x \sim \Phi(x)}{\sim \forall x \Phi(x)}$

     (NE) $\dfrac{\sim \exists x \Phi(x)}{\forall x \sim \Phi(x)}$ (AN) $\dfrac{\forall x \sim \Phi(x)}{\sim \exists x \Phi(x)}$

We can use these rule to derive more complicated rules such as

$$\frac{\sim \forall x \exists y \forall z \Phi(x, y, z)}{\exists x \forall y \exists z \sim \Phi(x, y, z)}$$

The derivation of this rule is as follows:

(1) SUBDERIVATION
  (1.1) $\sim \exists x \forall y \exists z \sim \Phi(x, y, z)$ (assumption)
  (1.2) $\forall x \sim \forall y \exists z \sim \Phi(x, y, z)$ (1.1, NE)
  (1.3) $\sim \forall y \exists z \sim \Phi(a, y, z)$ (1.2, $\forall$-elim: x/a)
  (1.4) $\exists y \sim \exists z \sim \Phi(a, y, z)$ (1.3, NA)
  (1.5) SUBDERIVATION
    (1.5.1) $\sim \exists z \sim \Phi(a, b, z)$ (assumption)
    (1.5.2) $\forall z \sim \sim \Phi(a, b, z)$ (1.5.1, NE)
    (1.5.3) $\sim \sim \Phi(a, b, c)$ (1.5.2, $\forall$-elim: z/c)
    (1.5.4) $\Phi(a, b, c)$ (1.5.3, $\sim$-elim)
    (1.5.5) $\forall z \Phi(a, b, z)$ (1.5.4, $\forall$-intro: z/c)
    (1.5.6) $\exists y \forall z \Phi(a, y, z)$ (1.5.5, $\exists$-intro: y/b)
  (1.6) $\exists y \forall z \Phi(a, y, z)$ (1.4, 1.5, $\exists$-elim: y/b)
  (1.7) $\forall x \exists y \forall z \Phi(x, y, z)$ (1.6, $\forall$-intro: x/a)
  (1.8) $\sim \forall x \exists y \forall z \Phi(x, y, z)$ (premiss)
(2) $\sim \sim \exists x \forall y \exists z \sim \Phi(x, y, z)$ (1, $\sim$-intro)
(3) $\exists x \forall y \exists z \sim \Phi(x, y, z)$ (2, $\sim$-elim)

Generalizing, we can bring together any number of applications of NA, NE, AN, and EN into the two **quantifier-swapping** rules:

$$(\text{q-swap}) \quad \frac{Q_1 x_1 Q_2 x_2 \cdots Q_n x_n \sim \Phi(x_1, x_2, \ldots, x_n)}{\sim Q'_1 x_1 Q'_2 x_2 \cdots Q'_n x_n \Phi(x_1, x_2, \ldots, x_n)}$$

$$(\sim \text{q-swap}) \quad \frac{\sim Q_1 x_1 Q_2 x_2 \cdots Q_n x_n \Phi(x_1, x_2, \ldots, x_n)}{Q'_1 x_1 Q'_2 x_2 \cdots Q'_n x_n \Phi \sim (x_1, x_2, \ldots, x_n)}$$

Here each of $Q_1, \ldots, Q_n$ is either $\forall$ or $\exists$, and for each i, Q'_i is $\forall$ if Q_i is $\exists$, and vice versa. The next example illustrates the use of the q-swap rules.

Example 5.1.7 Show that $\sim \exists x (P(x) \wedge \sim P(x))$ is logically true.

(1) SUBDERIVATION
  (1.1) $P(a) \wedge \sim P(a)$ (assumption)
  (1.2) $P(a)$ (1.1, $\wedge$-elim-left)
  (1.3) $\sim P(a)$ (1.1, $\wedge$-elim-right)
(2) $\sim (P(a) \wedge \sim P(a))$ (1, $\sim$-intro)
(3) $\forall x \sim (P(x) \wedge \sim P(x))$ (2, $\forall$-intro: x/a)
(4) $\sim \exists x (P(x) \wedge \sim P(x))$ (3, q-swap)

Example 5.1.8 Sometimes the success of a derivation depends critically on which term is chosen to instantiate the bound variable when applying $\forall$-elimination. This is particularly true when function symbols are involved, for then the range of available terms is that much greater. To illustrate this point, we derive $\forall x \exists y P(y, f(x))$ from $\forall x P(f(x), x)$:

(1) $\forall x P(f(x), x)$ (premiss)
(2) $P(f(f(a)), f(a))$ (1, $\forall$-elim: f(a)/x)

(3) $\exists y P(y, f(a))$ $(2, \exists\text{-intro: } y/f(f(a)))$
(4) $\forall x \exists y P(y, f(x))$ $(3, \forall\text{-intro: } x/a)$

Example 5.1.9   Validate the inference:

$$\forall x \forall y \forall z (R(x, y) \land R(y, z) \to R(x, z))$$
$$\sim \exists x R(x, x)$$

$$\overline{\sim \exists x \exists y (R(x, y) \land R(y, x))}$$

(1)   SUBDERIVATION
    (1.1) $\exists x \exists y (R(x, y) \land R(y, x))$ (premiss)
    (1.2)   SUBDERIVATION
        (1.2.1) $\exists y (R(a, y) \land R(y, a))$ (assumption)
        (1.2.2)   SUBDERIVATION
            (1.2.2.1) $R(a, b) \land R(b, a)$ (assumption)
            (1.2.2.2) $\forall x \forall y \forall z (R(x, y)$
                    $\land R(y, z) \to R(x, z))$ (premiss)
            (1.2.2.3) $R(a, b) \land R(b, a) \to R(a, a)$ $(1.2.2.2, \forall\text{-elim: } a/x, b/y, a/z)$
            (1.2.2.4) $R(a, a)$ $(1.2.2.1, 1.2.2.3, \to\text{-elim})$
            (1.2.2.5) $\exists x R(x, x)$ $(1.2.2.4, \exists\text{-intro: } x/a)$
        (1.2.3) $\exists x R(x, x)$ $(1.2.1, 1.2.2, \exists\text{-elim})$
    (1.3) $\exists x R(x, x)$ $(1.1, 1.2, \exists\text{-elim})$
    (1.4) $\sim \exists x R(x, x)$ (premiss)
(2) $\sim \exists x \exists y (R(x, y) \land R(y, x))$ $(1, \sim\text{-intro})$

This derivation looks more formidable than it really is. Essentially the argument goes as follows: if there are elements x and y such that $R(x, y)$ and $R(y, x)$, then, from the first premiss (which states, in effect, that R denotes a transitive relation), we can infer that we must have $R(x, x)$. But this contradicts the second premiss, so our assumption must be false, i.e. there are *no* elements x and y such that $R(x, y)$ and $R(y, x)$.

Exercise 5.1

  *(1)   Prove the converse of Example 5.1.6, i.e. show that $\exists x \sim F(x)$ implies $\sim \forall x F(x)$.

  (2)   Each of the following derivations is defective because one of the conditions for applying $\forall$-introduction or $\exists$-elimination is not met. State which condition is violated in each case, and explain how it is violated.

  *(a)   (1) $\exists x F(x)$ (premiss)
        (2)   SUBDERIVATION
            (2.1) $F(a)$ (assumption)
            (2.2) $\forall x F(x)$ $(2.1, \forall\text{-intro: } x/a)$
        (3) $\forall x F(x)$ $(1, 2, \exists\text{-elim: } x/a)$

  (b)   (1) $\forall x R(x, a)$ (premiss)
        (2) $R(a, a)$ $(1, \forall\text{-elim: } a/x)$
        (3) $\forall x R(x, x)$ $(2, \forall\text{-intro: } x/a)$

  *(c)   (1) $R(a, b)$ (premiss)
        (2) $\exists x R(x, c)$ (premiss)
        (3)   SUBDERIVATION
            (3.1) $R(a, c)$ (assumption)
            (3.2) $R(a, b) \land R(a, c)$ $(1, 3.1, \land\text{-intro})$
            (3.3) $\exists x (R(x, b) \land R(x, c))$ $(3.2, \exists\text{-intro: } x/a)$
        (4) $\exists x (R(x, b) \land R(x, c))$ $(2, 3, \exists\text{-elim: } x/a)$

(d) (1) SUBDERIVATION
      (1.1) $\forall x R(x, a)$ (premiss)
      (1.2) $R(a, a)$ (1.1, $\forall$-elim: a/x)
  (2) $\forall x R(x, a) \to R(a, a)$ (1, $\to$-intro)
  (3) $\forall y(\forall x R(x, a) \to R(y, y))$ (2, $\forall$-intro: y/a)

(e) (1) $\exists x(R(x, a) \land R(x, b))$ (premiss)
  (2) SUBDERIVATION
      (2.1) $R(a, a) \land R(a, b)$ (assumption)
      (2.2) $\exists x(R(x, x) \land R(x, c))$ (2.1, $\exists$-intro: a/x)
  (3) $\exists x(R(x, x) \land R(x, c))$ (1, 2, $\exists$-elim: x/a)

(3) Use natural deduction to validate the following inferences:

*(a) $\forall x(P(x) \to Q(x))$
     $P(a)$
     _____
     $\exists x Q(x)$

*(b) $\exists x(P(x) \land Q(x))$
     $\forall x(Q(x) \to R(x))$
     _____
     $\exists x(P(x) \land R(x))$

(c) $P(a)$
     _____
     $\forall x(R(x, a) \to \exists y(P(y) \land R(x, y)))$

*(d) $\exists x P(x) \to Q(a)$
     _____
     $\forall x(P(x) \to Q(a))$

(e) $\forall x(P(x) \to Q(a))$
     _____
     $\exists x P(x) \to Q(a)$

(f) $\forall x S(x, a, x)$
     $\forall x \forall y \forall z(S(x, y, z) \to S(x, f(y), f(z)))$
     _____
     $\forall x S(f(x), f(a), f(f(x)))$

*(g) $\forall x \exists y(P(x) \to R(x, y))$
     _____
     $\forall x(P(x) \to \exists y R(x, y))$

*(h) $\forall x(P(x) \to \exists y R(x, y))$
     _____
     $\forall x \exists y(P(x) \to R(x, y))$

(Hint: One way of doing this is to assert $P(a) \lor \sim P(a)$ using LEM, then derive $\exists y(P(a) \to R(a, y))$ for each disjunct and use $\lor$-elimination. Finish off with $\forall$-introduction.)

(i) $\forall x S(x, x)$
    $\forall y(T(a, y) \leftrightarrow \forall x(S(x, y) \to D(a, x)))$
    _____
    $\forall x(T(a, x) \to D(a, x))$

(j) $\forall x \forall y(P(x) \land P(y) \to S(x, y))$
    $\forall x(Q(x) \to P(f(x)))$

$$\frac{\begin{array}{l} \forall x(R(x) \to P(g(x))) \\ \forall x R(x) \\ \exists x Q(x) \end{array}}{\exists x(S(f(x), g(x)))}$$

(k) $\dfrac{\begin{array}{l} \exists x \forall y(P(y) \to R(x, y)) \\ P(a) \end{array}}{\exists x R(x, a)}$

(l) $\dfrac{\forall x \exists y \forall z P(x, y, z)}{\forall x \exists y P(x, y, x)}$

(4)   Write the schemas (c) and (d) of Exercise 4.1 in predicate calculus notation, and validate them using natural deduction.

(5)   Use natural deduction to show that each of the following schemas is logically true:

*(a) $\forall y(\forall x R(x, y) \to R(y, y))$

(b) $\forall x \exists y(P(x) \to P(y))$

(c) $\sim \exists x \forall y(S(x, y) \leftrightarrow \sim S(y, y))$

*(d) $\forall x \exists y(Q(x, y) \lor \forall z \sim Q(x, z))$

5.2   MODEL BUILDING

The replacement rules for the propositional calculus given in Section 3.3 can be supplemented by four rules for handling schemas containing quantifiers. The resulting system enables us to use model building for checking sets of predicate calculus schemas for satisfiability, and hence to check inferences for validity.

The additional rules are

Rule	X	Y
$[\forall]$	$S \cup \{\forall x \Phi(x)\}$	$S \cup \{\Phi(t), \forall x \Phi(x)\}$ (where t is any term constructed from constants occurring in the current set)
$[\exists]$	$S \cup \{\exists x \Phi(x)\}$	$S \cup \{\Phi(a)\}$ (where a is a constant which does not occur in the current set)
$[\sim \forall]$	$S \cup \{\sim \forall x \Phi(x)\}$	$S \cup \{\exists x \sim \Phi(x)\}$
$[\sim \exists]$	$S \cup \{\sim \exists x \Phi(x)\}$	$S \cup \{\forall x \sim \Phi(x)\}$

The rules $[\sim \forall]$ and $[\sim \exists]$ straightforwardly exploit the equivalences of $\sim \forall x$ with $\exists x \sim$ and $\sim \exists x$ with $\forall x \sim$ which we have already met. The other two rules are less straightforward.

In the rule $[\forall]$, the quantified schema $\forall x \Phi(x)$ is retained alongside its instantiation $\Phi(t)$. The reason for this is that $\Phi(t)$ does not exhaust the information which can be extracted from $\forall x \Phi(x)$. It may be necessary at a later stage of a refutation to instantiate $\forall x \Phi(x)$ to some other schema $\Phi(t')$, say, and this will be possible only if we retain the quantified schema in the current set. This is illustrated in Example 5.2.8 below.

This consideration does not apply to $\exists x\Phi(x)$, since this schema only guarantees the existence of *one* individual having the property denoted by Φ; there may be more, but this cannot be counted on. Hence in the rule $[\exists]$, an existentially quantified schema is actually *replaced* by its instantiation.

Note carefully the conditions relating to the choice of the instantiating term t. The universally quantified schema $\forall x\Phi(x)$ tells us that everything has the property denoted by Φ. It does not, however, vouch for the existence of anything at all: what it says is that if we are given any object, then that object will have the property denoted by Φ, but it does not itself *give* us any object. That is why we can only instantiate the bound variable to a term that has already been given to us: a term is given to us once its constituent constants are given to us, either by being already present in the original set of schemas, or by being introduced in an application of the rule $[\exists]$.

The schema $\exists x\Phi(x)$ *does* vouch for the existence of something: it tells us that something has the property denoted by Φ. We cannot, of course, assume that any object already given to us has that property, which is why we have to introduce a *new* constant to denote the object whose existence is guaranteed by the existentially quantified schema. That is why the term a introduced in rule $[\exists]$ must be a new one, i.e. a constant not occurring in the refutation up to that point.

Here are some examples to illustrate the use of the new rules.

Example 5.2.1   Show that $\exists x\forall yR(x, y)$ implies $\exists xR(x, a)$.

$$\begin{array}{lll}
\mathbf{C}_0 = \{\{\exists x\forall yR(x, y), \sim\exists xR(x, a)\}\} & & \\
\mathbf{C}_1 = \{\{\forall yR(b, y), \sim\exists xR(x, a)\}\} & & [\exists: b/x] \\
\mathbf{C}_2 = \{\{R(b, a), \forall yR(b, y), \sim\exists xR(x, a)\}\} & & [\forall: a/y] \\
\mathbf{C}_3 = \{\{R(b, a), \forall yR(b, y), \forall x \sim R(x, a)\}\} & & [\sim\exists] \\
\mathbf{C}_4 = \{\{R(b, a), \forall yR(b, y), \sim R(b, a), \forall x \sim R(x, a)\}\} & & [\forall: b/x] \\
\mathbf{C}_5 = \{\ \} & & [\text{del}]
\end{array}$$

Note that in the annotations for $[\forall]$ and $[\exists]$ we indicate which term is chosen to instantiate the bound variable. The term b is first introduced in $\mathbf{C}_1$ by the application of $[\exists]$; thereafter it is available for use by $[\forall]$, as happens in $\mathbf{C}_4$. The term a, which is used by $[\forall]$ in $\mathbf{C}_2$, is present in the original set.

If we write this refutation in tree form, we do not need to repeat the universally quantified schema when we apply the rule $[\forall]$; it suffices for us not to mark the schema as 'used up'. When we apply $[\exists]$, however, we do mark the existentially quantified schema as 'used up'. Thus the tree corresponding to the above refutation looks like this:

$$\begin{array}{lll}
\times\ (1) & \exists x\forall yR(x, y) & \\
\times\ (2) & \sim\exists xR(x, a) & \\
\hline
 & & 1[b/x] \\
(3) & \forall yR(b, y) & \\
\hline
 & & 2 \\
(4) & \forall x \sim R(x, a) & \\
\hline
 & & 3[a/y] \\
(5) & R(b, a) & \\
\hline
 & & 4[b/x] \\
(6) & \sim R(b, a) & \\
\hline\hline
 & & 5
\end{array}$$

(This 'tree' has no branches, only a trunk!)

Example 5.2.2   Validate the inference

$$\forall x P(x) \lor \forall x \sim P(x)$$
$$\sim P(a)$$
$$\overline{}$$
$$\sim P(b)$$

This time we give only the tree form of the refutation:

$$\times (1) \quad \forall x P(x) \lor \forall x \sim P(x)$$
$$(2) \quad \sim P(a)$$
$$(3) \quad \sim \sim P(b)$$
$$\underline{\hspace{8cm}} 1$$

(4) $\forall x P(x)$	(5) $\forall x \sim P(x)$
$\overline{}$ 4[a/x]	$\overline{}$ 5[b/x]
(6) $P(a)$	(7) $\sim P(b)$
$\overline{\overline{}}$ 2	$\overline{\overline{}}$ 3

In this example, one would naturally think of applying the rule [$\sim \sim$] at line (3), to give $P(b)$. As our tree shows, this is not necessary, since lines (3) and (7) are already contradictory.

Example 5.2.3   Prove that $\forall x(F(x) \to \exists y F(y))$ is logically true.

$$\times (1) \quad \sim \forall x(F(x) \to \exists y F(y))$$
$$\overline{} 1$$
$$\times (2) \quad \exists x \sim (F(x) \to \exists y F(y))$$
$$\overline{} 2[a/x]$$
$$\times (3) \quad \sim (F(a) \to \exists y F(y))$$
$$\overline{} 3$$
$$(4) \quad F(a)$$
$$\times (5) \quad \sim \exists y F(y)$$
$$\overline{} 5$$
$$(6) \quad \forall y \sim F(y)$$
$$\overline{} 6[a/y]$$
$$(7) \quad \sim F(a)$$
$$\overline{\overline{}} 4$$

Example 5.2.4   Show that $\forall x P(f(x), x)$ implies $\forall x \exists y P(y, f(x))$.

$$(1) \quad \forall x P(f(x), x)$$
$$\times (2) \quad \sim \forall x \exists y P(y, f(x))$$
$$\overline{} 2$$
$$\times (3) \quad \exists x \sim \exists y P(y, f(x))$$
$$\overline{} 3[a/x]$$
$$\times (4) \quad \sim \exists y P(y, f(a))$$
$$\overline{} 4$$
$$(5) \quad \forall y \sim P(y, f(a))$$
$$\overline{} 5[f(f(a))/y]$$
$$(6) \quad \sim P(f(f(a)), f(a))$$
$$\overline{} 1[f(a)/x]$$
$$(7) \quad P(f(f(a)), f(a))$$
$$\overline{\overline{}} 6$$

Note how the success of this refutation depends critically on a good choice of instantiation for the bound variables. A refutation like this cannot easily be constructed blindly: it helps to have some idea of where one is going, what contradiction one is going to produce.

Example 5.2.5   We shall validate the inference

$$\forall xS(x, a, x)$$
$$\forall x\forall y\forall z(S(x, y, z) \rightarrow S(x, f(y), f(z)))$$

$$S(f(a), f(a), f(f(a)))$$

The refutation is

    (1) $\forall xS(x, a, x)$
    (2) $\forall x\forall y\forall z(S(x, y, z) \rightarrow S(x, f(y), f(z)))$
    (3) $\sim S(f(a), f(a), f(f(a)))$
                                                                               $2[f(a)/x, a/y, f(a)/z]$
  $\times$ (4) $S(f(a), a, f(a)) \rightarrow S(f(a), f(a), f(f(a)))$
                                                                                    4
  (5) $\sim S(f(a), a, f(a))$ (6) $S(f(a), f(a), f(f(a)))$
                            $1[f(a)/x]$ 3
  (7) $S(f(a), a, f(a))$
                           5

Note that to derive line (4) from line (2) we have condensed three successive applications of $[\forall]$ into a single step. This is quite legitimate both for this rule and for $[\exists]$.

The strategy of this refutation repays close examination. The only specific fact we have initially is line (3). Examining lines (1) and (2), we note that the consequent of the conditional in (2) has the same form as the schema that appears negated in (3), that is, we can instantiate the variables in (2) so that the two expressions become identical. This gives us line (4). The rest of the derivation is straightforward.

The language of this example has a natural interpretation in terms of the natural numbers. If we interpret a to mean zero, f the successor function, and S(____, ..., ---) to mean 'the sum of ____ and ... is ---', then the two premisses of the inference represent true facts about the natural numbers, namely

(1) for every natural number x, $x + 0 = x$, and
(2) for all natural numbers x, y, z if $x + y = z$ then $x + (y + 1) = z + 1$.

These two rules in effect define the meaning of addition in terms of the successor function. The conclusion of the inference says that $1 + 1 = 2$. The refutation we presented can be interpreted informally as follows: assume that

(3) $1 + 1 \neq 2$.

From rule (2), we have that

(4) if $1 + 0 = 1$ then $1 + 1 = 2$;

in other words, *either*

(5) $1 + 0 \neq 1$,

or

(6) $1 + 1 = 2$.

But the former case contradicts

(7) $1 + 0 = 1$,

derived from the first rule; and the latter case contradicts our assumption. We thus have a contradiction, and the proof is complete.

So far we have applied model building only to *valid* inferences in the predicate calculus; what happens if we try to use the technique to test an invalid one? There are two possibilities, which are illustrated by the next two examples.

Example 5.2.6 Determine whether or not the inference

$$\forall x(P(x) \rightarrow Q(x))$$
$$\exists x(Q(x) \wedge R(x))$$
$$\overline{\exists x(P(x) \wedge R(x))}$$

is valid.

The model building procedure goes as follows:

```
        (1)   ∀x(P(x) → Q(x))
    × (2)   ∃x(Q(x) ∧ R(x))
    × (3)   ~∃x(P(x) ∧ R(x))
   ─────────────────────────── 3
        (4)   ∀x ~(P(x) ∧ R(x))
   ─────────────────────────── 2[a/x]
    × (5)   Q(a) ∧ R(a)
   ─────────────────────────── 5
        (6)   Q(a)
        (7)   R(a)
   ─────────────────────────── 1[a/x]
    × (8)   P(a) → Q(a)
   ─────────────────────────── 4[a/x]
    × (9)   ~(P(a) ∧ R(a))
 ───────────────────────────────────── 9
     (10)  ~P(a)              (11)  ~R(a)
 ─────────────────────── 8           ══════ 7
  (12)  ~P(a)    (13)  Q(a)
```

At this point, the tree is complete, since no more replacement rules can be applied (the universally quantified schemas 1 and 4 have not, admittedly, been used up, but there are no more constants available for instantiating the bound variables to). But the tree remains open, since the branches ending at (12) and (13) have not yielded any contradiction. We have, in fact, succeeded in building a model for the schemas 1, 2, and 3: this model can be represented by the set of schemas $\{\sim P(a), Q(a), R(a)\}$. To construct it explicitly, we put $D = \{I(a)\}$, $I(P) = \emptyset$, $I(Q) = I(R) = D$. The interpretation (D, I) satisfies schemas 1, 2, and 3, and hence shows that the inference we began with is invalid. (Note that in this interpretation we have not said what $I(a)$ actually is, only that it is the single element of D; of course it does not matter what it is!)

Example 5.2.7 Show that the inference

$$\forall x \exists y P(x, y)$$
$$\overline{P(a, a)}$$

is invalid.

Our tree begins as follows:

(1) $\forall x \exists y P(x, y)$
(2) $\sim P(a, a)$
——————————— $1[x/a]$
× (3) $\exists y P(a, y)$
——————————— $2[b/y]$
(4) $P(a, b)$
——————————— $1[x/b]$
× (5) $\exists y P(b, y)$
——————————— $2[c/y]$
(6) $P(b, c)$
——————————— $1[x/c]$
× (7) $\exists y P(c, y)$
——————————— $2[d/y]$
(8) $P(c, d)$
$\vdots$

and continues in this fashion *ad infinitum*—the development of the tree is totally determined, since at each stage there is only one choice of what to do next. And the tree never closes, since each time a new constant is introduced using the rule $[\exists]$, we can then immediately use that constant in another application of $[\forall]$, which gives rise to a new existentially quantified schema, allowing the introduction of yet another new constant. Thus the first schema, 1, spawns an infinite sequence of instantiations of itself, and at no point is a contradiction reached. Thus although the inference is indeed invalid, the model building method does not in itself show us this.

This failure is not a defect in the model building method; rather, it is an inescapable feature of the predicate calculus itself. For further discussion of this, see Section 5.4 below.

Example 5.2.8   Show that the inference

$$\forall x(A(x, b) \to \forall(a, x))$$
$$A(b, b)$$
——————————
$$A(a, a)$$

is valid (cf. Exercise 1.5, Question 1(h)). This example illustrates why it is necessary to retain a universally quantified schema even when the rule $[\forall]$ has been applied to it. The first premiss has to be used twice, with different instantiations of the variable.

(1) $\forall x(A(x, b) \to A(a, x))$
(2) $A(b, b)$
(3) $\sim A(a, a)$
————————————————————————— $1[x/b]$
× (4) $A(b, b) \to A(a, b)$
————————————————————————— 4
(5) $\sim A(b, b)$ (6) $A(a, b)$
=========== 2   ——————————————————— $1[x/a]$
(7) $A(a, b) \to A(a, a)$
——————————————————— 7
(7) $\sim A(a, b)$ (8) $A(a, a)$
=========== 6   =========== 3

Exercise 5.2

*(1) Use model building to validate each of the inferences given in Question (3) of Exercise 5.1.

(2) Use model building to establish the logical truth of the schemas in Question (5) of Exercise 5.1.

(3) Investigate the performance of model building when applied to each of the following invalid inferences:

*(a) $\dfrac{\forall x \exists y R(x, y)}{\exists x \forall y R(x, y)}$

(b) $\dfrac{\begin{array}{l}\exists x(P(x) \wedge Q(x)) \\ \exists x(Q(x) \wedge R(x))\end{array}}{\exists x(P(x) \wedge R(x))}$

*(c) $\dfrac{\begin{array}{l}\forall x(P(x) \rightarrow \exists y Q(x, y)) \\ \exists y \sim Q(a, y)\end{array}}{\sim P(a)}$

(d) $\dfrac{\begin{array}{l}\forall x(P(x) \rightarrow \exists y Q(y)) \\ \exists x P(x)\end{array}}{\exists x(P(x) \wedge Q(x))}$

(e) $\dfrac{\exists x \forall y Q(x, y, y)}{\exists x \exists z \forall y Q(x, y, z)}$

5.3 RESOLUTION

Resolution for the propositional calculus was introduced in Section 3.6. There we noted that this proof technique is applied to schemas of a special kind, called 'clauses'. In order to generalize the technique to cover the predicate calculus, we must first define what is meant by a clause in a first-order language.

We begin by listing a few examples. The following schemas are all first-order clauses:

$$R(a, b, c)$$
$$\forall x(\sim P(x) \vee Q(x, a) \vee R(x, a, f(x)))$$
$$\forall x \forall y(P(x) \vee Q(x, y))$$
$$\forall x \forall y \forall z(\sim P(x) \vee \sim P(y) \vee Q(a, z) \vee \sim R(x, f(a), z))$$

There are two points to notice about these schemas:

(i)   All the quantifiers are universal, and they occur at the beginning of the schemas.
(ii)   The body of each schema is a disjunction, and each of its disjuncts contains a single predicate letter, which may or may not be negated, together with its arguments.

We call the disjuncts occurring in the body of a clause *literals*; a literal is *negative* if it contains the negation sign, otherwise it is *positive*.

Our definition of a first-order clause now goes as follows:

A **positive literal** is a simple schema in which zero or more constants have been replaced by variables.

A **negative literal** is the negation of a positive literal.

A positive literal and its negation are called **complementary** literals (e.g. $P(x, a, f(y))$ and $\sim P(x, a, f(y))$).

A **clause** is a schema of the form $\forall x_1 \forall x_2 \cdots \forall x_m (L_1 \vee L_2 \cdots \vee L_n)$, where $L_1, \ldots, L_n$ are literals, and $x_1, \ldots, x_m$ are all the variables occurring in $L_1, \ldots, L_n$.

As with the propositional calculus, it is customary to use a shorthand notation whereby a clause is presented simply by giving the set of its component literals; so the four examples given above can be written as

$$\{ R(a, b, c) \}$$
$$\{ \sim P(x), Q(x, a), R(x, a, f(x)) \}$$
$$\{ P(x), Q(x, y) \}$$
$$\{ \sim P(x), \sim P(y), Q(a, z), \sim R(x, f(a), z) \}$$

No ambiguity is introduced by omitting the quantifiers; we know that in any clause the quantifiers must all be universal and must occur at the beginning. The order of a string of universal quantifiers is not significant, neither is the order of the disjuncts within a disjunction. Thus our third example above could equally well be taken as representing any of the four equivalent schemas

$$\forall x \forall y (P(x) \vee Q(x, y))$$
$$\forall y \forall x (P(x) \vee Q(x, y))$$
$$\forall x \forall y (Q(x, y) \vee P(x))$$
$$\forall y \forall x (Q(x, y) \vee P(x))$$

The resolution technique that we shall employ for handling first-order clauses is essentially the same as what we used for the propositional calculus, but with the added complication that now the resolution may take place within the scope of a universal quantifier. We present below a series of increasingly complicated examples to prepare the reader gently for the fully general definition of resolution in the predicate calculus.

Example 5.3.1 Resolve the clauses

$$\{ P(x), \sim Q(x, y) \} \qquad \text{and} \qquad \{ Q(x, y), R(y) \}.$$

If we 'blindly' apply the resolution rule in the form already given for the propositional calculus, we obtain the resolvent

$$\{ P(x), R(y) \}.$$

To justify this, we shall work with the full form of the clauses, namely

$$\forall x \forall y (P(x) \vee \sim Q(x, y)) \qquad \text{and} \qquad \forall x \forall y (Q(x, y) \vee R(y)).$$

Let (D, I) be any model for these two schemas, and let d and d' be any elements of D. The first clause says that

(1)   either $d \in I(P)$ or $(d, d') \notin I(Q)$

while the second clause says that

(2)   either $(d, d') \in I(Q)$ or $d' \in I(R)$.

Suppose $d \notin I(P)$; in that case we must have $(d, d') \notin I(Q)$, from (1), and hence, from (2), $d \in I(R)$. So either $d \in I(P)$ or $d' \in I(R)$; which means that (D, I) satisfies the clause

$$\forall x \forall y (P(x) \vee R(y)).$$

Thus this clause is a logical consequence of the two clauses we started out with. This is what justifies us in giving $\{P(x), R(y)\}$ as the resolvent of $\{P(x), \sim Q(x, y)\}$ and $\{Q(x, y), R(y)\}$.

Example 5.3.2   Resolve the clauses

$$\{P(x), \sim Q(x)\} \quad \text{and} \quad \{Q(a), R(b)\}.$$

This time the propositional calculus resolution rule does not apply, since neither clause contains a literal which is exactly complementary to a literal in the other. However, the first clause contains $\sim Q(x)$ which is *nearly* complementary to $Q(a)$ occurring in the second.

In full form, the clauses are

$$\forall x (P(x) \vee \sim Q(x)) \quad \text{and} \quad Q(a) \vee R(b).$$

Let (D, I) be any model for these two schemas. From the first schema, we know that for any domain element d,

(1)   either $d \in I(P)$ or $d \notin I(Q)$.

The second schema tells us that

(2)   either $I(a) \in I(Q)$ or $I(b) \in I(R)$.

Suppose $I(b) \notin I(R)$. Then from (2) we have that $I(a) \in I(Q)$, and since (1) applies to any domain element, we can in particular put $d = I(a)$, which tells us that $I(a) \in I(P)$. We thus have

(3)   either $I(a) \in I(P)$ or $I(b) \in I(R)$,

so (D, I) satisfies the clause

$$P(a) \vee R(b).$$

We can thus resolve the clauses $\{P(x), \sim Q(x)\}$ and $\{Q(a), R(b)\}$ to give $\{P(a), R(b)\}$.

What we have done in the last example can be summed up as follows. The clause $\{P(x), \sim Q(x)\}$, because it contains a variable, represents a *general* fact; this general fact entails as a consequence many *more concrete* facts, among them the facts represented by the clauses

$$\{P(a), \sim Q(a)\} \quad \text{and} \quad \{P(b), \sim Q(b)\}.$$

From amongst these *instantiations* of the original clause, we can select one, namely $\{P(a), \sim Q(a)\}$, which can be resolved directly with our other original clause $\{Q(a), R(b)\}$ to give us the resolvent $\{P(a), R(b)\}$.

This suggests the following strategy for resolving pairs of clauses: to resolve the clauses C_1 and C_2, look for *instantiations* C'_1 and C'_2 of these clauses such that C'_1 and C'_2 can be resolved together using the propositional calculus resolution rule.

In Example 5.3.2, only one of the two clauses needed to be instantiated to a new clause; we can, if we like, think of the other clause as having been instantiated to itself. In the next example, both clauses need to be instantiated further before resolution is possible.

Example 5.3.3   Resolve the clauses

$$\{\sim P(x, y), Q(x, y, a)\}$$
$$\{\sim Q(g(w), z, z), R(w, z)\}$$

We want instantiations of the two clauses so that the literals $Q(x, y, a)$ and $\sim Q(g(w), z, z)$ become complementary. The first argument in the negative literal has the form $g(\_)$, so we must instantiate the corresponding argument of the positive literal to this form, say x to $g(w)$. Note that this is not a *full* instantiation, since $g(w)$ contains a variable and hence could, if need be, be instantiated further (e.g. to $g(a)$). Next, we observe that the third argument of the positive literal is already fully instantiated to a, hence the corresponding argument z of the negative literal must be instantiated to a as well. This now determines that the second argument of both literals must also be instantiated to a.

Summarizing, we need to make the following substitutions:

$$x/g(w), y/a, z/a,$$

where the notation X/Y means 'instantiate X to Y'. This gives us the following instantiations of our clauses:

$$\{\sim P(g(w), a), Q(g(w), a, a)\},$$
$$\{\sim Q(g(w), a, a), R(w, a)\}.$$

These clauses have the resolvent

$$\{\sim P(g(w), a), R(w, a)\},$$

which we therefore give as the resolvent of the original two clauses.

In full first-order notation, what we have done is to infer from

$$\forall x \forall y (\sim P(x, y) \vee Q(x, y, a))$$

and

$$\forall w \forall z (\sim Q(g(w), z, z) \vee R(w, z))$$

the two less general clauses

$$\forall w (\sim P(g(w), a) \vee Q(g(w), a, a))$$

and

$$\forall w (\sim Q(g(w), a, a) \vee R(w, a))$$

which then jointly entail the resolvent clause

$$\forall w (\sim P(g(w), a) \vee R(w, a)).$$

Example 5.3.4   Resolve the clauses

$$\{\sim P(x), Q(f(x), a)\}$$
$$\{\sim Q(y, y), R(f(y), z)\}$$

We want to instantiate $Q(f(x), a)$ and $\sim Q(y, y)$ to a pair of complementary literals. Looking at the second argument in each case, we see that we have no choice but to instantiate y to a. But to get the first argument right we have to instantiate y to the form $f(\_)$. Since a does not have this form, no suitable instantiation is possible, and the two clauses cannot be resolved.

It will be seen from these examples that a crucial step in resolving predicate calculus clauses is to find instantiations of them containing complementary literals. Suppose C_1 contains a literal L_1 and C_2 contains L_2; then we are looking for substitutions for the variables occurring in the two clauses such that, after substitution, L_1 and L_2 become complementary. (For example, in Example 5.3.2 we put x/a, thereby making $\sim Q(x)$ from the first clause complementary to $Q(a)$ from the second, and in Example 5.3.3 we had $\{x/g(w), y/a, z/a\}$ in order to make $Q(x, y, a)$ complementary to $\sim Q(g(w), z, z)$.)

Making one literal complementary to another is achieved by making the positive member of the pair *identical* to the complement of the negative member. The process of making two literals identical is known as **unification**; the substitution by which it is achieved is called a **unifier** of the two literals. Thus

$$\{x/a\} \text{ is a unifier of } Q(x) \text{ and } Q(a);$$
$$\{x/g(w), y/a, z/a\} \text{ is a unifier of } Q(x, y, a) \text{ and } Q(g(w), z, z).$$

Note that there is not always a unique choice of unifier. In the second example just given,

$$\{x/g(b), w/b, y/a, z/a\}$$

is also a unifier of the two literals, since it makes both of them into $Q(g(b), a, a)$. But this unifier is unnecessarily specific: we do not need to instantiate w to anything in order to unify the two literals. The unifier we had originally does just the amount of work needed to achieve unification and no more: it is a **most general unifier** of the two clauses.

As we saw in Section 2.2, it is customary to use lower-case Greek letters such as θ, ϕ, ψ to denote sets of substitutions. In this notation, if X is any clause or literal, then the result of applying the substitution θ to X is written $X\theta$. And ϕ is a unifier of the literals L and L' just so long as $L\phi = L'\phi$.

We are now ready for a general statement of the resolution procedure for the predicate calculus:

Let $C = \{L_1, \ldots, L_m\}$ and $C' = \{L'_1, \ldots, L'_n\}$ be any two first-order clauses such that none of the variables occurring in C also occurs in C'. If there exist in C and C' two literals L_i and L'_j and a substitution θ such that $L_i\theta$ and $L'_j\theta$ are complementary, then the resolvent of C and C' is the clause

$$\text{res}(C, C') = (C - L_i)\theta \cup (C' - L'_j)\theta$$

$$= \{L_1\theta, \ldots, L_{i-1}\theta, L_{i+1}\theta, \ldots, L_m\theta, L'_1\theta, \ldots, L'_{j-1}\theta, L'_{j+1}\theta, \ldots, L'_n\theta\}$$

obtained by applying θ to the two clauses, taking their union, and omitting the complementary pair $L_i\theta$ and $L'_j\theta$.

Note carefully the stipulation that C and C' should not share any variables. The following example shows any this is required.

Example 5.3.5 Resolve the clauses

$$\{\sim P(x), Q(x, a)\}$$
$$\{\sim Q(b, x), R(x)\}$$

In order to unify $Q(b, x)$ with $Q(x, a)$, we should have to put both x/a *and* x/b, which of course are inconsistent requirements. Nonetheless, the two clauses do have an obvious resolvent, since the first implies

$$\{\sim P(b), Q(b, a)\}$$

and the second implies

$$\{\sim Q(b, a), R(a)\},$$

and these resolve together to give

$$\{\sim P(b), R(a)\}.$$

In order to get this result using our unification procedure, we must first *rename* the variables in one of the clauses. Variables are only dummy symbols after all; so our two original clauses are equivalent to

$$\{\sim P(x), Q(x, a)\}$$
$$\{\sim Q(b, y), R(y)\}.$$

We now have the unifier $\{x/b, y/a\}$, giving the same resolvent as before.

This example shows that the requirement that the two clauses must have no common variables is not a substantial restriction: if we wish to resolve two clauses which *do* have one or more common variables, all we have to do is to relabel enough of the variables so that they do not.

The unification algorithm

In order to resolve two clauses, we need to unify two literals: how can we set about doing that, other than by trial and error? It turns out that there is a systematic procedure for finding a unifier; it is called the unification algorithm. We present our version of this algorithm below, and then illustrate how it works with some examples.

The unification algorithm. *To unify two literals* L_1 *and* L_2 *having no shared variables.*

(1) Let $E_1 \leftarrow L_1$, $E_2 \leftarrow L_2$, $\theta \leftarrow \{\ \}$.
(2) If E_1 and E_2 are both null strings, halt and output the most general unifier θ; otherwise put $E_1 \leftarrow E_1\theta$, $E_2 \leftarrow E_2\theta$, and let X, Y be the first symbols in E_1 and E_2 respectively.
(3) If $X = Y$, delete X and Y from the beginning of E_1 and E_2 respectively, and go back to Step (2); otherwise proceed to Step (4). If the symbol deleted is either a function symbol or a predicate letter, then delete also the left parenthesis immediately following, together with the corresponding right parenthesis.

(4) If one or both of X and Y is a variable, go to (5), swapping X and Y if necessary so that X is a variable; otherwise halt and fail.

(5) If Y is either a variable or a constant, let $\theta \leftarrow \theta \cup \{X/Y\}$, delete X and Y from the beginning of E_1 and E_2, together with the commas immediately following them, if present, and go to step (2); otherwise go to (6).

(6) Y must be a function symbol; let Z be the expression consisting of Y together with its arguments.

(7) If Z contains X, then halt and fail; otherwise, put $\theta \leftarrow \theta \cup \{X/Z\}$, delete X and Z from the beginning of E_1 and E_2, together with the commas, if present, immediately following them, and go to Step (2).

In the next example, we illustrate the algorithm by tracing its step-by-step execution on a simple example. We use the Greek letter Λ (lambda) to denote the null (or empty) string.

Example 5.3.6 Unify the literals Q(x, f(a)) and Q(b, y).

(1) $E_1 \leftarrow$ Q(x, f(a))
$E_2 \leftarrow$ Q(b, y)
$\theta \leftarrow \{\ \}$

(2) $X \leftarrow$ Q
$Y \leftarrow$ Q

(3) $E_1 \leftarrow$ x, f(a)
$E_2 \leftarrow$ b, y

(2) $X \leftarrow$ x
$Y \leftarrow$ b

(3) $X \neq Y$, so:

(4) X is a variable, so:

(5) Y is a constant, so:
$\theta \leftarrow \{x/b\}$
$E_1 \leftarrow$ f(a)
$E_2 \leftarrow$ y

(2) $X \leftarrow$ f
$Y \leftarrow$ y

(3) $X \neq Y$, so:

(4) Y is a variable, so:
$X \leftarrow$ y
$Y \leftarrow$ f

(5) Y is a function symbol, so:

(6) $Z \leftarrow$ f(a)

(7) f(a) does not contain y, so:
$\theta \leftarrow \{x/b, y/f(a)\}$
$E_1 \leftarrow \Lambda$
$E_2 \leftarrow \Lambda$

(2) $E_1 = E_2 = \Lambda$, so halt and output $\{x/b, y/f(a)\}$.

Example 5.3.7 Unify the literals

$$P(f(x, y), y) \quad \text{and} \quad P(f(u, g(a)), b)$$

In this example, we do not give so much detail.
   On the first pass through the algorithm we delete P and its associated parentheses, to give

$$f(x, y), y \quad \text{and} \quad f(u, g(a)), b.$$

Next time round we delete f and the parentheses to give

$$x, y, y \quad \text{and} \quad u, g(a), b.$$

Next we put $\theta \leftarrow \{x/u\}$ or $\theta \leftarrow \{u/x\}$ (the choice is immaterial—assume the former), and our expressions become

$$y, y \quad \text{and} \quad g(a), b.$$

We now have $\theta \leftarrow \{x/u, y/g(a)\}$, and the expressions to compare become

$$g(a) \quad \text{and} \quad b.$$

(the remaining y has become $g(a)$ because of the application of the current substitution θ at step 2). Since neither g nor b is a variable, the algorithm halts and fails at this point: we conclude, correctly, that the original literals cannot be unified.

We now give some examples of resolution refutations in the predicate calculus.

Example 5.3.8 Validate the inference

$$\forall x \forall y (\sim P(x) \vee \sim Q(x, y) \vee R(y))$$
$$\forall x Q(f(x), g(x))$$
$$\underline{P(f(a))}$$
$$R(g(a))$$

We must show that the set consisting of the premisses together with the negation of the conclusion is unsatisfiable. In set notation, with variables renamed so that no two clauses share variables (as required by the unification algorithm), this set consists of the clauses

(1) $\{\sim P(x), \sim Q(x, y), R(y)\}$
(2) $\{Q(f(z), g(z))\}$
(3) $\{P(f(a))\}$
(4) $\{\sim R(g(a))\}$

Since these are all Horn clauses, it suffices to look for a resolution sequence starting with the negated conclusion,

(4) $\{\sim R(g(a))\}$.

The literal $R(g(a))$ which appears negated in (4) can be unified with the positive literal $R(y)$ from clause (1) by applying the unifier $\{y/g(a)\}$. The two clauses can then be resolved to give

(5) $\{\sim P(x), \sim Q(x, g(a))\}$.

We now have a choice: we can resolve clause (5) with either clause (2) or clause (3). Adopting the 'first come, first served' policy of Section 3.6, we choose the former alternative. A most general unifier for $Q(f(z), g(z))$ and $Q(x, g(a))$ is the substitution-set $\{x/f(a), z/a\}$; the resolvent of (2) and (5) is thus

(6) $\{\sim P(f(a))\}$.

We can now resolve (3) and (6) directly (i.e. without unification) to give the empty clause, and the refutation is complete.

The full resolution sequence for this example is thus:

$$\{\sim R(g(a))\}$$
$$\Big|\ 1, \{y/g(a)\}$$
$$\{\sim P(x),\ \sim Q(x, g(a))\}$$
$$\Big|\ 2, \{x/f(a), z/a\}$$
$$\{\sim P(f(a))\}$$
$$\Big|\ 3, \{\ \}$$
$$\mathbf{F}$$

Example 5.3.9   Validate the inference

$$\forall x(P(x) \vee Q(x))$$
$$\forall x(\sim P(f(x)) \vee R(x))$$
$$\forall x(\sim Q(f(x)) \vee S(g(x)))$$
$$\overline{\qquad R(a) \vee S(g(a)) \qquad}$$

In set notation, we have

(1) $\{\sim P(x), Q(x)\}$
(2) $\{\sim P(f(y)), R(y)\}$
(3) $\{\sim Q(f(z)), S(g(z))\}$
(4) $\sim R(a)$
(5) $\sim S(g(a))$

Note that the negation of the conclusion is a conjunction, since

$$\sim (R(a) \vee S(g(a))) \cong\ \sim R(a) \wedge \sim S(g(a)),$$

and hence contributes *two* clauses to the set. A suitable refutation sequence is

$$\sim S(g(a))$$
$$\Big|\ 3, \{z/a\}$$
$$\sim Q(f(a))$$
$$\Big|\ 1, \{x/f(a)\}$$
$$\sim P(f(a))$$
$$\Big|\ 2, \{y/a\}$$
$$R(a)$$
$$\Big|\ 4, \{\ \}$$
$$\mathbf{F}$$

In Example 5.3.9, a only occurs in the conclusion, so the inference cannot depend for its validity on its being a, rather than any other term, that occurs in the conclusion. We could just as well validate the conclusion $R(b) \lor S(g(b))$, using the same resolution sequence but with b replacing a at each of its occurrences. We can, if fact, validly infer from the premisses of our inference the conclusion

$$\forall x(R(x) \lor S(g(x)))$$

(cf. the rule of $\forall$-introduction in natural deduction). If we wish to validate this more general inference using resolution, though, we seem to encounter a problem. For although this new conclusion is a clause, $\{R(x), S(g(x))\}$, its *negation* is not. In fact

$$\sim \forall x(R(x) \lor S(g(x)))$$

is equivalent to

$$\exists x \sim (R(x) \lor S(g(x))),$$

and hence to

$$\exists x(\sim R(x) \land \sim S(g(x))).$$

The presence of an unnegated existential quantifier here makes this schema very unclauselike.

In order to get round this difficulty, we have to find a way of eliminating the existential quantifier. The process by which this is done is called

Skolemization

We begin with the observation that the schema

$$(1) \quad \exists x\Phi(x),$$

where Φ is any unary predicate, is satisfiable if and only if the schema

$$(2) \quad \Phi(a)$$

is satisfiable, where a is any individual constant not occurring in Φ. For suppose that (1) is satisfiable. Then there is an interpretation (D, I) for which $I'(\Phi)$ is non-empty; let d be any element of $I'(\Phi)$. Now consider the interpretation (D, J), defined so that

(a) $J(a) = d$, and
(b) $J(E) = I(E)$ for every basic expression E other than a

(where by 'basic expression' is meant any constant, predicate letter, or function symbol). Since a does not occur in Φ, we have $J'(\Phi) = I'(\Phi)$, by (b), and so in particular $d \in J'(\Phi)$. By (a), this means that $J(a) \in J'(\Phi)$, and hence (D, J) satisfies (2). We have proved that if (1) is satisfiable then so is (2). The converse of this result follows immediately from the fact that (2) logically implies (1). We may thus conclude that each of these two schemas is satisfiable just so long as the other is.

Note that this is not to say that the two schemas are equivalent: plainly they are not, since it is easy to construct a model for (1) which does not satisfy (2)—we just have to make sure that $I(a) \notin I'(\Phi)$. The relationship between (1) and (2) is weaker than logical equivalence; but so long as we are only interested in satisfiability of a set of schemas, and not in specific models for them, (1) and (2) are in effect equivalent for our purposes.

More generally, we can formulate the following rule:

Let **S** be any set of schemas, and let Φ by any unary predicate. Then the set

$$\mathbf{S} \cup \{\exists x \Phi(x)\}$$

is satisfiable if and only if the set

$$\mathbf{S} \cup \{\Phi(a)\}$$

is satisfiable, where a is any individual constant not occurring in either **S** or Φ.

We require a not to occur in **S** since if it did we could not be sure that (D, J) satisfied **S** simply because (D, I) did (thus, for example, the set $\{P(a), \exists x \sim P(x)\}$ is satisfiable, but $\{P(a), \sim P(a)\}$ is not—on the other hand $\{P(a), \sim P(b)\}$ *is* satisfiable, as predicted by our rule). The constant a introduced in the rule is called a **Skolem constant** after the Norwegian logician Thoralf Skolem, and the process of eliminating an existential quantifier by introducing a Skolem constant is called **skolemization**.

To illustrate, we return to the case discussed above, in which the premises of Example 5.3.9 are combined with a universally quantified version of the conclusion:

Example 5.3.10   Validate the inference

$$\forall x(P(x) \vee Q(x))$$
$$\forall x(\sim P(f(x)) \vee R(x))$$
$$\underline{\forall x(\sim Q(f(x)) \vee S(g(x)))}$$
$$\forall x(R(x) \vee S(g(x)))$$

Let **P** be the set consisting of the three premises. We must show that the set $\mathbf{P}' = \mathbf{P} \cup \{\exists x(\sim R(x) \wedge \sim S(g(x)))\}$ is unsatisfiable. Since a does not occur either in **P** or in $R(\_) \vee S(g(\_))$, we have, from our rule, that $\mathbf{P}'$ is unsatisfiable if and only if the skolemized form $\mathbf{P} \cup \{\sim R(a) \wedge \sim S(g(a))\}$ is; but this is just what we have already shown, in Example 5.3.9.

Skolemization need not, of course, be confined to the conclusion of an inference. This is illustrated by the next example.

Example 5.3.11   Validate the inference

$$\frac{\exists x \forall y P(x, y)}{\exists x P(x, x)}.$$

We must show that the set $\{\exists x \forall y P(x, y), \sim \exists x P(x, x)\}$ is unsatisfiable. The first schema, after skolemization, yields the clause $\{P(a, y)\}$; the second schema is equivalent to $\forall x \sim P(x, x)$, and hence yields the clause $\{\sim P(x, x)\}$. The refutation is now straightforward:

$$\sim P(x, x) \frac{P(a, y), \{x/a, y/a\}}{} \textbf{F}.$$

Not all existential quantifiers can be eliminated by the introduction of a Skolem constant. Things go wrong if the existential quantifier falls within the scope of a universal quantifier. Consider, for example, the set consisting of the schemas

$$\forall x \forall y (\sim R(x, y) \vee \sim R(f(x), y))$$
$$\forall z \exists w R(z, w).$$

This set is satisfiable: a suitable model is (D, I), where

$$D = \{0, 1\},$$
$$I(R) = \{(0, 1), (1, 0)\},$$
$$I(f)[0] = 1, I(f)[1] = 0.$$

If we try eliminating the existential quantifier by means of a Skolem constant, we obtain the schemas

$$\forall x \forall y (\sim R(x, y) \vee \sim R(f(x), y))$$
$$\forall z R(z, a).$$

This set is *unsatisfiable*: for the second schema entails both $R(a, a)$ and $R(f(a), a)$, whereas according to the first schema, these cannot both be true. So our attempted Skolemization has destroyed satisfiability, and hence cannot be used in the context of resolution refutation.

The problem arose because the existential quantifier occurred within the scope of a universal quantifier. The schema $\exists x P(x)$ says that something has the property denoted by P; we can legitimately invent a name for that 'something' (or one of them, if there are more than one), and this new name corresponds to the Skolem constant a in the schema P(a). But $\forall x \exists y R(x, y)$ does not say in an unqualified way that there is some one thing with a given property; rather, it says that for whatever x you choose I can find a y such that your x stands to my y in the relation denoted by R. In general, different xs will have different ys associated with them, so that the y that I select must be a *function* of the x that you name.

This suggests that the correct skolemization of $\forall x \exists y P(x, y)$ is not $\forall x P(x, a)$ but something like $\forall x P(x, f(x))$, where the functional dependence of the second argument on the first is made explicit. Our rule for introducing a Skolem constant can now be supplemented with a rule for introducing a (unary) *Skolem function symbol* as follows:

Let **S** be any set of schemas, and let Ψ by any binary predicate. Then the set

$$S \cup \{\forall x \exists y \Psi(x, y)\}$$

is satisfiable if and only if the set

$$S \cup \{\forall x \Psi(x, f(x))\}$$

is satisfiable, where f is any function symbol not occurring in either **S** or Ψ.

The 'if' part of this rule follows immediately from the fact that $\forall x \exists y \Psi(x, y)$ is a logical consequence of $\forall x \Psi(x, f(x))$. For the 'only if' part, suppose (D, I) satisfies the set

$$S \cup \{\forall x \exists y \Psi(x, y)\}.$$

Then for each d in D, there is an element d' of D such that $(d, d') \in I'(\Psi)$. Define a function $f: D \to D$ by choosing $f(d)$ to be any element of D for which $(d, f(d)) \in I'(\Psi)$. (*Note:* The function f is not the same as the function symbol f: notice the typographical difference.) Next, define an interpretation function J by the rules

(a) $J(f) = \{(d, f(d)) | d \in D\}$,
(b) $J(E) = I(E)$ for every basic expression E other than f.

Then (D, J) satisfies **S** (since f does not occur in **S**), and it also satisfies $\forall x \Psi(x, f(x))$ (since for each d in D, $(d, f(d))$ is in $J'(\Psi)$). Hence

$$\mathbf{S} \cup \{\forall x \Psi(x, f(x))\}$$

is satisfiable, as required.

Example 5.3.12   Validate the inference

$$\frac{\forall x \exists y (\sim P(x) \vee Q(x, y))}{\exists x P(x)}$$
$$\overline{\exists x \exists y Q(x, y)}$$

Skolemizing the premises, we obtain

$$\forall x (\sim P(x) \vee Q(x, f(x)))$$
$$P(a)$$

The conclusion, when negated, contains only universal quantifiers, so no skolemization is needed. The set of clauses to refute is thus

(1) $\{\sim P(x), Q(x, f(x))\}$
(2) $\{P(a)\}$
(3) $\{\sim Q(x, y)\}$

The refutation itself is simple:

$$\{\sim Q(x, y)\} \xrightarrow{\;1, \{y/f(x)\}\;} \{\sim P(x)\} \xrightarrow{\;2, \{x/a\}\;} \mathbf{F}$$

Of course, an existential quantifier may occur within the scope of more than one universal quantifier. The obvious generalization holds: *the arity of the Skolem function symbol is equal to the number of universal quantifiers within whose scope the existential quantifier falls.* We can regard a constant as a function symbol of arity zero, in which case the generalization covers Skolem constants too. Thus the fully general rule for skolemization is:

Let **S** be any set of schemas, and let Ψ be any n-ary predicate. Then the set

$$\mathbf{S} \cup \{\forall x_1 \forall x_2 \cdots \forall x_{n-1} \exists x_n \Psi(x_1, x_2, \ldots, x_{n-1}, x_n)\}$$

is satisfiable if and only if the set

$$\mathbf{S} \cup \{\forall x_1 \forall x_2 \cdots \forall x_{n-1} \Psi(x_1, x_2, \ldots, x_{n-1}, f(x_1, x_2, \ldots, x_{n-1}))\}$$

is satisfiable, where f is any $(n-1)$-ary function symbol not occurring in either **S** or Ψ.

The function symbol f introduced in this rule is called a **Skolem function symbol**; in the case $n = 1$, this is of course a Skolem constant.

Using skolemization, we can convert *any* first-order schema into a set of clauses, which is satisfiable if and only if the original schema is. The procedure for doing this is as follows:

Algorithm.   Conversion to clausal form

(1)   Eliminate all occurrences of $\rightarrow$ and $\leftrightarrow$ by means of the equivalences

$$A \rightarrow B \cong \sim A \vee B$$
$$A \leftrightarrow B \cong (A \wedge B) \vee (\sim A \wedge \sim B)$$

(2)   Reduce the scope of each occurrence of $\sim$ as far as possible, using the equivalences

$$\sim (A \wedge B) \cong \sim A \vee \sim B$$
$$\sim (A \vee B) \cong \sim A \wedge \sim B$$
$$\sim \sim A \cong A$$
$$\sim \forall x A \cong \exists x \sim A$$
$$\sim \exists x A \cong \forall x \sim A$$

(3)   Eliminate all existential quantifiers by skolemization.
(4)   If necessary, rename variables so that each universal quantifier binds a different variable. (For example, replace

$$\forall x P(x) \vee \forall x Q(x)$$

by the equivalent

$$\forall x P(x) \vee \forall y Q(y).)$$

(5)   Move all the universal quantifiers to the front. (For example, replace

$$\forall x P(x) \vee \forall y Q(y)$$

by the equivalent

$$\forall x \forall y (P(x) \vee Q(y)).)$$

(6)   Distribute disjunction over conjunction using the equivalences

$$A \vee (B \wedge C) \cong (A \vee B) \wedge (A \vee C)$$
$$(A \wedge B) \vee C \cong (A \vee C) \wedge (B \vee C)$$

(7)   Separate into clauses by distributing universal quantifiers over conjunction, using the equivalences

$$\forall x (\Phi(x) \wedge \Psi(x)) \cong \forall x \Phi(x) \wedge \forall x \Psi(x)$$
$$\forall x (\Phi(x) \wedge A) \cong \forall x \Phi(x) \wedge A$$
$$\forall x (A \wedge \Psi(x)) \cong A \wedge \forall x \Phi(x)$$

(where A is any schema not containing x), and delete any clause containing a complementary pair of literals.

The following example illustrates each of these steps.

Example 5.3.13   Convert into clausal form the schema

$$\exists x P(x) \rightarrow (\forall x (Q(x) \lor S(x)) \rightarrow \forall x \exists y R(x, y)).$$

Following the algorithm step by step, we obtain

(1) $\sim \exists x P(x) \lor (\sim \forall x (Q(x) \lor S(x)) \lor \forall x \exists y R(x, y))$
(2) $\forall x \sim P(x) \lor \exists x (\sim Q(x) \land \sim S(x)) \lor \forall x \exists y R(x, y)$
(3) $\forall x \sim P(x) \lor (\sim Q(a) \land \sim S(a) \lor \forall x R(x, f(x)))$
(4) $\forall x \sim P(x) \lor (\sim Q(a) \land \sim S(a)) \lor \forall y R(y, f(y))$
(5) $\forall x \forall y (\sim P(x) \lor (\sim Q(a) \land \sim S(a)) \lor R(y, f(y)))$
(6) $\forall x \forall y ((\sim P(x) \lor \sim Q(a) \lor R(y, f(y))) \land (\sim P(x) \lor \sim S(a) \lor R(y, f(y))))$
(7) $\forall x \forall y (\sim P(x) \lor \sim Q(a) \lor R(y, f(y))) \land \forall x \forall y (\sim P(x) \lor \sim S(a) \lor R(y, f(y))))$

In set notation, our set of clauses is

$$\{\{\sim P(x), \sim Q(a), R(y, f(y))\}, \{\sim P(x), \sim S(a), R(y, f(y))\}\}.$$

Note that in passing from (1) to (2) we 'silently' dropped the parentheses from around the second and third disjuncts, in effect rewriting $A \lor (B \lor C)$ in the form $A \lor B \lor C$. This is of course quite legitimate since disjunction is associative.

Putting together everything we have introduced in this section, we now know how to set about using resolution to validate any inference in the predicate calculus. First we convert the premises and the negation of the conclusion into clausal form, then we look for a resolution refutation, using unification to assist us in finding resolvable pairs of clauses.

Example 5.3.14   Validate the inference

$$\forall x (P(x) \rightarrow \exists y Q(x, y))$$
$$\frac{\forall x \forall y (Q(x, y) \rightarrow P(f(y)))}{\forall x (P(x) \rightarrow \exists y P(f(y)))}$$

After conversion to clausal form, we have the following set of clauses:

(1) $\{\sim P(x), Q(x, g(x))\}$
(2) $\{\sim Q(y, z), P(f(z))\}$
(3) $\{P(a)\}$
(4) $\{\sim P(f(w))\}$

Note that the Skolem function symbol in clause (1) is g: we could not use f since this symbol already occurs in the inference.

A refutation sequence for this set is

$$\{P(a)\}$$
$$\Big| 1, \{x/a\}$$
$$\{Q(a, g(a))\}$$
$$\Big| 2, \{y/a, z/g(a)\}$$

$$\{P(f(g(a)))\}$$
$$\Big| \ 4, \{w/g(a)\}$$
$$\textbf{F}$$

Note that most of the work in validating the inference comes from the conversion to clausal form. Once we have a set of clauses, especially if it is a set of Horn clauses, the actual refutation is comparatively quick. For this reason, resolution, as presented here, is recommended as a practical proof method only for schemas which are either in clausal form already or are sufficiently close to clausal form that the conversion process does not involve too much work.

Exercise 5.3

(1) For each of the following pairs of literals, determine whether or not they are unifiable, and give a most general unifier if they are:

* *(a) $F(x), F(g(a))$
* *(b) $F(a), F(g(x))$
* *(c) $F(g(a)), G(f(a))$
* (d) $F(a, x), F(y, b)$
* (e) $F(a, g(x)), F(y, b)$
* *(f) $F(x, f(a, x)), F(f(a, y), z)$
* (g) $G(x, x, y), G(x, y, a)$
* *(h) $H(f(a), g(a, y)), H(x, g(x, f(a)))$
* (i) $F(g(h(a, x), x), x), F(y, g(a, z))$
* (j) $F(g(x), x), F(y, g(y))$

(2) Use resolution to validate the following inferences

* *(a) $\forall x(\sim P(x) \vee Q(x))$
  $\forall x(\sim Q(x) \vee R(x))$
  $$\overline{\forall x(\sim P(x) \vee R(x))}$$

* *(b) $\exists x \forall y P(x, y)$
  $\forall x \forall y(\sim P(x, y) \vee Q(x, y))$
  $$\overline{\exists x \forall y Q(x, y)}$$

* (c) $\forall x(P(x) \vee Q(x))$
  $\exists x \sim P(x)$
  $$\overline{\exists x Q(x)}$$

* *(d) $\forall x \forall y \forall z(\sim R(x, y) \vee \sim R(y, z) \vee R(x, z))$
  $\forall x \forall y(\sim R(x, y) \vee R(y, x))$
  $\forall x \exists y R(x, y)$
  $$\overline{\forall x R(x, x)}$$

* (e) $\forall x P(x, a, x)$
  $\forall x \forall y \forall z \forall w(\sim P(x, y, z) \vee P(f(w, x), y, f(w, z)))$
  $$\overline{\forall x \forall y(P(f(x, a), f(y, a), f(x, f(y, a))))}$$

*(3) Use conversion to clausal form followed by resolution to validate the inferences in Exercise 5.1, Question (3).

5.4 SOUNDNESS, COMPLETENESS, AND DECIDABILITY

As with the propositional calculus, the three proof methods we have described for the predicate calculus can all be shown to be both sound and complete; but the proofs of soundness and completeness are more complicated than in the case of the propositional calculus. We shall not give the proofs in this book, but will content ourselves with a discussion of what the results mean.

The fact that the three methods are sound and complete amounts to the following:

(1) *Soundness*

   (a) If a schema A can be deduced from a set of schemas S using natural deduction, then A is a logical consequence of S.
   (b) If a set of schemas S is such that {S} can be reduced to the empty set using the model building replacement rules, then S is unsatisfiable.
   (c) If a set of schemas S is such that, after conversion to clausal form, it has a resolution refutation tree, then S is unsatisfiable.

(2) *Completeness*

   (a) If a schema A logically follows from a set of schemas S, then there exists a natural deduction derivation of A from S.
   (b) If a set of schemas S is unsatisfiable, then {S} can be reduced to the empty set using the model building replacement rules.
   (c) If a set of schemas S is unsatisfiable then, after conversion to clausal form, it has a resolution refutation tree.

The completeness of a given sound proof system for the predicate calculus was first shown by Kurt Gödel in 1930. This result is generally referred to as Gödel's completeness theorem. It is not to be confused with his celebrated incompleteness theorems of 1931, which we discuss later (see Section 6.3). A radically different form of completeness proof for the predicate calculus was subsequently given by Leon Henkin in 1949—this is the proof on which our completeness proof for the propositional calculus in Section 3.4 is based.

Although it is comforting to know that a given proof system for the predicate calculus is complete, this fact alone does not guarantee that the proof system is going to be useful in practice. It is one thing to know that, if an inference is valid, there is a proof of this in the proof system; it is quite another thing actually to *find* a proof and thereby determine that the inference *is* valid; and it is another thing still to show that there is *no* proof, and thereby determine that the inference is *in*valid. This it would be most satisfactory if a proof system could be supplemented with a **decision procedure** which would enable one, in every case, to decide whether or not a given inference was valid.

The search for a decision procedure for the predicate calculus occupied mathematical logicians for some twenty years starting from 1917, when the mathematician David Hilbert proposed the problem as one of a set of 23 so far unsolved problems which he urged the mathematicians of the day to devote their energies to resolving. The problem was known as the **decision problem** or, in the original German, the *Entscheidungsproblem*. Had the search for a solution been successful, then completeness would have followed, since if you have a guaranteed method of determining whether or not any inference is valid then this in itself constitutes a sound and complete proof system. As it happens, Gödel was able to prove completeness indirectly, without having access to a decision procedure, but the fact that the predicate calculus was shown to have a complete proof system must have strengthened people's hopes of eventually finding a decision procedure.

However, it was not to be. In 1936, Alonzo Church proved that *there cannot exist a complete decision procedure for the predicate calculus*. That is, any procedure that is suggested will at some point break down, either by coming to a wrong decision (e.g. identifying as valid some inference which in reality is invalid), or by simply failing to come to a decision at all. This result is now generally known as Church's theorem. In brief, Church's theorem states that the predicate calculus is **undecidable**.

In the same year, a completely different solution to the decision problem was discovered by Alan Turing. His work is particularly relevant to computer science, since it involved the creation of an abstract computational device, the Turing machine, which has been of central importance in the development of the theory of computation. Turing showed that the decision problem for the predicate calculus was equivalent to a seemingly quite different problem concerning Turing machines: namely, is there an effective procedure for determining whether or not an arbitrary computation on an arbitrary Turing machine will terminate? This problem is called the halting problem, and Turing was able to show that there cannot be such a procedure. In showing this, he in effect also solved the decision problem for the predicate calculus.

Warning: The above account is somewhat anachronistic in that the version of the decision problem which was being discussed at that time was to find a procedure not for determining whether or not an arbitrary *inference* was valid, but for determining whether or not an arbitrary *schema* was logically true. These two formulations amount to the same thing, since the inference from $P_1, \ldots, P_n$ to C is valid if and only if the schema $(P_1 \wedge \cdots \wedge P_n) \rightarrow C$ is logically true. A seemingly more general statement of the decision problem is: find a procedure for determining whether an arbitrary set of schemas S entails an arbitrary schema A. This could have turned out to be truly more general than the other versions of the problem, since S is allowed to contain infinitely many members, whereas an inference must have finitely many premisses. However, an important result in mathematical logic, called the compactness theorem, states that even if S is infinite, only finitely many of its members can be involved in any one entailment, so that if S entails A, then some finite subset of S also entails A.

The picture painted above is perhaps unduly gloomy, for although the predicate calculus is undecidable, i.e. does not possess a complete decision procedure, it is in fact **semi-decidable.** By this is meant that there exist procedures with the following property: given any inference of the predicate calculus, if it is valid than the procedure will discover that it is valid. This is not the same as full decidability, since there will always be some *invalid* inferences which the procedure will fail to reveal as such.

To demonstrate the semi-decidability of the predicate calculus, it is enough to have a sound and complete proof system. Our proof procedure then consists in systematically listing the proofs constructible in the system (in order of length, proofs of equal length being taken alphabetically) until we arrive at a proof of the inference under scrutiny. If the inference is valid, then, since the proof system is complete, there must be a proof of the inference in the system, and this proof will eventually get listed. Thus a valid inference can always be discovered to be valid by this method. But suppose our inference is invalid. Since the proof system is sound, no 'proof' of the inference exists. Since the number of possible proofs is infinite, our enumeration of the proofs will never come to an end (there is no end), so we can never say definitely that there does *not* exist a proof of our inference. Thus this method is no good for establishing invalidity.

The method of the last paragraph is of no practical value, even for establishing validity: it simply takes far too long to enumerate the proofs. We can get a better idea of the meaning of semi-decidability by recalling Example 5.2.7, in which we saw how the Model Building method failed to conclude that a certain inference was invalid, when in fact it was. It was not that the method incorrectly told us that the inference was valid; rather, it never came to a conclusion at all, generating instead an infinite succession of new schemas.

Of course, reasoning at a higher level, as it were from *outside* the model building method, we can *see* that the tree in that example is never going to close, and thus can say we have a conclusive proof of the invalidity of the inference. But this proof is not a proof *in* the proof system, it is 'higher-level' reasoning *about* the proof system; and Church's theorem shows us that if we tried to incorporate this higher-level reasoning into a new, improved proof system, we still wouldn't achieve a complete decision procedure. The trouble is that different trees will require different patterns of higher-level reasoning to enable us to see that they will never close, and no single finitely defined procedure can capture all the patterns of reasoning that will be required.

Thus while it may be true that

> for every inference I there is a method M such that M can be used to determine whether or not I is valid,

it is false, by Church's theorem, that

> there is a method M such that, for every inference I, M can be used to determine whether or not I is valid.

The second statement does not, of course, follow from the first, since the inference

$$\frac{\forall x(I(x) \to \exists y(M(y) \wedge W(x, y)))}{\exists y(M(y) \wedge \forall x(I(x) \to W(x, y)))}$$

is invalid (cf. Example 4.3.1).

The undecidability of the predicate calculus stands in stark contrast to the decidability of the propositional calculus. The propositional calculus is much more manageable than the predicate calculus, because any propositional schema has only finitely many interpretations (the rows of its truth table), whereas for a first-order schema the number of possible interpretations is infinite. Logical properties like satisfiability, entailment and logical truth have to do with the full range of possible interpretations of a schema or set of schemas. For example, A entails B just so long as every interpretation that satisfies A also satisfies B. So it is not surprising that, in a system like the propositional calculus, in which the full range of models is finite, there should be cut-and-dried methods for determining whether these logical properties hold in any particular case, whereas in the predicate calculus, with its infinitely many possible interpretations, such methods should forever elude us.

6 First-order Theories

6.1 FIRST-ORDER THEORIES AND AXIOM SYSTEMS

In Chapter 1, we discussed the inference schema

$$(1) \qquad \frac{\begin{array}{l} A \text{ is in } B \\ B \text{ is in } C \end{array}}{A \text{ is in } C}$$

and we decided that this was a valid schema on the grounds that none of its instances had true premises and false conclusion.

To translate this inference into the predicate calculus, we introduce a binary predicate R(____,...) to stand for '____ is in ...', and replace the schematic letters A, B, and C by individual constants a, b, and c respectively, giving

$$(2) \qquad \frac{\begin{array}{l} \mathsf{R(a, b)} \\ \mathsf{R(b, c)} \end{array}}{\mathsf{R(a, c)}}$$

Unfortunately, this inference is *invalid*! To see this, consider the interpretation (D, I), in which $D = \{1, 2, 3\}$, $I(\mathsf{a}) = 1$, $I(\mathsf{b}) = 2$, $I(\mathsf{c}) = 3$, and $I(\mathsf{R}) = \{(1, 2), (2, 3)\}$. Then the premises of (2) are both true in (D, I), whereas the conclusion is false. What has gone wrong?

The discrepancy between our original inference schema (1) and the predicate calculus version (2) arises from the fact that we naturally interpreted the predicate '____ is in ...' as having its normal English meaning. But in so interpreting it, we are narrowing down the range of possible ways in which the schema as a whole can be interpreted. Thus while the predicate calculus version (2) could equally well be taken as formalizing the invalid inference

$$\frac{\begin{array}{l} \text{Josie lives next door to Maggie} \\ \text{Maggie lives next door to Sue} \end{array}}{\text{Josie lives next door to Sue}}$$

our schema (1) does not include this inference amongst its instances.

Suppose now that we confine our attention to just those interpretations in which the predicate R denotes a *transitive* relation, that is, a relation which contains (d, d'') whenever it contains both (d, d') and (d', d''), for all triples of domain elements d, d', d''. Let us call such interpretations R-*transitive* interpretations. Then the inference (2) has the following *restricted* validity: every R-transitive interpretation which satisfies the premises also satisfies the conclusion.

The restricted validity of (2) can be seen as the logical basis for the validity of any English inference obtainable from (2) by interpreting the predicate letter R as denoting a transitive relation. Thus since the relation expressed by '____ is older than ...' is transitive, the inference

> Socrates is older than Plato
> Plato is older than Aristotle
> ―――――――――――――――――――
> Socrates is older than Aristotle

is valid. Similarly, '____ is a multiple of ...' is transitive, so the inference

> 1952 is a multiple of 478
> 478 is a multiple of 61
> ――――――――――――――――
> 1952 is a multiple of 61

is valid too—though this particular argument is not sound, since 1952 is *not* a multiple of 478 (but it *is* a multiple of 61—if this puzzles you, revise carefully the meaning of soundness and validity!).

Suppose we were to collect together all the predicate calculus inferences which are valid in this restricted sense. Corresponding to each such inference $P_1, \ldots, P_n/C$, there is a schema $P_1 \wedge \cdots \wedge P_n \rightarrow C$, and the restricted validity of the inference means that this schema must be true in all R-transitive interpretations. Thus corresponding to our inference (2), we have the schema

$$(3) \quad R(a, b) \wedge R(b, c) \rightarrow R(a, c).$$

Other schemas, not of this form, can also be true in all R-transitive interpretations. An example is the universal generalization of (3),

$$(4) \quad \forall x \forall y \forall z (R(x, y) \wedge R(y, z) \rightarrow R(x, z)),$$

another example is

$$(5) \quad \forall x \forall y \forall z \forall w (R(x, y) \wedge R(y, z) \wedge R(z, w) \rightarrow R(x, w)).$$

None of the schemas (3), (4), or (5) is logically true; that is, they are not true in all *possible* interpretations of their language; but they *are* true in all R-*transitive* interpretations.

The set of first-order schemas which are true in all R-transitive interpretations is called the **first-order theory** of a transitive relation (denoted by R), or of transitivity. In fact schema (4) alone is sufficient to *define* the first-order theory of transitivity, because this theory consists precisely of those schemas which are logical consequences of (4). For example, both (3) and (5) are in the theory, and they are both logical consequences of (4). So is

$$(6) \quad \forall x \forall y (R(x, y) \wedge R(y, x) \rightarrow R(x, x)).$$

Again, the fact that the inference (2) above is valid in the first-order theory of transitivity corresponds to the fact that the inference

$$\forall x \forall y \forall z (R(x, y) \wedge R(y, z) \rightarrow R(x, z))$$
$$R(a, b)$$
$$R(b, c)$$

$$R(a, c)$$

is universally valid, i.e. valid in the predicate calculus.

We say that (4) is an **axiom** for the theory of transitivity. It gives rise to a **sound and complete axiomatization** of the theory: sound because every logical consequence of the axiom is in the theory, and complete because every schema in the theory is a logical consequence of the axiom. We say that the schema (4) **completely axiomatizes** the theory.

As another example, consider **equivalence relations**. Let us call a relation on a set D an equivalence relation on D if there is a way of exhaustively partitioning D into non-overlapping blocks (called equivalence classes) such that d bears the given relation to d' just so long as d and d' belong in the same block. For example, in the domain consisting of the natural numbers, the relation

_____ differs from ... by a multiple of 3

is an equivalence relation, since we can partition the natural numbers into the three blocks

$$B_1 = \{0, 3, 6, 9, 12, 15, 18, \ldots\}$$
$$B_2 = \{1, 4, 7, 10, 13, 16, 19, \ldots\}$$
$$B_3 = \{2, 5, 8, 11, 14, 17, 20, \ldots\}$$

and two natural numbers differ from each other by a multiple of 3 if and only if they both belong to the same block (e.g., $17-8=9$, $16-1=15$, both multiples of 3, but $11-4=7$).

The first-order theory of an equivalence relation is completely determined by the set **E** consisting of the three schemas

    (R) $\forall x R(x, x)$
    (S) $\forall x \forall y (R(x, y) \rightarrow R(y, x))$
    (T) $\forall x \forall y \forall z (R(x, y) \wedge R(y, z) \rightarrow R(x, z))$

The third of these, (T), is the transitivity axiom we have already met. The second determines that the relation denoted by R is *symmetric* (i.e. d' stands in relation R to d whenever d stands in relation R to d'), while the first determines that it is *reflexive* (i.e. every domain element stands in relation R to itself). Thus an equivalence relation can be defined to be a reflexive, symmetric, transitive relation, and this is the usual way of defining it: the property that relates it to a partitioning of the domain, which we used to define it earlier, is usually proved to follow from the axioms.

How can we show that the set **E** completely axiomatizes the first-order theory of an equivalence relation? We have to show that the axiomatization defined by **E** is both sound and complete with respect to the theory. For soundness, it is enough that the axioms themselves belong to the theory, for if *they* are true in every interpretation (D, I) in which $I(R)$ is an equivalence relation, then so must all their logical consequences, from the definition of logical consequence. But the axioms belong to the theory just so long as every equivalence relation is reflexive, symmetric, and transitive. This is easy to verify, for the equivalence relation corresponding to a given partition can be expressed in the form

_____ belongs to the same block of the partition as ...

and this evidently has the three properties required.

For completeness, we require that every schema in the theory is a logical consequence of the axioms. This can be shown as follows.

Let (D, I) be any interpretation satisfying $\mathbf{E}$, so that $I(\mathsf{R})$ is reflexive, symmetric, and transitive. We shall show that $I(\mathsf{R})$ is an equivalence relation. For each d in D, define a subset B_d of D by the rule

$$d' \in B_d \text{ if and only if } (d, d') \in I(\mathsf{R}).$$

We shall show that the set $B = \{B_d \mid d \in D\}$ is a partition of D.

First, suppose B_d and B_e have an element in common, say f. Then (d, f) and (e, f) are in $I(\mathsf{R})$. Since $I(\mathsf{R})$ is symmetric, it also contains (f, e), and hence, by transitivity, it contains (d, e) as well. Now let g be any element of B_e. Then $I(\mathsf{R})$ contains both (e, g) and (d, e), so, by transitivity again, it contains (d, g), which means that g is in B_d. We have thus shown that B_e is a subset of B_d; and by reversing the roles of e and d in this argument, we also get that B_d is a subset of B_e. This shows us that $B_d = B_e$. Thus so long as two elements of B have an element in common, they are equal: put differently, this means that *distinct* elements of B are *disjoint*. Moreover, every element of D belongs to a member of B, since by reflexivity, $I(\mathsf{R})$ contains (d, d) for each d in D, and hence d is in B_d. From these two results it follows that B partitions D.

We now show that two elements of D are related by $I(R)$ if and only if they are in the same block of the partition. Suppose (e, f) is in $I(\mathsf{R})$. Then f is in B_e, and since e is in B_e (by reflexivity of $I(\mathsf{R})$), e and f belong in the same block of the partition. Suppose conversely that e and f belong in the same block. Since e is in block B_e, that must be the block in question. Hence f is in B_e, which means that (e, f) is in $I(\mathsf{R})$.

We have now shown that $I(\mathsf{R})$ is an equivalence relation on D. So any schema A in the theory of an equivalence relation must be satisfied by (D, I). But since (D, I) was any interpretation satisfying $\mathbf{E}$, it follows that A is a logical consequence of $\mathbf{E}$. Hence $\mathbf{E}$ determines a complete axiomatization of the theory of an equivalence relation.

Some definitions

(1) A **first-order theory** is any satisfiable set of first-order schemas closed under logical consequence.

In other words, a set **T** of first-order schemas is a first-order theory so long as

   (a) **T** is satisfiable;
   (b) every logical consequence of **T** is in **T**.

Now let C be any non-empty class of interpretations of a given first-order language, and consider the set **Th**(C) consisting of just those first-order schemas satisfied by every member of C. Then

(a) **Th**(C) is satisfiable, since any member of C is a model for **Th**(C);

(b) let A be any logical consequence of **Th**(C). Then A must be satisfied by any interpretation which satisfies **Th**(C). In particular, therefore, A is satisfied by every member of C. Hence A is a member of **Th**(C).

We have shown that **Th**(C) is a first-order theory. We define

(2) If C is any class of interpretations, the *first-order theory of C* is the set of schemas satisfied by every member of C.

A set of first-order schemas is **finitely definable** so long as it can be specified in a finite number of symbols. Thus any finite set of schemas is finitely definable, since it can be listed completely in a finite number of symbols. But in some cases an infinite set of schemas may be definable by means of a finite set of *higher-level schemas* in which functors or predicates are represented by schematic letters. Thus for example the single higher-level schema

$$\forall x \forall y (E(x, y) \to (\Phi(x) \to \Phi(y)))$$

finitely defines an infinite set of schemas which includes, for example, the schemas

$$\forall x \forall y (E(x, y) \to (P(x) \to P(y)))$$
$$\forall x \forall y (E(x, y) \to (\exists z R(x, z) \to \exists z R(y, z)))$$
$$\forall x \forall y (E(x, y) \to ((P(x) \to P(a)) \to (P(y) \to P(a))))$$

in which the predicate $\Phi(\_)$ is instantiated as $P(\_)$, $\exists z R(\_, z)$, and $P(\_) \to P(a)$ respectively.

(3) A (first-order) **axiom system** is any finitely definable set of first-order schemas.

Given a first-order theory **T** and an axiom system **S**, we say that

(4) **S** is **sound** with respect to **T** if and only if every logical consequence of **S** is in **T**;

(5) **S** is **complete** with respect to **T** if and only if every element of **T** is a logical consequence of **S**; and

(6) **S completely axiomatizes T** if and only if **S** is both sound and complete with respect to **T**, i.e. if and only **T** is equal to the set of logical consequences of **S**.

Moreover, if C is any class of interpretations, we say that **S characterizes** C if and only if **S** completely axiomatizes the first-order theory of C.

One curious consequence of our definitions is that an unsatisfiable set of axioms is complete with respect to any first-order theory whatever. For if **S** is unsatisfiable, then every schema is a logical consequence of **S**, so in particular every member of the first-order theory **T** is a logical consequence of **S**. This shows that completeness on its own is not of much interest: what we require is completeness *and* soundness together.

Exercise 6.1

(1) For each of the following relations, on the stated sets, decide whether it is (i) reflexive, (ii) symmetric, (iii) transitive:

*(a) ____ is greater than ...   (on the set of integers)

*(b) ____ is greater than or equal to ...   (on the set of integers)

(c) ____ has the same number of letters as ...   (on the set of English words)

*(d) ____ has at least one letter in common with ...   (on the set of English words)

(e) ____ admires ...   (on the set of all people)

*(f) ____ is sitting next to ...   (on a set of students in a class)

*(g) ____ is sitting in the same row as ...   (on a set of students in a class)

*(h) ____ is ...'s brother   (on the set of all people)

(i) ____ is ...'s brother   (on the set of all men)

(j) ____ has the same parents as ...   (on the set of all people)

(k) ____ runs faster than ...   (on the set of all computer programs)

(l) ____ computes the same results as ...   (on the set of all computer programs)

(m) ____ has a common border with ...   (on the set of all countries)

(2) Which of the following inferences are valid in the first-order theory of transitivity (i.e., assume R denotes a transitive relation)?

*(a) $R(a, b), R(b, a)/R(b, b)$

*(b) $R(a, a), R(b, a)/R(a, b)$

(c) $R(a, b)/\exists x(R(a, x) \wedge R(x, b))$

*(d) $\exists x(R(a, x) \wedge R(x, b))/R(a, b)$

(e) $\forall x R(a, x), \exists y R(y, b)/R(a, b)$

(f) $\exists x R(a, x), \exists y R(y, b)/R(a, b)$

*(g) $\forall x \forall y(R(x, y) \rightarrow R(y, x))/\forall x R(x, x)$

*(h) $\forall x \exists y(R(x, y) \wedge R(y, x))/\forall x R(x, x)$

(i) $R(a, b), \sim R(a, d), \exists x R(b, x), \forall y R(y, c)/ \sim R(c, d)$

(3) The first-order theory of a **strict partial order** relation (denoted by R) is defined by the axioms:

$$\text{(T)} \quad \forall x \forall y \forall z(R(x, y) \wedge R(y, z) \rightarrow R(x, z))$$
$$\text{(I)} \quad \forall x \sim R(x, x)$$

Show that the theory contains the following theorems:

*(a) $\forall x \forall y(R(x, y) \rightarrow \sim R(y, x))$

(b) $\forall x \forall y \forall z(R(x, y) \wedge R(y, z) \rightarrow \sim R(z, x))$

(c) $\forall x \forall y \forall z \forall w(R(w, x) \wedge R(y, z) \rightarrow \sim (R(x, y) \wedge R(z, w)))$

6.2   THE FIRST-ORDER THEORY OF IDENTITY

A special status in logic is accorded to the relation in which each object stands to itself but to no other object. This relation is referred to as the **identity** relation, and may be expressed in English in several different ways, among them

    ____is the same thing as...
    ____is identical to...
    ____is none other than...
    ____is equal to...
    ____equals...

or simply

    ____is...

A classical example, much discussed in the philosophical literature, is the statement

> The morning star is identical to the evening star.

The phrase 'the morning star' designates that planet which we sometimes see shining brightly in the eastern sky before sunrise, while the phrase 'the evening star' designates the planet which we sometimes see shining brightly in the western sky after sunset. And of course it is one and the same planet, Venus, which we see in these two circumstances: hence the morning star is identical to the evening star.

In mathematics, the symbol ' $=$ ' is used to represent the identity relation. When we write ' $2 \times 6 = 3 \times 4$ ', we are asserting that the number represented by the expression ' 2×6 ' is the *same* number as the one represented by the expression ' 3×4 ', namely 12.

Identity plays an important rôle in both mathematical and everyday reasoning. There are inferences which depend for their validity on the presence of a predicate which is interpreted to mean identity. An example is

> The capital of France is Paris
> The capital of any country is in that country
> ———————————————
> Paris is in France

This has the schema

$$E(f(a), b)$$
$$\forall x R(f(x), x)$$
$$\overline{\qquad\qquad}$$
$$R(b, a)$$

This is not generally valid, but it *is* valid, in the restricted sense discussed in Section 6.1, so long as the predicate $E(\_\_\_, ...)$ is interpreted to mean the identity relation.

In order to indicate that we require a predicate to be interpreted as referring to the identity relation, we use a special symbol for it. In accordance with usual mathematical practice, we use the 'equals' sign $=$, which we use as an *infix* predicate, i.e. we place it *between* its arguments rather than before them. When we supplement our first-order language with this special predicate, stipulated to be interpreted as the identity relation, we obtain a first-order theory known as the **predicate calculus with identity**. In this extended language, the inference discussed above looks like this:

$$f(a) = b$$
$$\forall x R(f(x), x)$$
$$\overline{\qquad\qquad}$$
$$R(b, a)$$

This inference is valid in the predicate calculus with identity.

The semantics for $=$ is quite straightforward. For *any* interpretation (D, I), we have

$$I(=) = \{(d,d)|d \in D\}.$$

Note that the meaning of $=$ is constant from one interpretation to another, and in this respect $=$ resembles the connectives and quantifiers more than it resembles run-of-the-mill predicate letters such as P, Q, and R, whose meanings vary across interpretations. Thus like the connectives and quantifiers, $=$ has the status of a **logical constant**.

The first-order theory of the identity relation consists of all logically true schemas in the predicate calculus with identity, It is completely axiomatized by the two axioms

> (I1) $\forall x(x = x)$
> (I2) $\forall x \forall y(x = y \wedge \Phi(x) \rightarrow \Phi(y))$

(I1) simply says that everything is identical to itself. (I2) is a higher-level schema: it has an instance for each choice of unary predicate to instantiate Φ. It says, in effect, that if A is the same object as B, then B has the same properties as A.

To illustrate, we shall use these axioms to validate the inference we gave earlier. From the second premiss, $\forall x R(f(x), x)$, we can infer the instance $R(f(a), a)$. This tells us that the object denoted by $f(a)$ has the property denoted by the predicate $R(\_, a)$. By the first premiss, this object is the same as the object denoted by b. Hence, by (E2), we can assert $R(b, a)$ as required.

In natural deduction proofs, we can implement the identity axioms as additional rules of inference:

$$(=\text{-intro}) \qquad \dfrac{}{t = t}$$

$$(=\text{-elim}) \qquad \dfrac{\begin{array}{l} t = t' \\ \Phi(t) \end{array}}{\Phi(t')}$$

in which t and t′ are any terms. Thus the argument by which we validated the inference just now can be formalized as follows:

> (1) $\forall x R(f(x), x)$ [Premiss]
> (2) $R(f(a), a)$ $[1, \forall\text{-elim: } a/x]$
> (3) $f(a) = b$ [Premiss]
> (4) $R(b, a)$ $[3, 2, =\text{-elim: } \Phi/R(\_, a)]$

Note how, in the application of the schematic rule ($=\text{-elim}$), we state explicitly in the annotation what complex predicate we are using to instantiate Φ.

The axioms (I1) and (I2) completely axiomatize the first-order theory of identity. As part of a proof of this, we shall show that these axioms can only be satisfied by an interpretation in (D, I) in which $I(=)$ is an equivalence relation on D. In other words, we show that the schemas (R), (S) and (T) of Section 6.1, with (R) replaced by $=$, are logical consequences of those axioms.

This is immediate in the case of (R), which just *is* (I1). For (S), we have the derivation

(1) SUBDERIVATION
      (1.1) $a = b$ [Assumption]
      (1.2) $a = a$ [$=$-intro: t/a]
      (1.3) $b = a$ [$1.1, 1.2, =$-elim: $\Phi/\_ = a$]
(2) $a = b \rightarrow b = a$ [$1, \rightarrow$-intro]
(3) $\forall x \forall y (x = y \rightarrow y = x)$ [$2, \forall$-intro: $x/a, y/b$]

Having established (S) as a logical consequence of the axioms for identity, we can supplement our natural deduction system with a derived rule of inference

$$(=\text{-symm}) \qquad \left|\begin{array}{l} t = t' \\ \hline t' = t \end{array}\right. .$$

The combination of ($=$-symm) with ($=$-elim) gives us a useful alternative form of the latter rule:

$$(=\text{-elim-right}) \qquad \left|\begin{array}{l} t = t' \\ \Phi(t') \\ \hline \Phi(t) \end{array}\right.$$

The proof (which is straightforward) is left as an exercise for the reader.

For transitivity, the derivation is

(1) SUBDERIVATION
      (1.1) $a = b \wedge b = c$ [Assumption]
      (1.2) $b = c$ [$1.1, \wedge$-elim-right]
      (1.3) $a = b$ [$1.1, \wedge$-elim-left]
      (1.4) $a = c$ [$1.2, 1.3, =$-elim: $\Phi/a = \_$]
(2) $a = b \wedge b = c \rightarrow a = c$ [$1, \rightarrow$-intro]
(3) $\forall x \forall y \forall z (x = y \wedge y = z \rightarrow x = z)$ [$2, \forall$-intro: $x/a, y/b, z/c$]

The corresponding derived rule of inference is

$$\left|\begin{array}{l} t = t' \\ t' = t'' \\ \hline t = t'' \end{array}\right. .$$

By combining together an arbitrary number of applications of this rule, we can generalize it to a chain of equalities of any length. For example, given $t_1 = t_2$, $t_2 = t_3$, $t_3 = t_4$, and $t_4 = t_5$, three successive applications of the rule give us

$$t_1 = t_3,\ t_3 = t_4,\ t_4 = t_5$$
$$t_1 = t_4,\ t_4 = t_5$$
$$t_1 = t_5.$$

Thus the general form of the rule is

$$(=\text{-trans}) \qquad \left|\begin{array}{l} t_1 = t_2 \\ t_2 = t_3 \\ \vdots \\ t_{n-1} = t_n \\ \hline t_1 = t_n \end{array}\right.$$

Function symbols and identity

Because of the way functions are defined, the value of any function is completely determined once its arguments are known. In particular, if two terms t and t′ denote the same object in a given interpretation, and ϕ is any unary functor, then the terms $\phi(t)$ and $\phi(t')$ must also denote the same object, so we have

$$\forall x \forall y (x = y \rightarrow \phi(x) = \phi(y)).$$

(For example, since $2 \times 6 = 3 \times 4$, we can be sure that $5 + (2 \times 6) = 5 + (3 \times 4)$.) This is the first of a series of **univalence theorems** ('univalent' means one-valued), one for each arity that a function symbol may have. The series continues:

$$\forall x_1 \forall x_2 \forall y_1 \forall y_2 (x_1 = y_1 \wedge x_2 = y_2 \rightarrow \phi(x_1, x_2) = \phi(y_1, y_2))$$
(for any binary functor ϕ);

$$\forall x_1 \forall x_2 \forall x_3 \forall y_1 \forall y_2 \forall y_3 (x_1 = y_1 \wedge x_2 = y_2 \wedge x_3 = y_3 \rightarrow \phi(x_1, x_2, x_3) = \phi(y_1, y_2, y_3))$$
(for any ternary functor ϕ);

and so on. Each of these is already a *theorem schema* since ϕ here plays the rôle of a schematic letter that can be replaced by any complex function symbol of the right arity. We can collect them together into a 'superschema':

(U) $\forall x_1 \ldots \forall x_n \forall y_1 \ldots \forall y_n (x_1 = y_1 \wedge \ldots \wedge x_n = y_n \rightarrow \phi(x_1, \ldots, x_n) = \phi(y_1, \ldots, y_n))$
(where ϕ is any *n*-ary functor).

Each instance of this superschema has its own proof. For the case of arity 1, the proof is:

(1) SUBDERIVATION
    (1.1) $a = b$ [assumption]
    (1.2) $\phi(a) = \phi(a)$ [$=$-intro: $x/\phi(a)$]
    (1.3) $\phi(a) = \phi(b)$ [1.1, 1.2, $=$-elim: $\Phi/\phi(a) = \phi(\_)$]
(2) $a = b \rightarrow \phi(a) = \phi(b)$ [1, $\rightarrow$-intro]
(3) $\forall x \forall y (x = y \rightarrow \phi(x) = \phi(y))$ [2, $\forall$-intro: x/a, y/b]

Having proved the result for 1, we can go on to prove it for arity 2 as follows:

(1) SUBDERIVATON
    (1.1) $a_1 = b_1 \wedge a_2 = b_2$ [assumption]
    (1.2) $a_1 = b_1$ [1.1, $\wedge$-elim-left]
    (1.3) $a_2 = b_2$ [1.1, $\wedge$-elim-right]
    (1.4) $\phi(a_1, a_2) = \phi(b_1, a_2)$ [1.2, U: $\phi/\phi(\_, a_2)$]
    (1.5) $\phi(a_1, a_2) = \phi(b_1, b_2)$ [1.3, 1.4, $=$-elim: $\Phi/\phi(a_1, a_2) = \phi(b_1, \_)$]
(2) $a_1 = b_1 \wedge a_2 = b_2 \rightarrow \phi(a_1, a_2) = \phi(b_1, b_2)$ [1, $\rightarrow$-intro]
(3) $\forall x_1 \forall x_2 \forall y_1 \forall y_2 (x_1 = y_1 \wedge x_2 = y_2 \rightarrow \phi(x_1, x_2) = \phi(y_1, y_2))$ [2, $\forall$-intro]

In a similar way, we can use the result for arity 2 to prove it for arity 3, and so on up through the positive integers (all we need do is to write out the general case—getting arity *n* from arity $n - 1$—in order to have a properly rigorous proof of the general result by mathematical induction).

Note that, strictly speaking, these proofs are *proof schemas*, since they contain the schematic letter ϕ standing for any functor. You only get an actual proof when you replace ϕ by a functor from the formal language: schematic letters for functors and predicates are not part of the formal language, they are part of the (relatively) informal symbolic apparatus we use for making general statements *about* the formal language.

We shall use the univalence theorems in the form of the following series of derived inference rules:

$$
\text{(U1)} \quad \frac{\begin{array}{|l} t = t' \end{array}}{\phi(t) = \phi(t')}
$$

$$
\text{(U2)} \quad \frac{\begin{array}{|l} t_1 = t'_1 \\ t_2 = t'_2 \end{array}}{\phi(t_1, t_2) = \phi(t'_1, t'_2)}
$$

$$
\vdots
$$

$$
\text{(Un)} \quad \frac{\begin{array}{|l} t_1 = t'_1 \\ \vdots \\ t_n = t'_n \end{array}}{\phi(t_1, \ldots, t_n) = \phi(t'_1, \ldots, t'_n)}
$$

Identity and indiscernibility

Although the axioms (I1) and (I2) completely axiomatize the first-order theory of identity, they do not *force* us to interpret $=$ to mean the identity predicate. Another possibility, consistent with the axioms, is to interpret $\ldots = \underline{\qquad}$ to mean '$\ldots$ and $\underline{\qquad}$ are indiscernible'. To say that two objects are **indiscernible** in an interpretation is to say that there is no way, within the resources provided by the interpretation itself, of telling them apart, i.e. that there is no predicate Φ in the language such that one of the objects is in $I'(\Phi)$ but the other not.

Example 6.2.1   Consider the first-order language with individual constants a, b, and c, unary predicate letter P, binary predicate letter Q, and unary function symbol f. Define the interpretation (D, I) as follows:

$$D = \{0, 1, 2, 3, 4\}$$
$$I(a) = 0$$
$$I(b) = 1$$
$$I(c) = 4$$
$$I(P) = \{0, 3, 4\}$$
$$I(Q) = \{(0, 4), (3, 4)\}$$

$I(f)$ maps 0, 1, 2, 3, 4 to 4, 2, 1, 4, 1 respectively

It is not hard to see that, for this interpretation, given any unary predicate Φ we have

$$0 \in I'(\Phi) \text{ if and only if } 3 \in I'(\Phi)$$
$$1 \in I'(\Phi) \text{ if and only if } 2 \notin I'(\Phi)$$

The reader should verify this for the case in which Φ is the predicate $P(\_) \rightarrow Q(\_, f(\_))$.

Thus in this interpretation, 0 and 3 are indiscernible, and so are 1 and 2. In addition, of course, each element is indiscernible with itself.

We now extend the language by adding the identity predicate $=$; but instead of interpreting it to *mean* identity, we shall interpret it to mean indiscernibility, so that

$$I(=) = \{(0,0),(0,3),(3,0),(3,3),(1,1),(1,2),(2,1),(2,2),(4.4)\}$$

It is easy to see that the identity axioms (I1) and (I2) are satisfied in this interpretation.

The first-order theory of identity consists of all those first-order schemas which are true in every interpretation in which $I(=)$ is the identity relation. Let us call such interpretations Type 1 interpretations. The first-order theory of indiscernibility, likewise, consists of all those schemas which are true in every interpretation in which $I(=)$ is the indiscernibility relation—we shall call these Type 2 interpretations. To say that (I1) and (I2) completely axiomatize the theory of identity is to say that the schemas true in every Type 1 interpretation are precisely the logical consequences of those axioms. We shall show that (a) *for every Type 2 interpretation there is a Type 1 interpretation which satisfies exactly the same schemas*. This means that if a schema is satisfied by every Type 1 interpretation then it must be satisfied by every Type 2 interpretation as well. And this implies that the first-order theory of identity is a subset of the first-order theory of indiscernibility. We shall also show that (b) *for every Type 1 interpretation there is a Type 2 interpretation which satisfies exactly the same schemas*. This means that the first-order theory of indiscernibility is a subset of the first-order theory of identity. The overall result of our endeavours will be to have shown that *identity and indiscernibility have the same first-order theory*.

To demonstrate the truth of the first italicized assertion above, we note that indiscernibility is an equivalence relation, and hence partitions its domain into blocks in the manner described in Section 6.1. Two domain elements belong to the same block if and only if they are indiscernible.

Now let (D, I) be any Type 2 interpretation, and let B be the set of blocks determined by $I(=)$. We shall define a Type 1 interpretation, with domain B, which satisfies *exactly the same schemas* as (D, I) does. Define an interpretation function J as follows:

(1) For each individual constant a, let $J(a)$ be the block to which $I(a)$ belongs;

(2) For each n-ary function symbol f, let $J(f)$ be the function which maps the n-tuple of blocks $(B_1,\dots,B_n)$ onto the block to which belongs the result of applying $I(f)$ to any n-tuple $(d_1,\dots,d_n)$, where $d_i \in B_i$ for $i = 1,\dots,n$. Note that it does not matter which d_i we choose from each B_i, since the elements of any one block are indiscernible and hence different choices cannot give rise to values of $I(f)$ in different blocks.

(3) For each n-ary predicate P, let $J(P)$ contain just those n-tuples $(B_1,\dots,B_n)$ for which, given any $d_1,\dots,d_n$ in $B_1,\dots,B_n$ respectively, $(d_1,\dots,d_n) \in I(P)$. Again, indiscernibility within a block ensures that this defines a unique value for $J(P)$.

It is not hard to see that the interpretation (B, J) satisfies exactly the same schemas as (D, I) does. And (B, J) is a Type 1 interpretation, since $J(=)$ contains a pair (B, B')

if and only if there are d, d' in B, B' respectively such that (d, d') is in $I(=)$. But this only happens if d and d' are indiscernible, and hence in the same block. So $J(=)$ contains (B, B') if and only if B and B' are the same block; in other words, $J(=)$ is the identity relation on B.

To illustrate, we return to our earlier example.

Example 6.2.1 (continued).   We have $B = (B_0, B_1, B_4)$, where $B_0 = \{0, 3\}$, $B_1 = \{1, 2\}$, and $B_4 = \{4\}$. Then

$$J(a) = B_0$$
$$J(b) = B_1$$
$$J(c) = B_4$$

$J(f)$ maps B_0, B_1, B_4 to B_4, B_1, B_1 respectively.

$$J(P) = \{B_0, B_4\}$$
$$J(Q) = \{(B_0, B_4)\}$$
$$J(=) = \{(B_0, B_0), (B_1, B_1), (B_4, B_4)\}.$$

Verify that the schema $\exists x(P(x) \land Q(x, f(x)))$ is true in both (D, I) and (B, J), and that the schema $\forall x(P(x) \to Q(x, f(x)))$ is false in both interpretations. In doing this you will come to understand the relationship between the two interpretations sufficiently well that it becomes obvious why they satisfy the same schemas. Determined readers should go on to produce a rigorous proof of this fact.

For the assertion (b) on p. 211, we take any Type 1 interpretation (D, I) and simply extend $I(=)$ so that it becomes the indiscernibility relation on D. No schema can change truth value as a result of this move, for this could only happen if there were some complex predicate which applied to one member of a pair of indiscernibles but not to the other, and this is ruled out by the definition of indiscernibility. We thus obtain a Type 2 interpretation which satisfies the same schemas as (D, I).

Counting

With the identity predicate at our disposal, we can generate a series of *complex existential quantifiers* which enable us not merely to assert that there exists an object with a given property but also to say *how many* such objects there are.

The simplest case is **uniqueness**, i.e. the existence of exactly one object with the given property. The schema

$$\Phi(a) \land \forall y(\Phi(y) \to y = a)$$

states that the object denoted by a has the property denoted by Φ, and that *anything* which has that property is identical to that object: in other words, that the object denoted by a is the *one and only* thing that has the property denoted by Φ. To say that there exists exactly one thing with that property, then, all we need do is to generalize this schema to

$$\exists x(\Phi(x) \land \forall y(\Phi(y) \to y = x)).$$

This says that there is something which has the property denoted by Φ and is identical to everything which has that property, i.e. that there is exactly one thing with that property. This is often abbreviated to

$$\exists!x\Phi(x).$$

but we shall use the more readily generalizable notation

$$\exists_1 x\Phi(x).$$

The fact that we can express uniqueness in the predicate calculus with identity allows us to dispense with function symbols. Recall from Section 4.4 our attempt to represent the statement 'Mary's father is a poet' by means of the schema $\exists x(F(x, a) \wedge P(x))$. We did not adopt this suggestion, because it failed to imply that Mary has only one father, and that was our original motivation for introducing function symbols. Now, though, we *can* express the uniqueness of Mary's father. We can regiment 'Mary's father is a poet' as:

> For some x, x is father to Mary, and nothing else is father to
> Mary, and x is a poet.

The phrase 'nothing else is father to Mary' can be regimented further to 'for every y, if y is father to Mary then $y = x$', giving us the predicate calculus representation

$$\exists x(F(x, a) \wedge \forall y(F(y, a) \rightarrow y = x) \wedge P(x))$$

for the full statement.

This way of analysing definite descriptions, which does not make use of function symbols, was introduced by Bertrand Russell in 1905, and hence is generally known as **Russell's theory of descriptions**. In principle it allows us to do all our predicate logic without using function symbols at all, and this has certain advantages from a theoretical point of view; in practice, though, the convenience of using function symbols tips the balance decisively in their favour.

Returning now to the subject of counting, the schema

$$\exists x\exists y(\Phi(x) \wedge \Phi(y) \wedge \sim x = y)$$

says that there exist *at least two* things with the property denoted by Φ. To say that there are *exactly* two such things, we must add that anything with that property is identical to one of the two already mentioned:

$$\exists x\exists y(\Phi(x) \wedge \Phi(y) \wedge \sim x = y \wedge \forall z(\Phi(z) \rightarrow z = x \vee z = y)).$$

This schema is abbreviated to

$$\exists_2 z\Phi(x).$$

The extension of this notation to higher numbers is obvious.

Example 6.2.2 We shall prove that $1 + 1 = 2$. This statement means that the union of two disjoint sets each containing just one element is a set containing exactly two elements. In terms of properties rather than sets, this means that if there is exactly one thing with property A, and exactly one thing with property B, and nothing with both properties together, then there are exactly two things with the property of having either property A or property B. Thus what we must prove is that the inference

$$\exists_1 x \Phi(x)$$
$$\exists_1 x \Psi(x)$$
$$\underline{\sim \exists x(\Phi(x) \wedge \Psi(x))}$$
$$\exists_2 x(\Phi(x) \vee \Psi(x))$$

is valid.

In the proof (which is a monster!), we use the full forms of the premisses and conclusion, obtained by expanding the abbreviations $\exists_1$ and $\exists_2$.

(1)	$\exists x(\Phi(x) \wedge \forall y(\Phi(y) \rightarrow y = x))$	[premiss]
(2)	SUBDERIVATION	
	(2.1) $\Phi(a) \wedge \forall y(\Phi(y) \rightarrow y = a)$	[assumption]
	(2.2) $\Phi(a)$	[2.1, $\wedge$-elim]
	(2.3) $\exists x(\Psi(x) \wedge \forall y(\Psi(y) \rightarrow y = x))$	[premiss]
	(2.4) SUBDERIVATION	
	(2.4.1) $\Psi(b) \wedge \forall y(\Psi(y) \rightarrow y = b)$	[assumption]
	(2.4.2) $\Psi(b)$	[2.4.1, $\wedge$-elim]
	(2.4.3) SUBDERIVATION	
	(2.4.3.1) $a = b$	[assumption]
	(2.4.3.2) $\Phi(b)$	[2.4.3.1, 2.2, =-elim]
	(2.4.3.3) $\Phi(b) \wedge \Psi(b)$	[2.4.3.2, 2.4.2, $\wedge$-intro]
	(2.4.3.4) $\exists x(\Phi(x) \wedge \Psi(x))$	[2.4.3.3, $\exists$-intro]
	(2.4.3.5) $\sim \exists x(\Phi(x) \wedge \Psi(x))$	[premiss]
	(2.4.4) $\sim a = b$	[2.4.3, $\sim$-intro]
	(2.4.5) $\Phi(a) \vee \Psi(a)$	[2.2, $\vee$-intro]
	(2.4.6) $\Phi(b) \vee \Psi(b)$	[2.4.2, $\vee$-intro]
	(2.4.7) $(\Phi(a) \vee \Psi(a)) \wedge (\Phi(b) \vee \Psi(b))$	[2.4.5, 2.4.6, $\wedge$-intro]
	(2.4.8) $(\Phi(a) \vee \Psi(a)) \wedge (\Phi(b) \vee \Psi(b)) \wedge \sim a = b$	[2.4.7, 2.4.4, $\wedge$-intro]
	(2.4.9) SUBDERIVATION	
	(2.4.9.1) $\Phi(c) \vee \Psi(c)$	[assumption]
	(2.4.9.2) SUBDERIVATION	
	(2.4.9.2.1) $\Phi(c)$	[assumption]
	(2.4.9.2.2) $\forall y(\Phi(y) \rightarrow y = a)$	[2.1, $\wedge$-elim]
	(2.4.9.2.3) $\Phi(c) \rightarrow c = a$	[2.4.9.2.2, $\forall$-elim]
	(2.4.9.2.4) $c = a$	[2.4.9.2.1, 2.4.9.2.3, $\rightarrow$-elim]
	(2.4.9.2.5) $c = a \vee c = b$	[2.4.9.2.4, $\vee$-intro]
	(2.4.9.3) SUBDERIVATION	
	(2.4.9.3.1) $\Psi(c)$	[assumption]
	(2.4.9.3.2) $\forall y(\Psi(y) \rightarrow y = b)$	[2.4.1, $\wedge$-elim]
	(2.4.9.3.3) $\Psi(c) \rightarrow c = b$	[2.4.9.3.2, $\forall$-elim]
	(2.4.9.3.4) $c = b$	[2.4.9.3.1, 2.4.9.3.3, $\rightarrow$-elim]
	(2.4.9.3.5) $c = a \vee c = b$	[2.4.9.3.4, $\vee$-intro]
	(2.4.9.4) $c = a \vee c = b$	[2.4.9.1, 2.4.9.2, 2.4.9.3, $\vee$-elim]
	(2.4.10) $\Phi(c) \vee \Psi(c) \rightarrow c = a \vee c = b$	[2.4.9, $\rightarrow$-intro]
	(2.4.11) $\forall z(\Phi(z) \vee \Psi(z) \rightarrow z = a \vee z = b)$	[2.4.10, $\forall$-intro]
	(2.4.12) $(\Phi(a) \vee \Psi(a)) \wedge (\Phi(b) \vee \Psi(b)) \wedge \sim a = b \wedge \forall z(\Phi(z) \vee \Psi(z) \rightarrow z = a \vee z = b)$	
		[2.4.8, 2.4.11, $\wedge$-intro]

(2.4.13) $\exists x \exists y((\Phi(x) \vee \Psi(x)) \wedge (\Phi(y) \vee \Psi(y)) \wedge \sim x = y \wedge \forall z(\Phi(z) \vee \Psi(z) \rightarrow z = x \vee z = y)$
[2.4.12, ∃-intro]

(2.5) $\exists x \exists y((\Phi(x) \vee \Psi(x)) \wedge (\Phi(y) \vee \Psi(y)) \wedge \sim x = y \wedge \forall z(\Phi(z) \vee \Psi(z) \rightarrow z = x \vee z = y))$
[2.3, 2.4, ∃-elim]

(3) $\exists x \exists y((\Phi(x) \vee \Psi(x)) \wedge (\Phi(y) \vee \Psi(y)) \wedge \sim x = y \wedge \forall z(\Phi(z) \vee \Psi(z) \rightarrow z = x \vee z = y))$
[1, 2, ∃-elim]

It would be possible, though extremely tedious, to construct similar proofs for other particular arithmetical facts such as $3 + 7 = 10$ or $100 + 11 = 111$. One can even express and prove facts involving multiplication. For example, to say that $2 \times 3 = 6$ is to say that if there are two things with the property A, and three things with the property B, then there are six ordered pairs in which the first element has the property A and the second element has the property B. Using $p(x, y)$ to represent the ordered pair whose first element is x and whose second element is y, and including an appropriate axiom for this function as our third premiss, the truth of $2 \times 3 = 6$ corresponds to the validity of the inference

$$\exists_2 x \Phi(x)$$
$$\exists_3 x \Psi(x)$$
$$\forall x \forall y \forall z \forall w(p(x, y) = p(w, z) \leftrightarrow x = w \wedge y = z)$$

$$\overline{\exists_6 x(x = p(y, z) \wedge \Phi(y) \wedge \Psi(z))}$$

The reader can judge from the magnitude of the proof of $1 + 1 = 2$ how much more complicated the validation of *this* inference is going to be!

Although the use of complex existential quantifiers enables us to express particular numerical facts, and to prove particular arithmetical results such as the ones we discussed above, it cannot be used to express *general* arithmetical laws such as the commutativity of addition, i.e. the rule that for every pair of numbers m and n, $m + n = n + m$. To say this, in full generality, we need to be able to quantify over numbers, and this cannot be done on the approach we are discussing, since numbers are not represented by terms.

This approach to arithmetic is therefore a blind alley. A more satisfactory approach is to use terms, rather than quantifiers, to represent numbers, for then we *can* quantify over numbers and hence make general arithmetical statements. This approach is what is understood by the expression 'first-order arithmetic', and is the subject of the next section.

Exercise 6.2

(1) Use natural deduction to validate the following inferences in the predicate calculus with identity:

*(a) $a = f(b)$
$b = f(c)$
$\forall x(f(f(x)) = x)$

$\overline{a = c}$

*(b) $\exists x \forall y(x = y)$

$\overline{\forall x \forall y(x = y)}$

(c) $\forall x(P(x) \to Q(x))$
$\exists y(P(y) \wedge a = f(y))$

———————————

$\exists z(Q(z) \wedge a = f(z))$

*(d) $\forall x \forall y \exists z(f(x) = z \wedge g(y) = z)$

———————————

$\exists x(f(x) = x)$

(e) $\exists x \exists y \forall z(x = z \vee y = z)$
$\sim \forall x(x = f(f(x)))$

———————————

$\forall x(x = f(x))$

(2) Use natural deduction to show that the following schemas are theorems of the predicate calculus with identity:

    (a) $\forall x \exists y(x = y)$
    (b) $\forall x \forall y \forall z(x = y \wedge \sim y = z \to \sim x = z)$
  *(c) $\forall x(F(x) \leftrightarrow \exists y(y = x \wedge F(y)))$

(3) In this question, you are to simulate the formal syntax and semantics of **Prop**, using predicate letters L, S, I, T, and function symbols n, k, a, c, e with the following intended meanings:

          L(____): '____is a schematic letter'
          P(____): '____ is a propositional schema'
          I(____): '____is an interpretation'
          S(____,...): '____is satisfied by...'
          If terms t and t' denote the schemas T, T' respectively, then n(t),
          k(t, t'), a(t, t'), c(t, t'), and e(t, t') denote the schemas $\sim T$, $T \wedge T'$,
          $T \vee T'$, $T \to T'$, and $T \leftrightarrow T'$ respectively.

Write down axioms which ensure that

  *(a) P(t) is satisfied if and only if t denotes a schema under the intended interpretation;
  *(b) there is an interpretation for each assignment of truth values to the schematic letters;
  (c) the satisfaction relation (denoted by S) conforms to the semantic rules for **Prop**.

Show that in the first-order theory thus constructed the following schemas are theorems:

    (d) $\forall x \forall y(S(x) \wedge I(y) \to T(a(x, n(x)), y))$
    (e) $\forall x \forall y \forall z(T(n(k(x, y)), z) \leftrightarrow T(a(n(x), n(y)), z))$
    (f) $\forall x \forall y(L(x) \wedge L(y) \to \exists z(T(n(c(x, y)), z)))$

6.3 ARITHMETIC

In this section we shall discuss the first-order theory of the arithmetic of the natural numbers. This is of fundamental importance in that it furnishes us with an exact illustration of both the scope and the limitations of first-order logic, as well as having been, historically, the chief stimulus to its development.

    As we saw in Example 4.4.2, we can represent the natural numbers using an individual constant to represent zero and a unary function symbol to represent the successor relation, since the natural numbers are, by definition, zero, the successor of zero, the successor of the successor of zero, and so on. In this section we shall use 0 as the individual constant, and s as the function symbol. Thus the terms

$$0, s(0), s(s(0)), s(s(s(0))), \ldots$$

represent 0, 1, 2, 3,..., and so on. Note carefully the distinction between 0 and 0: the former is a symbol in our formal theory, the latter is the number zero. The symbol 0 comes to denote the number 0 only by virtue of our setting up an interpretation (D, I) in which $I(0) = 0$.

The intended interpretation of 0 and s is given by the rules

$$D = \mathbf{N}$$
$$I(0) = 0$$

$I(\text{s})$ maps each natural number onto its successor.

Here **N** denotes the set of natural numbers.

The first-order theory of the successor function consists of all those schemas in this language which come out true under the interpretation $(\mathbf{N}, I)$. This theory may be axiomatized by means of the following three axioms:

(S1) $\forall x (\sim s(x) = 0)$
(S2) $\forall x \forall y (s(x) = s(y) \rightarrow x = y)$
(S3) $[\Phi(0) \wedge \forall x (\Phi(x) \rightarrow \Phi(s(x)))] \rightarrow \forall x \Phi(x)$

Axiom (S1) says that 0 is not the successor of any natural number, while Axiom (S2) says that two distinct natural numbers must have distinct successors. In natural deduction proofs, we shall use (S2) in the form of the rule of inference

$$(\text{s-elim}) \quad \left| \frac{s(t) = s(t')}{t = t'} \right. .$$

The converse of this rule,

$$(\text{s-intro}) \quad \left| \frac{t = t'}{s(t) = s(t')} \right. ,$$

does not need to be separately axiomatized, since it is an instance of the rule (U1) (see Section 6.2).

Our third axiom, (S3), is the principle of **mathematical induction**. It says that if some property (denoted by the complex predicate Φ) is possessed by 0, and is also possessed by the successor of each natural number which possesses it, then it is possessed by *every* natural number. Informally, we can argue as follows: since 0 has the property, so must its successor, 1. But then in that case so must the successor of 1, which is 2. And there is nothing to stop us from arguing like this all the way up through the natural numbers.

The rule of inference corresponding to (S3) is

$$(\text{ind}) \quad \left| \begin{array}{l} \Phi(0) \\ \left| \dfrac{\Phi(a)}{\Phi(s(a))} \right. \\ \hline \forall x \Phi(x) \end{array} \right.$$

(where a does not occur in Φ, in a premiss, or in the
assumption of a current subderivation)

The restrictions on a are the same as for $\forall$-intro: the reason for this is that the
induction axiom itself, (S3), contains a universally quantified form of the conditional
corresponding to the subderivation in (ind), so to justify (ind) on the basis of the
axiom we should have to apply $\forall$-intro to that conditional (see Exercise 6.3a, Question
(1)).

Arithmetic proper is concerned with the operations of addition and multiplication
defined on the natural numbers. We begin by considering just addition. To represent
this, we require a binary function symbol whose interpretation is the function that
maps each pair of natural numbers onto their sum. We could use a symbol such as
$f(\_\_\_\_, \ldots)$ for this, but in order to stress the intended interpretation we shall find it
more convenient to use the symbol $+$, which we shall write in the usual way as an
infix operator (i.e. we write $a+b$ rather than $+(a, b)$). The intended interpretation
of $+$ is given by the rule that

$$I(+) \text{ maps } (d, d') \text{ to } d + d'.$$

Note that, as with 0 and 0, we must distinguish carefully between the symbol $+$
appearing as a function symbol in our formal language and the symbol '$+$' used
informally to denote the operation of addition.

We axiomatize the first-order theory of addition by the following two axioms:

(A1) $\forall x(x + 0 = x)$
(A2) $\forall x \forall y(x + s(y) = s(x + y))$

To illustrate their use, we shall prove once again that $1 + 1 = 2$. In the present system,
this amounts to showing that the schema $s(0) + s(0) = s(s(0))$ is a logical consequence
of the axioms. The proof is

(1) $s(0) + s(0) = s(s(0) + 0)$ [A2: x/s(0), y/0]
(2) $s(0) + 0 = s(0)$ [A1: x/s(0)]
(3) $s(s(0) + 0) = s(s(0))$ [2, s-intro]
(4) $s(0) + s(0) = s(s(0))$ [1, 3, =-trans].

We now proceed to some more general theorems.

(AT1) $\forall x(0 + x = x)$

This is very like (A1), and with our familiarity with the properties of addition we
might be inclined to think that it is 'obviously' the same: but (A1) and (AT1) are
distinct schemas, and so if we are concerned with establishing a proof system for
arithmetic then we have to *prove* that (AT1) is a logical consequence of the axioms
for addition. This is, in fact, quite straightforward, and affords our first illustration
of the use of the induction axiom

(1) $0 + 0 = 0$ [A1: x/0]
(2)   SUBDERIVATION
     (2.1) $0 + a = a$ [assumption]
     (2.2) $0 + s(a) = s(0 + a)$ [A2: x/0, y/a]

(2.3) $0 + s(a) = s(a)$ [2.1, 2.2, =-intro: $\Phi/0 + s(a) = s(\_)$]

(3) $\forall x(0 + x = x)$ [1, 2, ind: $\Phi/0 + \_ = \_$]

(AT2) $\forall x \forall y(s(y) + x = s(y + x))$

This bears the same relation to (A2) as (AT1) bears to (A1).

(1) $s(a) + 0 = s(a)$ [A1: x/s(a)]
(2) $a + 0 = a$ [A1: x/a]
(3) $s(a + 0) = s(a)$ [2, s-intro]
(4) $s(a) + 0 = s(a + 0)$ [3, 1, =-elim-right]
(5) $\forall y(s(y) + 0 = s(y + 0))$ [4, $\forall$-intro: y/a]
(6)   SUBDERIVATION
    (6.1) $\forall y(s(y) + a = s(y + a))$ [assumption]
    (6.2) $s(b) + a = s(b + a)$ [6.1, $\forall$-elim: b/y]
    (6.3) $b + s(a) = s(b + a)$ [A2: x/b, y/a]
    (6.4) $s(b) + a = b + s(a)$ [6.2, 6.3, =-elim-right]
    (6.5) $s(s(b) + a) = s(b + s(a))$ [6.4, s-intro]
    (6.6) $s(b) + s(a) = s(s(b) + a)$ [A2: x/s(b), y/a]
    (6.7) $s(b) + s(a) = s(b + s(a))$ [6.6, 6.5, =-trans]
    (6.8) $\forall y(s(y) + s(a) = s(y + s(a)))$ [6.7, $\forall$-intro: y/b]
(7) $\forall x \forall y(s(y) + x = s(y + x))$ [5, 6, ind: $\Phi/\forall y(s(y) + \_ = s(y + \_))$]

We are now in a position to prove a result of fundamental importance in the theory of addition, namely the

Commutative law of addition

(AT3) $\forall x \forall y(x + y = y + x)$.

    (1) $0 + a = a$ [AT1: x/a]
    (2) $a + 0 = a$ [A1: x/a]
    (3) $0 + a = a + 0$ [2, 1, =-elim-right]
    (4) $\forall y(0 + y = y + 0)$ [3, $\forall$-intro: y/a]
    (5)   SUBDERIVATION
        (5.1) $\forall y(a + y = y + a)$ [assumption]
        (5.2) $a + b = b + a$ [5.1, $\forall$-elim: b/y]
        (5.3) $s(a) + b = s(a + b)$ [AT2: x/b, y/a]
        (5.4) $s(a + b) = s(b + a)$ [5.2, s-intro]
        (5.5) $s(b + a) = b + s(a)$ [A2: x/a, y/b]
        (5.6) $s(a) + b = b + s(a)$ [5.3, 5.4, 5.5, =-trans]
        (5.7) $\forall y(s(a) + y = y + s(a))$ [5.6, $\forall$-intro: y/b]
    (6) $\forall x \forall y(x + y = y + x)$ [4, 5, ind: $\Phi/\forall y(\_ + y = y + \_)$]

If you have followed these proofs attentively, you should by now be familiar with the standard pattern of a proof which makes use of the rule (ind). To prove $\forall x \Phi(x)$, we first establish $\Phi(0)$; then, in a subderivation, we derive $\Phi(s(a))$ from the assumption $\Phi(a)$. Lastly, we apply (ind) to this result and $\Phi(0)$ to give the required conclusion. Thus in every case the structure of the proof is

(1)
$\vdots$

(n) $\Phi(0)$
$(n+1)$ SUBDERIVATION
         $(n+1.1)\,\Phi(a)$ [assumption]
                $\vdots$
         $(n+1.m)$ $\Phi(s(a))$
$(n+2)$ $\forall x \Phi(x)$ $[n, n+1, \text{ind}]$

The section of the proof in which we establish $\Phi(0)$ is called the **base step** and the section in which we establish $\forall x(\Phi(x) \rightarrow \Phi(s(x)))$ is called the **induction step**.

The first-order theory of addition for the natural numbers, expressed in the language containing 0, $=$, s, and $+$, is called **Pressburger arithmetic**, and is completely axiomatized by the axioms we have given, namely

(I1) $\forall x(x = x)$
(I2) $\forall x \forall y(x = y \wedge \Phi(x) \rightarrow \Phi(y))$
(S1) $\forall x(\sim s(x) = 0)$
(S2) $\forall x \forall y(s(x) = s(y) \rightarrow x = y)$
(S3) $(\Phi(0) \wedge \forall x(\Phi(x) \rightarrow \Phi(s(x)))) \rightarrow \forall x \Phi(x)$
(A1) $\forall x(x + 0 = x)$
(A2) $\forall x \forall y(x + s(y) = s(x + y))$

That is, every schema in the language which comes out true when the symbols are interpreted in the standard way (i.e. so that 0 means zero, $=$ means equality, s means the successor function, and $+$ means addition) is a logical consequence of these axioms, and hence can be derived from them using a complete proof system for the predicate calculus with identity, such as natural deduction.

Exercise 6.3a
*(1) Show that the rule (ind) is justified on the basis of the axiom (S3).
(2) Prove the *associative law of addition*:,

(AT4) $\forall x \forall y \forall z(x + (y + z) = (x + y) + z)$.

Introducing multiplication

Multiplication can be introduced into first-order arithmetic by means of another binary function symbol, $*$, with the intended interpretation given by

$$I(*) \text{ maps } (d, d') \text{ to } dd'.$$

We shall use two axioms for $*$:

(M1) $\forall x(x * 0 = 0)$
(M2) $\forall x \forall y(x * s(y) = x * y + x)$.

Note that we read $*$ as binding more strongly than $+$, so that, for example, $x + y * z$ means $x + (y * z)$ and not $(x + y) * z$.

To illustrate, we show that $2 \times 2 = 4$, i.e. that the schema $s(s(0)) * s(s(0)) = s(s(s(s(0))))$ is a logical consequence of the axioms.

(1) $s(s(0))*0 = 0$ [M1: $x/s(s(0))$]
(2) $s(s(0))*s(0) = s(s(0))*0 + s(s(0))$ [M2: $x/s(s(0)), y/0$]
(3) $s(s(0))*s(0) = 0 + s(s(0))$ [1, 2, = -elim]
(4) $0 + s(s(0)) = s(s(0))$ [AT1: $x/s(s(0))$]
(5) $s(s(0))*s(0) = s(s(0))$ [3, 4, = -trans]
(6) $s(s(0))*s(s(0)) = s(s(0))*s(0) + s(s(0))$ [M2: $x/s(s(0)), y/s(0)$]
(7) $s(s(0))*s(s(0)) = s(s(0)) + s(s(0))$ [5, 6, = -elim]
(8) $s(s(0)) + s(s(0)) = s(s(0) + s(s(0)))$ [AT2: $x/s(0), y/s(s(0))$]
(9) $s(0) + s(s(0)) = s(0 + s(s(0)))$.   [AT2: $x/0, y/s(s(0))$]
(10) $s(0) + s(s(0)) = s(s(s(0)))$ [4, 9, = -elim]
(11) $s(s(0)) + s(s(0)) = s(s(s(s(0))))$ [10, 8, = -elim]
(12) $s(s(0))*s(s(0)) = s(s(s(s(0))))$ [7, 11, = -trans]

We can go on to prove a series of theorems for $*$ similar to the theorems we have already proved for $+$, namely

(MT1) $\forall x(0*x = 0)$
(MT2) $\forall x \forall y(s(x)*y = x*y + y)$
(MT3) $\forall x \forall y(x*y = y*x)$
(MT4) $\forall x \forall y \forall z(x*(y*z) = (x*y)*z).$

as well as the *distributive laws*

(MT5) $\forall x \forall y \forall z(x*(y + z) = x*y + x*z)$
(MT6) $\forall x \forall y \forall z((x + y)*z = x*z + y*z)$

which involve both addition and multiplication. We shall prove (MT3), assuming the previous results, and leave the rest as exercises for the reader.

(1) $a*0 = 0$ [M1: x/a]
(2) $0*a = 0$ [MT1: x/a]
(3) $0*a = a*0$ [1, 2, = -elim-right]
(4) $\forall y(0*y = y*0)$ [3, $\forall$-intro: y/a]
(5) **SUBDERIVATION**
    (5.1) $\forall y(a*y = y*a)$ [assumption]
    (5.2) $a*b = b*a$ [5.1, $\forall$-elim: b/y]
    (5.3) $s(a)*b = a*b + b$ [MT2: $x/a, y/b$]
    (5.4) $s(a)*b = b*a + b$ [5.2, 5.3, = -elim]
    (5.5) $b*s(a) = b*a + b$ [M2: $x/b, y/a$]
    (5.6) $s(a)*b = b*s(a)$ [5.5, 5.4, = -elim-right]
    (5.7) $\forall y(s(a)*y = y*s(a))$ [5.6, $\forall$-intro: y/b]
(6) $\forall x \forall y(x*y = y*x)$ [4, 5, ind: $\Phi/\forall y(\_ *y = y* \_)$]

Of course, it is not customary, when doing mathematics, to spell out in detail all the logical steps of one's reasoning: it would take far too long, and one would be in danger of 'not seeing the mathematics for the logic'. In practice, one takes for granted the most basic properties such as the symmetry and transitivity of equality, the commutativity and associativity of both addition and multiplication. But for the logician, what is of primary interest is whether these 'basic' properties can be derived from a very small number of even more basic principles—in short, whether mathematics as a whole can be finitely axiomatized.

It would be natural to suppose that the axioms for Pressburger arithmetic, together with (M1) and (M2), should completely axiomatize the first-order theory of addition and multiplication on the natural numbers. That this is not, in fact, the case, was first demonstrated by Gödel in his celebrated *incompleteness theorems* of 1931. In fact, Gödel showed more than that one particular axiom system failed to axiomatize arithmetic: he showed that the first-order theory of arithmetic is *not completely axiomatizable*, that is, that it is not even *possible* to construct an axiom system which completely axiomatizes the theory. He showed that in any sound axiomatization of first-order arithmetic, there exists a schema in the theory which is not derivable from the axioms. If one tries to get round this problem by adding that schema to one's set of axioms, then Gödel's result implies that for the new set of axioms there will still be a schema in the theory not derivable from them. The manner in which Gödel proved his results, though of the greatest interest, lies outside the scope of this book. It you want to follow up this topic, accessible starting points are Nagel and Newman (1958) and Hofstadter (1979).

Exercise 6.3b   In this exercise all the axioms (or their corresponding rules of inference) given in this section may be used.

(1) Construct natural deduction proofs for the theorems *(MT1), (MT2), *(MT4), (MT5) and (MT6) given above. You will find it helpful to prove (MT5) and (MT6) *before* proving (MT4).

(2) Prove the following theorems (*Note*: it is not necessary to use induction in (a) and (b)):

*(a) $\forall x(x*s(0) = x)$
(b) $\forall x(s(s(0))*x = x + x)$
(c) $\forall x(x = 0 \lor \exists y(x = s(y)))$
*(d) $\forall x(\sim x = s(x))$
(e) $\forall x \forall y(x + y = x \rightarrow y = 0)$
(f) $\forall x \forall y \forall z(x + y = x + z \rightarrow y = z)$
(g) $\forall x \forall y(x + y = 0 \rightarrow x = 0)$
*(h) $\forall x \forall y(x*y = 0 \rightarrow x = 0 \lor y = 0)$
(i) $\forall x \forall y(x*y = s(0) \rightarrow x = s(0))$
(j) $\forall x \exists y(y + y = x \lor y + y = s(x))$
(k) $\forall x \forall y(x + x = y + y \rightarrow x = y)$
(l) $\forall x \forall y \forall z(x*y + y = x*z + z \rightarrow y = z)$
(m) $\forall x \forall y \forall z(x*y = x*z \rightarrow x = 0 \lor y = z)$
(n) $\forall x \forall y(\sim y + y = s(x + x))$

Warning: these get more difficult as you go down the list; the later ones, in particular, are only recommended for the more dedicated readers.

7 Modal and Temporal Logic

7.1 POSSIBILITY AND NECESSITY

Modal logic is the logic of *possibility* and *necessity*, but it is notoriously hard to give a clear definition of what these terms really mean. In this section we shall introduce the main ideas of modal logic by way of two simple examples. These will prepare the way for a more general treatment in the next section.

Example 7.1.1   Noughts and Crosses   In the game of noughts and crosses (known in America as 'tic-tac-toe'), there are three possible outcomes, which we symbolize by schematic letters of the propositional calculus as follows:

P   Noughts wins
Q   Crosses wins
R   The game is drawn

At each stage in a game, each of these propositions is either true or false. For example, P is true in the state

W_5

×	0	×
	0	×
×	0	0

but Q and R are both false in this state. Similarly, R is true in the state

W_9

×	0	×
0	0	×
×	×	0

but P and Q are both false. In state

W_1

×	0	×
	0	
×		0

P, Q, and R are all false.

States W_5 and W_9 are *accessible* from state W_1, since if W_1 occurs at a certain stage in a game, then W_5 and W_9 are both possible later stages of the same game. Thus at W_1, it is possible for noughts to win, since the final outcome of the game may be the state W_5; but it is also possible that the game will be drawn, since W_9 is a possible final outcome too. Finally, it is obviously possible for crosses to win too!

Given any schema A, we shall write

$$\Diamond A$$

to mean that it is *possible* for A to be true. In our noughts and crosses game, this means that A *is* true in some state of play accessible from the current state. Thus what we said in the last paragraph can be expressed by saying that $\Diamond$P, $\Diamond$Q, and $\Diamond$R are all true in state W_1.

The schema $\Diamond$P may be read in any of the following ways:

It is possible for noughts to win
Noughts may win
Noughts might win
Noughts can win
Possibly, noughts will win.

The symbol $\Diamond$ is called a **modal operator**. It is usually read 'possibly', corresponding to the last of the readings of $\Diamond$P just given.

We can apply $\Diamond$ to any schema, including schemas which themselves contain $\Diamond$. For example, we know that P and Q cannot both be true at any stage of the game. This means that the schema $\Diamond$(P $\wedge$ Q), which says that P $\wedge$ Q is true in some state accessible from the current one, must be false at every stage; and this in turn means that the schema $\sim \Diamond$(P $\wedge$ Q) is *true* at every stage. (You can think of this as a *theorem* of noughts and crosses.)

Now consider the state

	×	0	×
W_3		0	×
	×		0

Satisfy yourself that it is *not* possible for crosses to win from this position. In fact the only possible outcomes are the states W_5 and W_9 that we met above. So in state W_3, the schema $\Diamond$Q is *false*, and hence the schema $\sim \Diamond$Q is true.

Another way of saying this is that the negation $\sim$Q ('Crosses does not win') is *inevitable* or *necessary*. For this purpose, we introduce another modal operator $\Box$, read 'necessarily'. A schema

$$\Box A$$

is true in a given state W just so long as A is true in every state accessible from W. Thus in state W_3, the schema $\Box \sim$Q is true, since, as we saw, in this state it is inevitable that crosses will not win.

The schemas $\sim \Diamond$Q and $\Box \sim$Q are thus both true in state W_3; as we noted, they are merely two ways of saying the same thing. They are, in fact, *equivalent* schemas. We can demonstrate this easily. For $\sim \Diamond$A is true in state W if and only if it is not the case that there is a state accessible from W in which A is true; writing $R(\underline{\quad},\ldots)$ for '$\underline{\quad}$ is accessible from ...', and $T(\underline{\quad},\ldots)$ for '$\underline{\quad}$ is true in state...', we can write this in predicate calculus notation[1] as

$$\sim \exists x (R(x, W) \wedge T(A, x)).$$

But this is equivalent to the schema

$$\forall x (R(x, W) \rightarrow \sim T(A, x))$$

which says that A is false in every state accessible from W. This in turn means that $\sim$ A is *true* is every state accessible from W; and this is just what we express in modal logic by saying that the schema $\square \sim$ A is true in W. Thus in any state, $\sim \diamond$ A is true if and only is $\square \sim$ A is true, and hence these two schemas are equivalent. Using the properties of negation, we can rephrase this by saying that $\square$A is equivalent to $\sim \diamond \sim$ A; this gives rise to the possibility of *defining* the necessity operator $\square$ in terms of the possibility operator $\diamond$, in a way that is strictly analogous to the way in which we can define the universal quantifier in terms of the existential (see Section 4.3, p. 135). The same thing can be seen informally by noting that the statement schema

It is not possible for $\cdots$ not to happen

is merely a long-winded way of saying.

It is inevitable that $\cdots$ will happen.

Returning to our game of noughts and crosses, the diagram on the next page shows all the positions accessible from W_1. Below the diagram are tabulated the truth-values of P, Q, R and the schemas obtained by prefixing each of these with one of the modal operators.

For the sake of clarity, only the Ts have been shown in table, the empty boxes corresponding to Fs. You may be surprised to see that the schemas $\square$P, $\square$Q, and $\square$R are labelled as true in states W_2, W_5, W_9, and W_{10}. It may seem strange to say that in state W_2, for example, in which crosses has already won, that noughts will inevitably win. This is a point at which the technical meaning we have given to $\square$ diverges from the intuitive meaning of necessity. We have said that $\square$A is true at a state W so long as A is true in every state accessible from W: that is, that for every state W', if W' is accessible from W then A is true in W'. But W_2 is an *end state*: there are no states accessible from it. So it is vacuously true of any state that if it is accessible from W_2 then A is true at it—the antecedent of the conditional is always false, so the conditional itself is always true. Hence *any* schema of the form $\square$A is true in any end state.

We could avoid saying that $\square$P is true in W_2 by stipulating that each state is to count as accessible from itself: this amounts to regarding accessibility as a *reflexive*

---

[1] Note here a slight departure from our usual conventions: we are using the capital letters W and A as individual constants, the former denoting a state of play in noughts and crosses, the latter denoting a propositional schema.

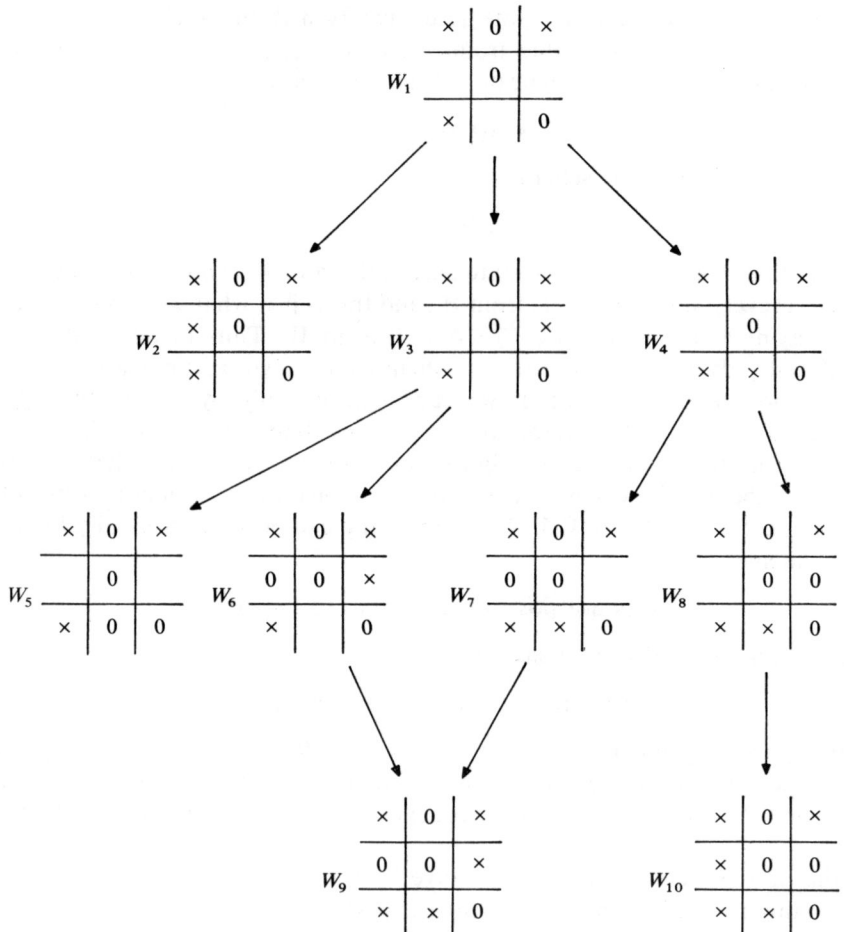

	P	◇P	□P	Q	◇Q	□Q	R	◇R	□R
W_1		T			T			T	
W_2			T	T		T			T
W_3		T						T	
W_4					T			T	
W_5	T		T			T			T
W_6								T	T
W_7								T	T
W_8					T	T			
W_9			T			T	T		T
W_{10}			T	T		T			T

relation. In that case $\Box Q$ would still be true at W_2, but $\Box P$ and $\Box R$ would no longer be. But if we do this, other consequences ensue. For example, $\Diamond A$ would now be true in any state in which A itself is true, and $\Box A$ could not be true in a state unless A itself were true in that state. So on this understanding of accessibility, our table would have to be greatly modified (see Exercise 7.1a, Question (1)).

Complex modalities

Since the operator $\Diamond$ can be prefixed to any schema, it can in particular be prefixed to a schema such as $\Diamond Q$ which already contains a model operator. In terms of our noughts and crosses game, how should the schema $\Diamond \Diamond Q$ be interpreted? Well, from the rule for interpreting $\Diamond$,

$$\Diamond \Diamond Q \text{ is true in state } W$$

means that

$$\Diamond Q \text{ is true in a state accessible from } W$$

which in turn means that

$$Q \text{ is true in a state accessible from a state accessible from } W.$$

For example, Q is true in W_{10}, which is accessible from W_8, which in turn is accessible from W_4. Hence $\Diamond Q$ is true in W_8, and $\Diamond \Diamond Q$ is true in W_4.

Of course, $\Diamond Q$ is also true in W_4, since W_{10} is accessible from W_4 too: indirect accessibility (via an intermediate state) counts as accessibility just as much as direct accessibility does. Put differently, we are treating accessibility as a *transitive* relation. It is not hard to see that so long as accessibility is transitive, we must have $\Diamond A$ true whenever we have $\Diamond \Diamond A$ true. This in turn means that the schema $\Diamond \Diamond A \rightarrow \Diamond A$ is true in every state, and hence that $\Box (\Diamond \Diamond A \rightarrow \Diamond A)$ is true in every state as well. Complex schemas like this can get quite hard to translate back into everyday language; this schema says something like.

It is necessary that if A is possibly possible, then A is possible,

which is hardly 'everyday' language! Nonetheless, this kind of statement has its uses, not just in philosophical discussions of the meaning of possibility and necessity, but also in more down-to-earth technical matters directly relevant to computing.

Suppose now that we try changing the rules again, and consider the *intransitive* accessibility relation that we get if we stipulate that state W' is to count as accessible from state W only if state W' can be reached from state W in a single move, with no intervening states. With this interpretation of accessibility, the schema $\Diamond Q$ says that crosses can win next move, while the schema $\Box Q$ says that crosses must win next move. So $\Diamond Q$ is now still true at W_8, but it is no longer true at W_4, although $\Diamond \Diamond Q$ is. As with the case when we made accessibility reflexive, intransitive accessibility also substantially alters the table of truth values we compiled above (see Exercise 7.1a, Question (3)).

Exercise 7.1a The following questions all refer to the noughts and crosses example.

(1) For each of the following schemas determine which of the states $W_1, \ldots, W_{10}$ it is true at (use the original version, in which accessibility is transitive but not reflexive):

*(a) $\Diamond P \wedge \Diamond Q$
*(b) $\Diamond (P \wedge Q)$
 (c) $\Box (P \vee Q)$
*(d) $\Box Q \vee \Box R$
*(e) $\Box (Q \vee R)$
*(f) $P \rightarrow \Box Q$
 (g) $\Diamond \Diamond P$
 (h) $\Diamond \Box P$
 (i) $\Diamond P \wedge \sim \Diamond Q$
*(j) $\Box Q \rightarrow \Box P$
 (k) $\Diamond (\Diamond Q \wedge \sim \Diamond \Diamond Q)$
 (l) $\Diamond (\Diamond R \wedge \Box R)$

(2) (a) Show that the schema $\Diamond R \wedge \Diamond Q \wedge \sim \Diamond P$ is true at W_4 and at no other state.
  *(b) Write down schemas that are true (i) at W_3 only, (ii) at W_8 only.
   (c) Show that a schema is true at W_6 if and only if it is true at W_7.
  *(d) Find another pair of states related in this way.

*(3) Write down the table (similar to the one on p. 226) that is obtained if each state is regarded as accessible from itself (reflexive transitive accessibility).

(4) Write down the table that is obtained if a state is only regarded as accessible from another if it can be reached from it in one move (intransitive accessibility).

Example 7.1.2 Logical truth and falsity   In the propositional calculus, the schema $P \wedge \sim P$ is logically true (a tautology) because it is true in every interpretation, i.e. in every line of the truth table. Suppose we think of the possible interpretations as states analogous to the states of play in noughts and crosses. From the standpoint of any one interpretation, all the others can be seen as possible alternative states of the world. So we can think of each interpretation as being accessible from all the others, and from itself. In this way we can express logical truth and falsity in terms of modal logic.

For example, the logical truth of $P \vee \sim P$ can be expressed by the schema $\Box (P \vee \sim P)$, which says that $P \vee \sim P$ is *necessarily* true, i.e. true in every interpretation. This schema, $\Box (P \vee \sim P)$, is itself true in every interpretation, since all the interpretations are mutually accessible. Likewise, $\Diamond (P \rightarrow \sim P)$ is true in every interpretation because there is an interpretation (accessible to all) in which $P \rightarrow \sim P$ is true, namely any interpretation in which P is false. So $P \rightarrow \sim P$ is *possibly* true. On the other hand, $\Diamond (P \wedge \sim P)$ is not true, since $P \wedge \sim P$ is false in every interpretation, and hence is not possible (a contradiction).

This way of understanding the modal operators is particularly simple. This becomes clear if we consider strings of two or more operators. Suppose $\Diamond A$ is true in some interpretation I. This means that A is true in some interpretation I': we don't need to specify that I' is accessible from I, since all the interpretations are mutually accessible. Now take any interpretation I''. Since A is true in I', $\Diamond A$ must be true in I''. The truth of A in an interpretation suffices to guarantee the truth of $\Diamond A$ in *every* interpretation. But this means that $\Box \Diamond A$ is true in I (or indeed in any other interpretation). We thus see that in this version of modal logic, $\Diamond A$ implies $\Box \Diamond A$.

Conversely, suppose $\Box \Diamond A$ is true in I. Then $\Diamond A$ must be true in every interpretation. In particular, $\Diamond A$ is true in I. Hence $\Box \Diamond A$ implies $\Diamond A$. Taking our two implications together, we see that, in this logic, $\Box \Diamond A$ is equivalent to $\Diamond A$.

It can similarly be shown that $\Diamond \Diamond A$ is equivalent to $\Diamond A$ and that both $\Box \Box A$ and

◇ □A are equivalent to □A. In short, each string of two consecutive modal operators can be replaced by the second operator alone. From this, it immediately follows that *any* string of consecutive modal operators can be replaced by its last element alone. For example, ◇ □ □ ◇ □A is equivalent to □A.

This modal logic is therefore much simpler than the logic we had in the noughts and crosses example, in which these reductions are not possible.

To summarize this section, then, modal logic provides us with a way of handling the ideas of necessity and possibility. It does this by postulating a system of 'states' endowed with a relation called 'accessibility'. Necessity in a state is then interpreted to mean truth in *all* states accessible from that state, and possibility as truth in *some* state accessible from it. The logical relationships amongst the modal schemas (equivalence, entailment, etc.) will depend, in general, on the nature of the accessibility relation, e.g. whether it is transitive or reflexive. Amongst the specific relationships of this kind, we noted that ◇ ◇A entails ◇A so long as accessibility is transitive, that □A entails A and A entails ◇A so long as accessibility is reflexive, and that any string of consecutive modal operators can be replaced by its last member if all states are mutually accessible.

Exercise 7.1b

(1) Investigate the properties of an accessibility relation which makes ◇A entail ◇ ◇A.

(2) Suppose that we divide the interpretations of propositional calculus into several groups, and declare that one interpretation is to count as accessible from another if and only if it is in the same group as it. Show that the special properties of the modal logic discussed in Example 7.1.2 remain unchanged by this.

7.2 FORMAL SYNTAX AND SEMANTICS OF PROPOSITIONAL MODAL LOGIC

Modal operators can be introduced either into the propositional calculus or into the predicate calculus. The latter case presents special difficulties of its own, so in this book we shall confine our attention to propositional modal logic (PML).

Formal syntax

The formal syntax of PML is quite straightforward. We start with the same rules as for **Prop**, and add in the modal operators □ and ◇:

Lexicon of PML	
Schematic letters:	$P_0 P_1 P_2 P_3 \cdots$
Connectives:	$\sim \wedge \vee \rightarrow \leftrightarrow$
Modal operators:	◇ □
Parentheses:	( )

Rules of formation for PML

(1) If P_i and P_j are schematic letters, then

$$P_i, \sim P_i, (P_i \wedge P_j), (P_i \vee P_j), (P_i \rightarrow P_j), (P_i \leftrightarrow P_j)$$
$$\Box P_i, \Diamond P_i$$

are all schemas.

(2) If S is a schema, then so is any substitution-instance of S obtained by substituting schemas for the schematic letters in S.

(3) Nothing is a schema unless it can be generated by the rules 1 and 2 above

---

As with the propositional calculus, we relax the rules for the sake of readability, allowing unsubscripted letters P, Q, R, S, etc., for the schematic letters, and introducing conventions for omitting parentheses.

Some examples of correctly formed modal schemas are

$$\Box P$$
$$\Box \sim P$$
$$\Box P \vee Q$$
$$\Box (P \vee Q)$$
$$\Box P \vee \Diamond Q$$
$$\Box (P \vee \Diamond Q)$$
$$\Box \Diamond \sim \Box (P \rightarrow \Diamond (Q \wedge \Box R))$$

Note that in $\Box P \vee Q$, the scope of $\Box$ is just P, whereas in $\Box (P \vee Q)$ it is P $\vee$ Q.

Formal semantics

An interpretation of PML has three ingredients. First, we require a set W of **states** (often called **possible worlds** or just **worlds**).

Second, we require an **accessibility relation** R on W. This can be presented either by giving a list of ordered pairs of elements of W, or by means of a rule which determines which pairs are to count as related by R. Together, W and R are known as a **modal frame** (or simply a **frame**), and we shall sometimes write things like

$$F = (W, R)$$

to establish a particular frame. We shall use R as an infix operator: thus wRw' means that state w' is accessible from state w.

Third, we require an **interpretation function** which maps each schematic letter onto a subset of W, so that for the schematic letter P, $I(P)$ is the set of states in which P is to count as true.

The interpretation function is extended to cover complex schemas by means of the following rules:

- For a schematic letter A, $I'(A) = I(A)$
- $I'(\sim A) = W - I'(A)$
- $I'(A \wedge B) = I'(A) \cap I'(B)$
- $I'(A \vee B) = I'(A) \cup I'(B)$
- $I'(A \rightarrow B) = I'(\sim A \vee B)$
- $I'(A \leftrightarrow B) = I'(A \rightarrow B) \cap I'(B \rightarrow A)$
- $I'(\Diamond A) = \{w \in W \mid \text{ for some } w' \in I'(A), wRw'\}$
- $I'(\Box A) = \{w \in W \mid \text{for every } w' \in I'(A), wRw'\}$

The three ingredients together, W, R, and I, make up a **formal interpretation** (W, R, I) for PML.

Satisfaction and validity

For a schema A and a state w in W, we write

$$\vDash_{W,R,I} A[w]$$

to mean that w is in $I'(A)$. In this case we say that A is **satisfied** by (W, R, I) at state w. If A is satisfied by (W, R, I) at *every* state in W, we write

$$\vDash_{W,R,I} A,$$

that is

$$\vDash_{W,R,I} A \text{ means that for every } w \in W, \vDash_{W,R,I} A[w].$$

In this case we say that (W, R, I) satisfies A.

It may happen that for some modal frame (W, R), whichever interpretation function I we choose, (W, R, I) satisfies A. In this case we say that A is *valid* with respect to (or in) the frame (W, R), and write

$$\vDash_{W,R} A.$$

Thus

$$\vDash_{W,R} A \text{ means that for every interpretation } I \text{ on } (W, R), \vDash_{W,R,I} A,$$

or equivalently

$$\vDash_{W,R} A \text{ means that for every state } w \text{ in } W \text{ and every interpretation } I$$
$$\text{on } (W, R), \vDash_{W,R,I} A[w].$$

If, finally, A is valid in *every* frame (and hence satisfied by every interpretation), then we say that A is **universally valid**, and write

$$\vDash A.$$

Thus

$$\vDash A \text{ means that for every frame } (W, R), \vDash_{W,R} A,$$

or equivalently

$\models A$ means that for every interpretation (W, R, I), $\models_{W,R,I} A$,

or again,

$\models A$ means that for every interpretation (W, R, I) and every state w in W, $\models_{W,R,I} A[w]$.

The noughts and crosses example (Example 7.1.1) can be regarded as an interpretation of PML with three schematic letters since it has all the right ingredients:

- There is a set $\{W_1, \ldots, W_{10}\}$ of states.
- There is an accessibility relation on the states.
- In each state it is determined which of the schematic letters P, Q, R are to count as true, and which as false.

Similarly, Example 7.1.2 provides another interpretation of PML, this time with an arbitrary number of schematic letters, since

- There is a set of states, i.e. the set of possible propositional calculus interpretations of the schematic letters.
- There is an accessibility relation on the states, in this case the *universal* relation which relates every pair of states to each other.
- In each state it is determined which schematic letters are true (i.e., the ones true in the propositional calculus interpretation identified with the state).

Universally valid modal schemas

The set of universally valid modal schemas can be characterized fairly simply; the following examples take us through the characterization one step at a time.

Example 7.2.1   If A is universally valid, so is $\Box A$.

To prove this, suppose that A is any universally valid modal schema. This means that in any interpretation (W, R, I), and for any state w in W, A is satisfied by I at w. In particular, for a given state w, A is satisfied by I at every state accessible from w, and hence, from the semantic rule for $\Box$, the schema $\Box A$ is satisfied by I at w. Since (W, R, I) was an arbitrary interpretation, and w an arbitrary state in W, it follows that $\Box A$ is universally valid, as required.

Example 7.2.2   For any modal schemas A and B, the schema $\Box(A \rightarrow B) \rightarrow (\Box A \rightarrow \Box B)$ is universally valid.

To see this, let (W, R, I) be any interpretation, and let w be any state in W. Abbreviate the schema above to S. Now either $\Box(A \rightarrow B)$, the antecedent of S, is satisfied by I at w, or it is not. If it is not, then from the semantic rule for $\rightarrow$, S *is* satisfied by I at w.

Suppose on the other hand that $\Box(A \to B)$ is satisfied at w, and consider the schema $\Box A$. If $\Box A$ is not satisfied at w, then $\Box A \to \Box B$ is satisfied at w, and hence so is S.

Suppose, then, that $\Box A$ *is* satisfied at w, and let w' be any state accessible from w in (W, R). Since both $\Box(A \to B)$ and $\Box A$ are satisfied at w, both $A \to B$ and A must be satisfied at w', by the semantics for $\Box$, and hence, by the semantics for $\to$, B must be satisfied at w'. Since w' is *any* state accessible from w, $\Box B$ must be satisfied at w, which means that $\Box A \to \Box B$, and hence also S, is satisfied at w.

We have now shown that S is always satisfied at w, and since w was an arbitrary state in an arbitrary interpretation it follows that S is universally valid.

Example 7.2.3 Let A be the result of uniformly substituting PML schemas for the schematic letters occurring in any tautology of the propositional calculus. Then A is universally valid.

(The sort of schema considered here is exemplified by $\Box A \to (\Diamond B \to \Box A)$, which is a modal substitution-instance of the propositional tautology $P \to (Q \to P)$.)

Let w be any state in any interpretation (W, R, I). Let A′ be the propositional tautology of which A is a substitution-instance. Let A′ contain the schematic letters $P_1, \ldots, P_n$, and let $A_1, \ldots, A_n$ be the modal schemas substituted for them to give A. Let the truth values of $A_1, \ldots, A_n$ at w in (W, R, I) be $V_1, \ldots, V_n$ respectively, and consider the propositional calculus interpretation J in which each P_i receives the truth-value V_i. The truth-value of A at w in (W, R, I) must be the same as the truth value of A′ in J, since both are built up in the same way out of ingredients with corresponding truth values. But since A′ is a tautology, it is true in J; hence A is true at w in (W, R, I), and since w was arbitrary, A is universally valid.

Example 7.2.4 If A and $A \to B$ are universally valid, so is B.

This follows from the semantic rule for $\to$, which implies that B is true in any state in which A and $A \to B$ are both true. Since these two schemas are, by hypothesis, universally valid, it follows that B is also universally valid.

The four results of Examples 7.2.1–4 form the basis of the natural deduction system for modal logic which we present in the next section, and which can be used for proving the universal validity of other modal schemas.

It is also generally possible to prove universal validity by arguing directly in terms of interpretations, as we did in the examples above. Here are two more examples to illustrate this.

Example 7.2.5 If A and B are any PML schemas, then the schema

$$\Diamond(A \vee B) \to \Diamond A \vee \Diamond B$$

is universally valid.

Suppose the antecedent, $\Diamond(A \vee B)$, is satisfied by (W, R, I) at state w. Then for some w' accessible from w in (W, R), $A \vee B$ is satisfied at w', and hence either A is satisfied at w' or B is satisfied at w'. If A is satisfied at w', then since w' is accessible from w, $\Diamond A$ is satisfied at w. Likewise, if B is satisfied at w', $\Diamond B$ is satisfied at w. Hence either $\Diamond A$ or $\Diamond B$ is satisfied at w, i.e., $\Diamond A \vee \Diamond B$ is satisfied at w. It follows that $\Diamond(A \vee B) \to \Diamond A \vee \Diamond B$ is universally valid.

Example 7.2.6   Prove that $\Diamond(A \to B) \to (\Box A \to \Diamond B)$ is universally valid.

Let C abbreviate the theorem to be proved, let (W, R, I) be any interpretation, and let w be any element of W.
   Suppose $\Diamond(A \to B)$ is satisfied by (W, R, I) at w. This means that there is a state w', accessible from w, at which $A \to B$ is satisfied:

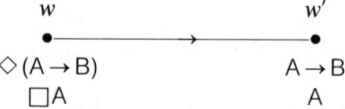

$$
\begin{array}{cc}
w & w' \\
\bullet\!\!\longrightarrow\!\!\bullet & \\
\Diamond(A \to B) & A \to B
\end{array}
$$

Now suppose that $\Box A$ is satisfied at w. Then A must be satisfied at every state accessible from w; in particular, A is satisfied at w':

$$
\begin{array}{cc}
w & w' \\
\bullet\!\!\longrightarrow\!\!\bullet & \\
\Diamond(A \to B) & A \to B \\
\Box A & A
\end{array}
$$

Since both $A \to B$ and A are satisfied at w', B must be too, from the semantic rule for $\to$. Hence $\Diamond B$ is satisfied at w:

$$
\begin{array}{cc}
w & w' \\
\bullet\!\!\longrightarrow\!\!\bullet & \\
\Diamond(A \to B) & A \to B \\
\Box A & A \\
\Diamond B & B
\end{array}
$$

We have now shown that $\Box A \to \Diamond B$ is satisfied at any state where $\Diamond(A \to B)$ is satisfied; and hence that C is universally valid.

   In order to show that a schema is *not* universally valid, we must find a counter-example, i.e. an interpretation containing a state at which the schema is not satisfied. The next example illustrates this.

Example 7.2.7   Show that the schema $\Box(A \lor B) \to \Box A \lor \Box B$ is not universally valid.

A suitable counter-example is provided by the interpretation illustrated below:

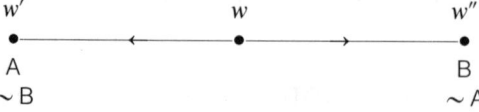

$$
\begin{array}{ccc}
w' & w & w'' \\
\bullet\!\!\longleftarrow\!\!\bullet\!\!\longrightarrow\!\!\bullet & & \\
A & & B \\
\sim B & & \sim A
\end{array}
$$

Since $A \lor B$ is satisfied at both w' and w'', and these are the only states accessible from w, $\Box(A \lor B)$ is satisfied at w. On the other hand, neither $\Box A$ nor $\Box B$ is satisfied at w.

Exercise 7.2
   (1) Prove that the following schemas are universally valid.

   *(a) $\Box(A \land B) \to \Box A \land \Box B$
   (b) $\Box A \land \Box B \to \Box(A \land B)$
   (c) $\Diamond(A \land B) \to \Diamond A \land \Diamond B$

(d) $\Diamond A \vee \Diamond B \to \Diamond (A \vee B)$
(e) $\Box (A \to B) \to (\Diamond A \to \Diamond B)$

(2) Show that the following schemas are not universally valid.

*(a) $\Box A \to A$
(b) $\Box A \to \Diamond A$
(c) $\Box A \to \Box \Box A$
(d) $\Diamond A \vee \Diamond B \to \Diamond (A \wedge B)$
(e) $A \to \Diamond A$

7.3 A PROOF SYSTEM FOR PROPOSITIONAL MODAL LOGIC

We can use the results proved in Examples 7.2.1–4, together with the definition of $\Diamond$ in terms of $\Box$, to set up a natural deduction system for PML. This system contains all the rules for **Prop**, as given in Section 3.2, together with the rules:

$$(\Box\text{-intro}) \quad \begin{array}{|l} A \\ \hline \Box A \end{array} \quad \begin{array}{l} \text{(so long as A does not} \\ \text{depend on a premiss or} \\ \text{an undischarged assumption)} \end{array}$$

$$(K) \quad \begin{array}{|l} \Box (A \to B) \\ \Box A \\ \hline \Box B \end{array}$$

$$(\Diamond\text{-intro}) \quad \begin{array}{|l} \sim \Box \sim A \\ \hline \Diamond A \end{array}$$

$$(\Diamond\text{-elim}) \quad \begin{array}{|l} \Diamond A \\ \hline \sim \Box \sim A \end{array}$$

In proofs, we shall often replace a sequence of purely propositional rules by a single rule

$$(\text{tautology}) \quad \begin{array}{|l} \\ \hline A \end{array} \quad \begin{array}{l} \text{(so long as A is a} \\ \text{substitution-instance} \\ \text{of a tautology)} \end{array}$$

This enables us to take the purely propositional part of our logic for granted, and focus on the specifically modal part.

The condition attached to the $\Box$-introduction rule is needed because Example 7.2.1 only licenses us to infer $\Box A$ if we know that A is universally valid—so we need to be in a position to assert A unconditionally in order to deduce $\Box A$. If A is a premiss or an assumption, or is derived from one of these, then this is precisely to say that A is *not* being asserted unconditionally.

We illustrate the system by proving two theorems.

Example 7.3.1 $\Box A \wedge \Box B \to \Box (A \wedge B)$

(1)   SUBDERIVATION

     (1.1) $\Box A \wedge \Box B$ (assumption)
     (1.2) $\Box A$ (1.1, $\wedge$-elim-left)
     (1.3) $\Box B$ (1.1, $\wedge$-elim-right)
     (1.4) $A \to (B \to A \wedge B)$ (tautology)
     (1.5) $\Box (A \to (B \to A \wedge B)))$ (1.4, $\Box$-intro)
     (1.6) $\Box (B \to A \wedge B)$ (1.2, 1.5, K)
     (1.7) $\Box (A \wedge B)$ (1.3, 1.6, K)
(2) $\Box A \wedge \Box B \to \Box (A \wedge B)$ (1, $\to$-intro)

Example 7.3.2 $\Diamond (A \to B) \to (\Box A \to \Diamond B)$

(1)   SUBDERIVATION

     (1.1) $\Diamond (A \to B)$ (assumption)
     (1.2)   SUBDERIVATION

         (1.2.1) $\Box A$ (assumption)
         (1.2.2) $A \to (\sim B \to \sim (A \to B))$ (tautology)
         (1.2.3) $\Box (A \to (\sim B \to \sim (A \to B)))$ (1.2.2, $\Box$-intro)
         (1.2.4) $\Box (\sim B \to \sim (A \to B))$ (1.2.1, 1.2.3, K)
         (1.2.5)   SUBDERIVATION

             (1.2.5.1) $\Box \sim B$ (assumption)
             (1.2.5.2) $\Box \sim (A \to B)$ (1.2.4, 1.2.5.1, K)
             (1.2.5.3) $\sim \Box \sim (A \to B)$ (1.1, $\Diamond$-elim)
         (1.2.6) $\sim \Box \sim B$ (1.2.5, $\sim$-intro)
         (1.2.7) $\Diamond B$ (1.2.6, $\Diamond$-intro)
     (1.3) $\Box A \to \Diamond B$ (1.2, $\to$-intro)
(2) $\Diamond (A \to B) \to (\Box A \to \Diamond B)$ (1, $\to$-intro)

Two very important theorems, which are closely related to the introduction and elimination rules for $\Diamond$, are

$$\Box \sim A \leftrightarrow \sim \Diamond A$$

and

$$\Diamond \sim A \leftrightarrow \sim \Box A.$$

We prove the second of these, leaving the first (which is somewhat easier), to the reader.

(1)   SUBDERIVATION

     (1.1) $\sim \Box A$ (assumption)
     (1.2) $\sim \sim A \to A$ (tautology)
     (1.3) $\Box (\sim \sim A \to A)$ (1.2, $\Box$-intro)
     (1.4)   SUBDERIVATION

         (1.4.1) $\Box \sim \sim A$ (assumption)
         (1.4.2) $\Box A$ (1.3, 1.4.1, K)
         (1.4.3) $\sim \Box A$ (1.1, rep)
     (1.5) $\sim \Box \sim \sim A$ (1.4, $\sim$-intro)
     (1.6) $\Diamond \sim A$ (1.5, $\Diamond$-intro)
(2)   SUBDERIVATION

     (2.1) $\Diamond \sim A$ (assumption)
     (2.2) $A \to \sim \sim A$ (tautology)

(2.3) $\square(A \rightarrow \sim \sim A)$ (2.2, $\square$-intro)
(2.4)   SUBDERIVATION
      (2.4.1) $\square A$ (assumption)
      (2.4.2) $\square \sim \sim A$ (2.3, 2.4.1, K)
      (2.4.3) $\sim \square \sim \sim A$ (2.1, $\diamond$-elim)
  (2.5) $\sim \square A$ (2.4, $\sim$-intro)
(3) $\diamond \sim A \leftrightarrow \sim \square A$ (1, 2, $\leftrightarrow$-intro)

These theorems give rise to four derived rules of inference, as follows:

$$(\square \sim) \quad \begin{array}{|l} \square \sim A \\ \hline \\ \sim \diamond A \end{array} \qquad (\sim \square) \quad \begin{array}{|l} \sim \square A \\ \hline \\ \diamond \sim A \end{array}$$

$$(\diamond \sim) \quad \begin{array}{|l} \diamond \sim A \\ \hline \\ \sim \square A \end{array} \qquad (\sim \diamond) \quad \begin{array}{|l} \sim \diamond A \\ \hline \\ \square \sim A \end{array}$$

Another useful derived rule of inference is

$$(K \diamond) \quad \begin{array}{|l} \square(A \rightarrow B) \\ \diamond A \\ \hline \\ \diamond B \end{array}$$

This is proved as follows:

(1) $\square(A \rightarrow B)$ (premiss)
(2) $\diamond A$ (premiss)
(3) $(A \rightarrow B) \rightarrow (\sim B \rightarrow \sim A)$ (tautology)
(4) $\square((A \rightarrow B) \rightarrow (\sim B \rightarrow \sim A))$ (3, $\square$-intro)
(5) $\square(\sim B \rightarrow \sim A)$ (1, 4, K)
(6)   SUBDERIVATION
      (6.1) $\square \sim B$ (assumption)
      (6.2) $\square \sim A$ (6.1, 5, K)
      (6.3) $\sim \square \sim A$ (2, $\diamond$-elim)
(7) $\sim \square \sim B$ (6, $\sim$-intro)
(8) $\diamond B$ (7, $\diamond$-intro)

These rules can be used to help prove more complicated results such as $\sim \square \diamond \diamond A \rightarrow \diamond \square \square \sim A$:

(1)   SUBDERIVATION
      (1.1) $\sim \square \diamond \diamond A$ (assumption)
      (1.2) $\diamond \sim \diamond \diamond A$ (1.1, $\sim \square$)
      (1.3)   SUBDERIVATION
            (1.3.1) $\sim \diamond \diamond A$ (assumption)
            (1.3.2) $\square \sim \diamond A$ (1.3.1, $\sim \diamond$)
            (1.3.3)   SUBDERIVATION
                  (1.3.3.1) $\sim \diamond A$ (assumption)
                  (1.3.3.2) $\square \sim A$ (1.3.3.1, $\sim \diamond$)

$$(1.3.4) \quad \sim \Diamond A \rightarrow \Box \sim A \qquad\qquad (1.3.3, \rightarrow\text{-intro})$$
$$(1.3.5) \quad \Box(\sim \Diamond A \rightarrow \Box \sim A) \qquad (1.3.4, \Box\text{-intro})$$
$$(1.3.6) \quad \Box\Box \sim A \qquad\qquad\qquad (1.3.2, 1.3.5, K)$$
$$(1.4) \quad \sim \Diamond\Diamond A \rightarrow \Box\Box \sim A \qquad (1.3, \rightarrow\text{-intro})$$
$$(1.5) \quad \Box(\sim \Diamond\Diamond A \rightarrow \Box\Box \sim A) \quad (1.4, \Box\text{-intro})$$
$$(1.6) \quad \Diamond\Box\Box \sim A \qquad\qquad\qquad (1.2, 1.5, K \Diamond)$$
$$(2) \quad \sim \Box\Diamond\Diamond A \rightarrow \Diamond\Box\Box \sim A \qquad (1, \rightarrow\text{-intro})$$

The pattern of this derivation can be generalized to yield proofs of all instances of the following metatheorem:

$$\sim O_1 O_2 \ldots O_n A \leftrightarrow O_1' O_2' \ldots O_n' \sim A$$

where each of O_i is either $\Box$ or $\Diamond$ and O_i' is the other one (i.e. if O_i is $\Box$ then O_i' is $\Diamond$, and *vice versa*). The corresponding generalized rules of inference are

(op-swap)
$$\frac{O_1 O_2 \ldots O_n \sim A}{\sim O_1' O_2' \ldots O_n' A}$$

($\sim$ op-swap)
$$\frac{\sim O_1 O_2 \ldots O_n A}{O_1' O_2' \ldots O_n' \sim A}$$

Note the close analogy with the quantifier-swapping rules of the predicate calculus. We shall make extensive use of these rules in the next section.

Exercise 7.3

(1) Construct natural deduction proofs for each of the following universally valid schemas:

*(a) $\Box(A \wedge B) \rightarrow \Box A \wedge \Box B$
(b) $\Diamond(A \wedge B) \rightarrow \Diamond A \wedge \Diamond B$
(c) $\Diamond(A \vee B) \leftrightarrow \Diamond A \vee \Diamond B$
(d) $\Box A \vee \Diamond \sim A$
*(e) $\Box A \vee \Diamond(B \vee \sim B)$
(f) $\Box(A \wedge \sim A) \vee \Diamond(A \vee \sim A)$
(g) $\Box A \wedge \Diamond B \rightarrow \Diamond(A \wedge B)$

(2) Without using the rules $(\sim \Box)$, $(\sim \Diamond)$, $(\Box \sim)$, $(\Diamond \sim)$, (op-swap) or ($\sim$ op-swap), prove the theorem $\Box \sim A \leftrightarrow \sim \Diamond A$.

*(3) Derive the inference rule

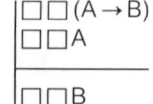

(4) Use mathematical induction to show that for each positive integer n the rule

$$\frac{\Box^n(A \rightarrow B) \quad \Box^n A}{\Box^n B}$$

can be derived using natural deduction. (Here $\Box^n$A means A preceded by n copies of the operator $\Box$, so that, e.g., $\Box^4$A is the schema $\Box\Box\Box\Box$A.)

7.4 MODAL THEORIES

We saw in Section 7.1 that the schema

$$(T) \quad \diamond\diamond A \rightarrow \diamond A$$

is satisfied at every state in an interpretation (W, R, I) so long as R is a transitive relation on W, in other words that T is valid in any transitive modal frame (W, R). The set of modal schemas which are valid in every transitive frame is called the **modal theory of transitivity** (cf. the first-order theory of transitivity discussed in Section 6.1). Thus every instance of the schema T is contained in the modal theory of transitivity.

There is a sense in which the schema T *exactly* characterizes the notion of transitivity. Suppose (W, R) is any modal frame in which T is valid. We shall show that R must be a transitive relation.

To see this, let $w_1 R w_2$ and $w_2 R w_3$; we must show that $w_1 R w_3$. To this end, let P be any schematic letter, and define the interpretation function I so that $I(P) = \{w_3\}$. Since P is satisfied at w_3, which is accessible from w_2, which in turn is accessible from w_1, it follows that $\diamond\diamond$P is satisfied at w_1. Since the schema T is valid in (W, R), it is satisfied by (W, R, I) at state w_1; at this state, therefore, we have $\diamond\diamond$P and $\diamond\diamond$P $\rightarrow$ $\diamond$P and hence also $\diamond$P. This means that P is satisfied at some state accessible from w_1. But the only state at which P is satisfied is w_3. It follows that w_3 is accessible from w_1, as required.

A proof system for the modal theory of transitivity can therefore be obtained from the proof system for general modal logic given in the last section by adding the schema $\diamond\diamond A \rightarrow \diamond A$ as an *axiom*, or, more conveniently, by introducing a new inference rule

$$(\text{trans-}\diamond) \quad \left| \begin{array}{c} \diamond\diamond A \\ \hline \diamond A \end{array} \right.$$

An equivalent formulation of this rule, using necessity rather than possibility, is

$$(\text{trans-}\Box) \quad \left| \begin{array}{c} \Box A \\ \hline \Box\Box A \end{array} \right.$$

Either of these rules can be derived from the other; for example, assuming (trans-$\diamond$), we have

(1) $\Box$A   (premiss)
(2) SUBDERIVATION
    (2.1) $\sim\Box\Box$A   (assumption)
    (2.2) $\diamond\diamond\sim$A   (2.1, $\sim$ op-swap)
    (2.3) $\diamond\sim$A   (2.2, trans-$\diamond$)

(2.4)	$\sim \Box A$	$(2.3, \diamond \sim)$
(2.5)	$\Box A$	$(1, \text{rep})$
(3)	$\sim \sim \Box \Box A$	$(2, \sim\text{-intro})$
(4)	$\Box \Box A$	$(3, \sim\text{-elim})$

The reverse derivation is left as an exercise (see Exercise 7.4, Question (1a)).

Certain modal theories have been singled out for study either because they have particularly interesting formal properties or because they confer on the modal operators some of the important properties we intuitively feel that possibility and necessity (or some related notions) ought to possess.

It seems—more or less—natural, for example, to say that if some statement P is necessarily true, then it is *necessarily* necessarily true, that if we can't avoid having P true, then we can't avoid the necessity of having P true.

Again, on most informal interpretations of 'necessary' and 'possible' it seems natural to insist that what is necessarily the case *is* the case, and that what is the case is *possible*. An exception would be if we understand 'necessary' and 'possible' to mean 'compulsory' and 'permissible' respectively: just because something is compulsory, it does not follow that it gets done, nor does the fact that something is done imply that it is permissible. But more commonly, there is good reason to accept the principles that what is necessary is true, and what is true is possible.

These principles correspond to the rules

$$(\text{refl-}\diamond) \qquad \begin{array}{c} A \\ \hline \diamond A \end{array}$$

$$(\text{refl-}\Box) \qquad \begin{array}{c} \Box A \\ \hline A \end{array}$$

As with the (trans) rules, each of these two rules can be derived from the other (see Exercise 7.4, Question (1b)), and hence it is not strictly necessary (though very convenient) to include both. These rules are called (refl) because they are characteristic of interpretations (W, R, I) in which the relation R is *reflexive*, i.e. in which each state is accessible from itself (see Exercise 7.4, Question 2).

The modal theory S4

The combination of (trans) with (refl) gives rise to modal theory—the theory of a reflexive, transitive relation—known as S4. This is one of a series of modal theories (S1–S5) first studies by C. I. Lewis in the early years of this century and known collectively as the *Lewis modal systems*.

An example of a schema which is valid in S4 but not universally is

$$\Box \Box A \rightarrow \Box A.$$

This follows immediately from (refl-$\Box$) by putting $\Box A$ in place of A. Note that this result and (trans-$\Box$) together imply that the schemas $\Box \Box A$ and $\Box A$ are equivalent in S4. In words, 'necessarily necessary' reduces to plain 'necessary'. Similarly, $\Diamond\Diamond$A is equivalent in S4 to $\Diamond$A, i.e. 'possibly possible' reduces to 'possible'.

Using these results, all occurrences of $\Diamond\Diamond$ or $\Box\Box$ can be replaced by $\Diamond$ and $\Box$ respectively. This means that the only strings of consecutive modal operators which cannot be reduced in this way are those in which $\Box$ and $\Diamond$ alternate, i.e.

$$\Diamond, \Diamond\,\Box, \Diamond\,\Box\,\Diamond, \Diamond\,\Box\,\Diamond\,\Box, \Diamond\,\Box\,\Diamond\,\Box\,\Diamond, \ldots$$
$$\Box, \Box\,\Diamond, \Box\,\Diamond\,\Box, \Box\,\Diamond\,\Box\,\Diamond, \Box\,\Diamond\,\Box\,\Diamond\,\Box, \ldots$$

It turns out that reductions are possible for most of these strings too. We have, for example:

Example 7.4.1 $\Box\,\Diamond\,\Box\,\Diamond$A $\to \Box\,\Diamond$A

(1)   SUBDERIVATION
     (1.1) $\Box\,\Diamond$A   (assumption)
     (1.2) $\Diamond$A   (1.1, refl-$\Box$)
(2) $\Box\,\Diamond$A $\to \Diamond$A   (1, $\to$-intro)
(3) $\Box(\Box\,\Diamond$A$\to\Diamond$A)   (2, $\Box$-intro)
(4)   SUBDERIVATION
     (4.1) $\Diamond\,\Box\,\Diamond$A   (assumption)
     (4.2) $\Diamond\,\Diamond$A   (3, 4.1, K $\Diamond$)
     (4.3) $\Diamond$A   (4.2, trans-$\Diamond$)
(5) $\Diamond\,\Box\,\Diamond$A $\to\Diamond$A   (4, $\to$-intro)
(6) $\Box(\Diamond\,\Box\,\Diamond$A$\to\Diamond$A)   (5, $\Box$-intro)
(7)   SUBDERIVATION
     (7.1) $\Box\,\Diamond\,\Box\,\Diamond$A   (assumption)
     (7.2) $\Box\,\Diamond$A   (6, 7.1, K)
(8) $\Box\,\Diamond\,\Box\,\Diamond$A $\to \Box\,\Diamond$A   (7, $\to$-intro)

Example 7.4.2 $\Box\,\Diamond$A $\to \Box\,\Diamond\,\Box\,\Diamond$A

(1)   SUBDERIVATION
     (1.1) $\Box\,\Diamond$A   (assumption)
     (1.2) $\Diamond\,\Box\,\Diamond$A   (refl-$\Diamond$)
(2) $\Box\,\Diamond$A $\to\Diamond\,\Box\,\Diamond$A   (1, $\to$-intro)
(3) $\Box(\Box\,\Diamond$A$\to\Diamond\,\Box\,\Diamond$A)   (2, $\Box$-intro)
(4)   SUBDERIVATION
     (4.1) $\Box\,\Diamond$A   (assumption)
     (4.2) $\Box\,\Box\,\Diamond$A   (4.1, trans-$\Box$)
     (4.3) $\Box\,\Diamond\,\Box\,\Diamond$A   (3, 4.2, K)
(5) $\Box\,\Diamond$A $\to\Box\,\Diamond\,\Box\,\Diamond$A   (4, $\to$-intro)

These two results together imply that $\Box\,\Diamond$A is equivalent to $\Box\,\Diamond\,\Box\,\Diamond$A in S4, and hence that the operator string $\Box\,\Diamond\,\Box\,\Diamond$ can be reduced to $\Box\,\Diamond$. Applying the op-swap rules, we also obtain that $\Diamond\,\Box$A is equivalent to $\Diamond\,\Box\,\Diamond\,\Box$A, so $\Diamond\,\Box\,\Diamond\,\Box$ can be reduced to $\Diamond\,\Box$.

The upshot of all this is that any string of four or more consecutive modal operators can be reduced in S4 to a string of length 3 or less. For example, the string $\Diamond\,\Box\,\Box\,\Diamond\,\Diamond\,\Box\,\Box\,\Box\,\Diamond$ can be reduced first to $\Diamond\,\Box\,\Diamond\,\Box\,\Diamond$ (by the equivalence of $\Diamond\,\Diamond$,

$\Box\Box$ to $\Diamond$, $\Box$ respectively), then the first four operators of this string can be reduced to $\Diamond\Box$ using the result just obtained, giving $\Diamond\Box\Diamond$ for the whole string.

Altogether there are just seven non-equivalent operator strings in S4: the empty string (as, e.g., in the schema P), $\Box$, $\Diamond$, $\Box\Diamond$, $\Diamond\Box$, $\Box\Diamond\Box$, and $\Diamond\Box\Diamond$. The logical relationships between these is as shown in the following diagram, in which the arrows represent implication:

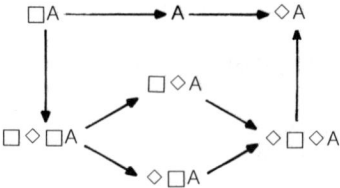

The proof of these relationships is left as an exercise for the reader (see Exercise 7.4, Question 3).

The modal theory S5

Of the other Lewis systems, the one most frequently encountered is S5, which is the modal theory of an equivalence relation: i.e. the theorems of S5 are precisely those schemas which are valid in any modal frame (W, R) for which R is an equivalence relation. We have already met an example of such an interpretation: Example 7.1.2. The relation in this example is a *universal* relation, i.e. all states are mutually accessible. This is a special case of the more general notion of an equivalence relation, and gives rise to the same modal theory.

Since an equivalence relation is both transitive and reflexive, S5 contains S4: that is, the rules of inference characteristic of S4 are also valid in S5, and hence every theorem of S4 is also a theorem of S5. But S5 also contains theorems *not* in S4, since an equivalence relation, as well as being both transitive and reflexive, is also *symmetric*, i.e. for every pair of states w, w', if wRw' then $w'Rw$.

To find a schema that is valid in any symmetric frame, let R be a symmetric relation on W, and suppose that (W, R, I) satisfies A at state w. Let w' be any state accessible from w. By symmetry, w is accessible from w', so $\Diamond$A is satisfied at w'. Since w' was *any* state accessible from w, it follows that $\Box\Diamond$A is satisfied at w. Hence the schema

$$(S) \qquad A \to \Box\Diamond A$$

is valid in any frame with a symmetric accessibility relation, and hence must be a theorem of S5.

To see that the schema S serves to characterize symmetry exactly, suppose that S is satisfied at every state in any interpretation based on the modal frame (W, R), and let $w_1 R w_2$. We must show that $w_2 R w_1$. Let P be any schematic letter, and define an interpretation function I so that $I(\text{P}) = \{w_1\}$. Then since S is satisfied throughout (W, R, I), it is satisfied at w_1, at which, therefore, the schema $\Box\Diamond$P is satisfied. Since w_2 is accessible from w_1, $\Diamond$P is satisfied at w_2, and hence P is satisfied at some state accessible from w_2. But the only state at which P is satisfied in this interpretation is w_1, and hence this state must be accessible from w_2, as required.

We may therefore axiomatize S5 by adding either S or an equivalent such as $\Diamond\Box A \to A$ as an axiom schema to S4; or, as we did with transitivity and reflexiveness, to add one or other of the rules of inference

$$(\text{symm-}\Box) \qquad \dfrac{\begin{array}{c} A \end{array}}{\Box\Diamond A} \qquad (\text{symm-}\Diamond) \qquad \dfrac{\begin{array}{c} \Diamond\Box A \end{array}}{A}$$

In S5, any string of consecutive modal operators can be reduced to a single operator, namely the last element of the string. This is because the schemas

$$\Box\Box A \leftrightarrow \Box A$$
$$\Diamond\Diamond A \leftrightarrow \Diamond A$$
$$\Box\Diamond A \leftrightarrow \Diamond A$$
$$\Diamond\Box A \leftrightarrow \Box A$$

are all theorems (or rather theorem schemas) of S5. We have already proved the first two in S4, and they carry over directly into S5. The third is proved in the next example, the fourth is left as an exercise.

Example 7.4.3 $\Box\Diamond A \leftrightarrow \Diamond A$ is valid in S5.

(1) SUBDERIVATION
     (1.1) $\Box\Diamond A$ (assumption)
     (1.2) $\Diamond A$ (1.1, refl-$\Box$)
(2) SUBDERIVATION
     (2.1) $\Diamond A$ (assumption)
     (2.2) $\Box\Diamond\Diamond A$ (2.1, symm-$\Box$)
     (2.3) SUBDERIVATION
          (2.3.1) $\Diamond\Diamond A$ (assumption)
          (2.3.2) $\Diamond A$ (2.3.1, trans-$\Diamond$)
     (2.4) $\Diamond\Diamond A \to \Diamond A$ (2.3, $\to$-intro)
     (2.5) $\Box(\Diamond\Diamond A \to \Diamond A)$ (2.4, $\Box$-intro)
     (2.6) $\Box\Diamond A$ (2.2, 2.5, K)
(3) $\Box\Diamond A \leftrightarrow \Diamond A$ (1, 2, $\leftrightarrow$-intro)

Two informal interpretations of the modal operators

The modal logic of knowledge

S4 and S5 have both been used to model *epistemic modality*, the logic of *knowledge*. We read $\Box A$ as, for example, 'It is known that A', and assume that

(a) if something is known, then it is known that it is known,

and

(b) if something is known then it is true

(a false proposition can be *believed*, with total conviction, but it cannot be *known*). Assumption (a) corresponds to the inference rule (trans-□), and (b) to (refl-□). Hence the logic of 'it is known that', on these assumptions, contains S4.

How is ◇ interpreted on this scheme? Since ◇A is equivalent to ∼ □ ∼ A, it may be read as

'It is not known that A is false'

and hence as

'A is compatible with what is known'

(since if A were *not* compatible with what is known, that would enable it to be known that A was false—assuming perfect reasoning ability on the part of the knowers).

Given this, the S5 rule (symm-□) comes out as the assumption

(c)   If A is true, then it is known to be compatible with what is known.

This may strike one as a much more doubtful assumption than either (a) or (b). After all, there are surely innumerable true statements which have not even been considered, so nothing is known about them at all, much less that they are compatible with what is known. And even for a truth which *has* been considered, we may simply not be in a position to say whether or not it is compatible with what we know.

Perhaps the best bet for an adequate logic of knowledge would be a modal theory intermediate between S4 and S5, i.e. a theory which contains all the theorems of S4, and some, but not all, of the theorems which are in S5 but not S4. Such intermediate systems have been included in the Lewis hierarchy under names such as S4.2 and S4.3. For more details, see Hughes and Cresswell (1968).

A modal theory of future time

There is another, quite different way of reading the modal operators which results in a theory intermediate between S4 and S5. Read □A as

'A is true now and always will be true'.

Then ◇A, i.e. ∼ □ ∼ A, must be read as 'It is not the case that A is and always will be false', i.e. as 'A is true now or will eventually be true'. With these readings, all the rules of S4 receive intuitively acceptable interpretations.

Thus (trans-◇) says that if it is or will be true that A is or will be true, then A is or will be true. That this is reasonable is readily seen: for if it *is* true that A is or will be true, then of course the conclusion follows; and if it *will be* true at some future time *t* that A is or will be true then either A will be true at *t* or A will be true at some time later than *t*, and in either case A will be true. All this merely amounts to saying that the temporal order is transitive: that what will be future is already future.

Similarly, (refl-◇) simply says that if it is now true that A, then it is or will be true that A, which is certainly the case since a disjunction does follow from each of its disjuncts (so 'A now or A in the future' follows from 'A now').

Thus the modal theory corresponding to this particular temporal intepretation at any rate *contains* S4: that is, every theorem of S4 is valid in this system. Is it the same as S4? To test this, we should have to check whether the seven distinct operator strings of S4 correspond to distinct meanings under the interpretation.

For example, is it possible to have

> 'It is and always will be the case that it is or
> will be the case that A is and always will be true'

(i.e. $\Box \Diamond \Box A$) without having

> 'A is and always will be true'

(i.e. $\Box A$)? If this is not possible, then the operator strings $\Box \Diamond \Box$ and $\Box$ must be equivalent in our interpretation, which means that the system in question is strictly stronger than S4; if, on the other hand, it *is* possible, then $\Box \Diamond \Box$ is not equivalent to $\Box$ in our interpretation, which means that the system is strictly weaker than S5. So either way, we obtain useful information about the system.

In fact it *is* possible. Suppose that time t' is later than time t, and that A is true at t' and at all later times, but false at all times earlier than t'. Then $\Box A$ is true at t' and at all times later than t'. Hence $\Diamond \Box A$ is true at *all* times, since at any time you choose, either $\Box A$ is true then (if the time you choose is t' or later) or $\Box A$ will be true at a later time (if the time you choose is earlier than t'—in which case t' itself is a later time). Hence in particular $\Diamond \Box A$ is true at t and all later times, which means that $\Box \Diamond \Box A$ is true at t. But $\Box A$ is false at t, and hence it *is* possible to have $\Box \Diamond \Box A$ without having $\Box A$.

This shows us that the modal theory determined by our interpretation is strictly weaker than S5, for in S5, $\Box \Diamond \Box A$ is equivalent to $\Box A$. It is therefore either S4 or intermediate between S4 and S5.

To determine which, we consider the two schemas $\Diamond \Box \Diamond A$ and $\Box \Diamond A$. We know that the first of these is implied by the second in S4, and hence also in our system, which contains S4. But in S4, the second schema is not implied by the first, i.e. the two schemas are not equivalent. In our system, on the other hand, they *are* equivalent.

To see this, let t be any time and suppose that $\Diamond \Box \Diamond A$ is true at time t. Then $\Box \Diamond A$ is true either at t or at some later time t'. Suppose the latter, i.e. $\Box \Diamond A$ true at t'. This means that $\Diamond A$ is true at t' and at every later time. So at any time you choose that is later than t', A is true either then or at some later time. This means that however far you go into the future, there will always be times at which A is true. Now take any time whatever, say t''. From what we have just said, there must be a time later than t'' at which A is true. Hence $\Diamond A$ is true at t''. In particular, $\Diamond A$ is true at t and every later time, and hence $\Box \Diamond A$ is true at t. We thus have that $\Diamond \Box \Diamond A \rightarrow \Box \Diamond A$ is true at t; and since t was arbitrary, it follows that this schema is valid in our system.

We have now shown that the modal theory determined by interpreting $\Box A$ as 'A is and always will be true' is intermediate between S4 and S5: it is in fact the system known as S4.3.

To summarize this section, we have seen that modal theories can be obtained by adding to PML various new rules of inference (or the corresponding axiom schemas).

Typically, a rule of inference will correspond to some condition on the accessibility relation in interpretations of the logic, so that the axiomatic theory thus obtained is exactly the modal theory of accessibility relations which satisfy that condition.

Thus the rules (trans-$\square$), (refl-$\square$), and (symm-$\square$) correspond to a transitive, reflexive, or symmetric accessibility relation, respectively. The modal theory S4 obtained by adding (trans-$\square$) and (refl-$\square$) to PML is thus the modal theory of a transitive, reflexive relation, while S5, the theory obtained by adding (symm-$\square$) to S4, is the modal theory of an equivalence relation.

In addition, we considered two intuitively plausible ways of interpreting the modal operators informally, in terms of knowledge and time, and we investigated which formal modal theories corresponded to these informal interpretations. In both cases, we concluded that what was required was a modal theory intermediate between S4 and S5.

Exercise 7.4

(1) Use natural deduction to prove that

*(a)　the rule (trans-$\diamond$) can be derived from the rule (trans-$\square$);
　(b)　the rules (refl-$\square$) and (refl-$\diamond$) can each be derived from the other;
　(c)　the rules (symm-$\square$) and (symm-$\diamond$) can each be derived from the other.

*(2) Show that the schema $\square A \rightarrow A$ is valid in a modal frame (W, R) if and only if the relation R is reflexive.

(3) Show that the following are theorems of S4:

*(a)　$\square A \rightarrow \square \diamond \square A$
　(b)　$\square \diamond \square A \rightarrow \square \diamond A \wedge \diamond \square A$
　(c)　$\square \diamond A \vee \diamond \square A \rightarrow \diamond \square \diamond A$
　(d)　$\diamond \square \diamond A \rightarrow \diamond A$

(4) Show that $\diamond \square A \leftrightarrow \square A$ is a theorem of S5.

(5) Show that the schema $\square A \rightarrow \diamond A$ is valid in a modal frame (W, R) if and only if the frame has the property that for every w in W, there is a w' in W accessible from w.

(6) For each of the following informal interpretations of the necessity operator $\square$, state the corresponding interpretation of $\diamond$, and determine, for each of the modal schemas below, whether the schema makes sense under the interpretation, and if it does, whether it is valid.

Schemas

(i)　$\square A \rightarrow A$,
(ii)　$\square A \rightarrow \diamond A$,
(iii)　$\square \diamond A \rightarrow \diamond A$,
(iv)　$\square \square A \rightarrow \square A$,
(v)　$\square A \rightarrow \square \square A$.

Interpretations (readings of $\square A$)

*(a)　'It is obligatory that A'.
　(b)　'It has always been the case that A'
　(c)　'John knows that A'

*(d)   'John believes that A'
 (e)   'It can be proved that A'
 (f)   'It is written in the Bible that A'

7.5   *TEMPORAL LOGIC*

In the last section we looked at a temporal interpretation of the modal operators, by which □A was read 'it is and always will be the case that A' and ◇A as 'it is or will be the case that A'. This is already the beginnings of a temporal logic, but for a fully fledged temporal logic we need to be able to talk about the *past* as well as the future.

In **tense logic**, there are *two* pairs of modal operators, called **tense operators**, F and G for talking about the future, P and H for talking about the past. Informally, we read them as follows:

> FA: 'It will be the case that A'
> GA: 'It will always be the case that A'
>
> PA: 'It has been the case that A'
> HA: 'It has always been the case that A'.

And just as the 'weak' modal operator ◇ can be defined in terms of the 'strong' operator □ and negation, so we stipulate that the equivalences

$$FA \cong \sim G \sim A$$
$$PA \cong \sim H \sim A$$

are to hold; equivalently, we include in our proof system the rules

$$(\text{F-intro}) \quad \frac{\sim G \sim A}{FA} \qquad (\text{F-elim}) \quad \frac{FA}{\sim G \sim A}$$

$$(\text{P-intro}) \quad \frac{\sim H \sim A}{PA} \qquad (\text{P-elim}) \quad \frac{PA}{\sim H \sim A}$$

The temporal interpretations of ◇ and □ discussed in the last section can now be obtained by treating the schemas ◇A and □A as abbreviations for the tense-logical schemas A ∨ FA and A ∧ GA respectively.

The formal semantics for tense logic makes use of **temporal frames**. These are essentially no different from modal frames, except that sometimes it is stipulated that they should always be transitive. In the interests of realism, one might reasonably insist that they be both irreflexive and asymmetric as well; for more on this, see below.

We write $(T, <)$ to denote a temporal frame, and refer to the elements of T as **times** rather than states. The relation $<$ may be read as 'precedes' or 'is earlier than', and we shall sometimes find it convenient to write $t < t'$ as $t' > t$, where $>$ is read 'follows' or 'is later than'.

The semantic rules for the tense operators are as follows.

$\vDash_{T,<,I}$FA[t] if and only if $\vDash_{T,<,I}$A[t'] for some $t' > t$;

$\vDash_{T,<,I}$GA[t] if and only if $\vDash_{T,<,I}$A[t'] for every $t' > t$;

$\vDash_{T,<,I}$PA[t] if and only if $\vDash_{T,<,I}$A[t'] for some $t' < t$;

$\vDash_{T,<,I}$HA[t] if and only if $\vDash_{T,<,I}$A[t'] for every $t' < t$;

Because of the formal similarity of these rules to the rules for the modal operators $\square$ and $\lozenge$, many of the results of modal logic carry over directly into tense logic. For example:

(1) If in any universally valid modal schema, $\lozenge$ and $\square$ are replaced by F and G respectively, then the result is a universally valid tense-logical schema.

(2) If in any universally valid modal schema, $\lozenge$ and $\square$ are replaced by P and H respectively, then the result is a universally valid tense-logical schema.

Thus we can say without further ado that the following schemas are tense-logically valid:

$$H(A \to B) \to (HA \to HB)$$
$$G(A \to B) \to (GA \to GB)$$

$$H(A \to B) \to (PA \to PB)$$
$$G(A \to B) \to (FA \to FB)$$

$$F(A \lor B) \leftrightarrow FA \lor FB$$
$$P(A \lor B) \leftrightarrow PA \lor PB$$

$$H(A \land B) \leftrightarrow HA \land HB$$
$$G(A \land B) \leftrightarrow GA \land GB$$

and we may also affirm the validity of the rules

(G-intro)
$$\dfrac{A}{GA}$$
(where A is any universally valid tense-logical schema)

(H-intro)
$$\dfrac{A}{HA}$$
(where A is any universally valid tense-logical schema)

(KG)
$$\dfrac{G(A \to B) \quad GA}{GB}$$
(KH)
$$\dfrac{H(A \to B) \quad HA}{HB}$$

as well, of course, as the tautology rule.

Derived rules such as $K\lozenge$, op-swap, and $\sim$ op-swap carry over, with appropriate changes, into tense logic too. In the op-swap rules, H is swapped with P, and G with F, so for example from the schema HHGPF $\sim$ A we can infer $\sim$ PPFHGA, and vice versa.

More interesting are the universally valid temporal schemas which essentially involve both past (P, H) and future (F, G) operators, for such schemas have no analogues in modal logic.

Example 7.5.1

The temporal schemas

$$A \rightarrow GPA$$
$$A \rightarrow HFA$$
$$PGA \rightarrow A$$
$$FHA \rightarrow A$$

are all universally valid.

The first schema, for example, says that if something is true now, then it will always be the case that it *has* been true. The fourth says that if it is ever going to be the case that A has always been true then A is true now. (What do the middle two schemas say?)

To prove the first schema, suppose A is true at time t in an interpretation $(T, <, I)$, and let t' be any time in T later than t. Then since $t < t'$, and A is true at t, we have that PA is true at t'. But t' is *any* time later than t, so GPA is true at t.

The other schemas are proved similarly (see Exercise 7.5, Question (1)).

In a natural deduction system for tense logic, the universal validity of these schemas can be represented by the rules

$$(\text{GP-intro}) \quad \frac{A}{GPA} \qquad (\text{PG-elim}) \quad \frac{PGA}{A}$$

$$(\text{HF-intro}) \quad \frac{A}{HFA} \qquad (\text{FH-elim}) \quad \frac{FHA}{A}$$

The elimination rules can be derived from the introduction rules, and vice versa, so only one set is needed—though of course it is convenient to have both at one's disposal in practice (see Exercise 7.5, Question (2)).

The universally valid temporal schemas are valid in all temporal frames, including frames which we should regard as quite inappropriate as models of time. As with modal logic, things become more interesting if we restrict our attention to certain classes of frames, either by stipulating certain additional rules of inference or by placing constraints directly on the 'earlier than' relation.

The latter approach forces us to address the questions of how we want to model time and exactly what, in such a model, we want 'earlier than' to mean. Some of the possibilities are discussed below. It is important to realize that an appropriate answer to the question 'What model of time should we use?' depends on what we want a model of time *for*; the question is therefore quite distinct from the question, 'What is time really like?', which is presumably a matter for physics or philosophy rather than for logic.

Transitive and intransitive models

It might seem that the 'earlier than' relation has to be *transitive*, that is, that if t_1 is earlier than t_2, and t_2 is earlier than t_3, then t_1 has to be earlier than t_3. Formally, a temporal frame $(T, <)$ is *transitive* so long as it has the property

TRANSITIVITY
For all t, t', t'' in T, if $t < t'$ and $t' < t''$ then $t < t''$.

However, it is possible to develop a coherent temporal logic in which time is *not* transitive.

Suppose we represent instants of time by real numbers, instant t being exactly t hours later than some 'zero hour', represented by 0. Now define $t \lhd t'$ to mean that t is between t' and $t' + 1$, i.e. t precedes t' by less than one hour. This relation is not transitive since, for example, we have $0 \lhd 0.6$ and $0.6 \lhd 1.2$, but *not* $0 \lhd 1.2$ (since 1.2 is *more* than one hour later than 0).

How should tense-logical schemas be interpreted in relation to this intransitive frame? The schema PA means 'A has been true within the last hour', HA means 'A was true throughout the last hour', with analogous readings (with 'next' instead of 'last') for FA and GA.

The schema PPA means that A was true at a time within the hour preceding a time within the last hour; an appropriate reading is therefore 'A has been true within the last two hours'. Similarly, GGGA means 'A will be true throughout the next three hours', and so on.

The schema PPA $\wedge$ $\sim$ PA means that A has been true within the last two hours but not within the last hour, i.e. A was true at some time between two hours ago and hour ago. Note that this schema cannot be true on a transitive frame: its negation is part of the temporal theory of transitivity.

The schema PFA means that A is true at some time within the hour following a time within the last hour. A little reflection should serve to convince you that this means 'A is true at some time within an hour on either side of now'.

Transitive models are characterized by the theorems

$$PPA \rightarrow PA$$
$$FFA \rightarrow FA$$
$$HA \rightarrow HHA$$
$$GA \rightarrow GGA$$

which recall the modal transitivity axioms. The corresponding rules of inference are

$$(\text{trans-P}) \quad \left|\frac{PPA}{PA}\right. \qquad (\text{trans-H}) \quad \left|\frac{HA}{HHA}\right.$$

$$(\text{trans-F}) \quad \left|\frac{FFA}{FA}\right. \qquad (\text{trans-G}) \quad \left|\frac{GA}{GGA}\right.$$

We don't need all of these rules, since any one of them suffices to derive the rest. The rules on the left are straightforwardly derivable from the ones on the right using op-swap. The tricky part is deriving the ones on the right from each other. Here we derive (trans-H) from (trans-G):

(1)	HA	(premiss)
(2)	SUBDERIVATION	
	(2.1) FFHA	(assumption)
	(2.2) SUBDERIVATION	
	(2.2.1) $\sim$ FHA	(assumption)
	(2.2.2) G $\sim$ HA	(2.2.1, $\sim$ op-swap)

	(2.2.3)	GG ~ HA	(2.2.2, trans-G)
	(2.2.4)	~ FFHA	(2.2.3, op-swap)
	(2.2.5)	FFHA	(2.1, rep)
(2.3)		~ ~FHA	(2.2, ~ -intro)
(2.4)		FHA	(2.3, ~ -elim)
(3)	FFHA → FHA		(2, → -intro)
(4)	H(FFHA → FHA)		(3, H-intro)
(5)	SUBDERIVATION		
	(5.1)	FHA	(assumption)
	(5.2)	HFFHA	(5.1, HF-intro)
	(5.3)	HFHA	(4, 5.2, KH)
	(5.4)	SUBDERIVATION	
		(5.4.1) FHA	(assumption)
		(5.4.2) A	(5.4.1, FH-elim)
	(5.5)	FHA → A	(5.4, → -intro)
	(5.6)	H(FHA → A)	(5.5, H-intro)
	(5.7)	HA	(5.3, 5.6, KH)
(6)	FHA → HA		(5, → -intro)
(7)	H(FHA → HA)		(6, H-intro)
(8)	HFHA		(1, HF-intro)
(9)	HHA		(7, 8, KH)

Reflexivity and symmetry

It is natural to want the 'earlier than' relation to be neither reflexive nor symmetric: that is, we do not want any time to be earlier than itself, nor do we want two times each of which is earlier than the other. In other words, we want our temporal frame $(T, <)$ to have the properties:

IRREFLEXIVITY
For any t in T, it is not the case that $t < t$.

ASYMMETRY
For no pair of times t, t' in T is it the case that both $t < t'$ and $t' < t$.

Although these are natural requirements for temporal frames, they turn out not to be axiomatizable in tense logic: that is, there are no tense-logical schemas such that the class of frames on which they are valid is precisely the class of irreflexive frames, or precisely the class of asymmetric ones.

The failure of tense logic to capture the notions of asymmetry and irreflexivity is perhaps not as drastic as it sounds, so long as we have transitivity. For consider a transitive frame $(T, <)$ in which there are times t, t' such that $t < t'$ and $t' < t$—so $<$ is neither asymmetric nor, by transitivity, irreflexive. Then we can replace this frame by a new frame $(T_a, <_a)$ as follows:

T_a is just like T except that the times t and t' are replaced by two both-ways infinite sequences $..., t_{-1}, t_0, t_1, ...$ and $t'_{-1}, t'_0, t'_1, ...$ of times, related by

$$\cdots <_a t_{-1} <_a t'_{-1} < t_0 <_a t'_0 < t_1 <_a t'_1 \cdots.$$

For any other time t'' in T, if $t'' < t$ then $t'' <_a t_i$ for each integer i; and likewise with $>$ instead of $<$ or with t', t_i' instead of t, t_i.

Given an interpretation $(T, <, I)$ on the old frame, we can define an interpretation $(T_a, <_a, I_a)$ on the new one by the rules

For each integer i, $t_i \in I_a(A)$ if and only if $t \in I(A)$, and $t_i' \in I_a(A)$ if and only if $t' \in I(A)$.

For every time t'' distinct from t and t', $t'' \in I_a(A)$ if and only if $t'' \in I(A)$.

Then it is not hard to see that $(T, <, I)$ satisfies exactly the same schemas as $(T_a, <_a, I_a)$; and moreover, the latter interpretation is both asymmetric and irreflexive.

Linear and branching time

Usually, we think of time as having a *linear* structure: this is implicit in the way we measure time using real numbers. A temporal frame $(T, <)$ is **linear** so long as it is transitive and has the property

LINEARITY
For all times t, t' in T, exactly one of $t < t', t = t', t' < t$ holds.

To see the force of linearity, observe how it *fails* to hold in a branching structure such as:

Here none of the relations $t < t', t = t'$ and $t' < t$ holds. The linearity condition breaks down similarly in a structure with *parallel* (i.e. disjoint) time lines:

Branching structures may contain both branching into the future, as in

and branching 'into' the past, as in

Branching temporal frames are sometimes used to explain the idea of *alternative possible histories*, for example to model the execution of a non-deterministic algorithm. At any given time, several different futures are possible, and these are represented by different branches into the future; equally, the state of the world at a given time may have arisen from any of several different pasts, which are represented by different branches into the past. The tree of possible moves in noughts and crosses which we looked at in Section 7.1 could be thought of in this way as a branching temporal frame. Most of the branching is into the future, but some lines do converge, as when W_6 and W_7 both lead to W_9.

A temporal frame which allows branching into the future but not into the past is called *left-linear*, the flow of time being conventionally represented in diagrams as moving from left to right across the page. A frame $(T, <)$ is **left-linear** so long as it is transitive and has the property

LEFT-LINEARITY
For all times t, t', t'' in T,
if $t' < t$ and $t'' < t$, then exactly one of $t' < t''$, $t' = t''$, $t'' < t'$ holds.

Left-linearity means that the set of times which precede any given time is linearly ordered: it constitutes the unique *past history* of that time. In the diagram below, for example, the past history of t_3 runs from t_0 through t_1 and t_2 to t_3:

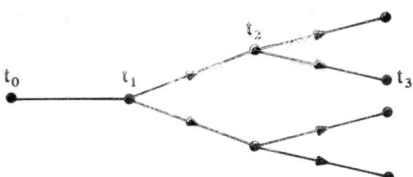

Right-linearity is defined analogously, using the property

RIGHT-LINEARITY
For all times t, t', t'' in T,
if $t' > t$ and $t'' > t$, then exactly one of $t' < t''$, $t' = t''$, $t'' < t'$ holds.

Left-linearity and right-linearity together just fail to imply full linearity, the exception being the case of parallel time lines:

This model is both left-linear and right-linear, but it is not linear, though each of the separate lines individually is.

A linear frame is necessarily both irreflexive and asymmetric, and this is true of left-linear and right-linear frames as well. This means that the tense-logical theory of linearity fails to characterize the irreflexive and asymmetric nature of linear frames; it is in fact identical with the theory of **semi-linearity**, which is defined by the condition

SEMI-LINEARITY
For every t, t' in T, at least one of $t < t', t = t', t > t'$ holds.

Semi-left-linearity and semi-right-linearity can be defined analogously.

Left-linearity, and hence semi-left-linearity, is completely axiomatized by the schema

$$(LL) \quad PA \land PB \to P(A \land B) \lor P(A \land PB) \lor P(PA \land B).$$

This says that if A was true at some past time and B was true at some past time, then either A and B were true together (at the same past time), or A was true at a time when B had already been true, or B was true at a time when A had already been true.

We shall show that a temporal frame on which (LL) is valid is necessarily semi-left-linear. Let $(T, <)$ be such a frame and let t be any time in T. Let t' and t'' be times for which $t' < t$ and $t'' < t$. We shall show that at least one of $t' < t'', t' = t''$, and $t'' < t'$ holds.

Define an interpretation I on $(T, <)$ such that $I(A) = \{t'\}$ and $I(B) = \{t''\}$, i.e. t' and t'' are the only times at which the interpretation satisfies A and B respectively. Since both t' and t'' precede t, we have

$$\vDash_{T, <, I} PA \land PB[t]$$

so by (LL), which must be satisfied at any time in any interpretation based on $(T, <)$, we have

$$\vDash_{T, <, I} P(A \land B) \lor P(A \land PB) \lor P(PA \land B)[t].$$

From the semantic rule for disjunction, we must have at least one of

(a) $\vDash_{T, <, I} P(A \land B)[t]$,
(b) $\vDash_{T, <, I} P(A \land PB)[t]$,
(c) $\vDash_{T, <, I} P(PA \land B)[t]$.

In case (a), we have A and B both satisfied at some time earlier than t. But the only times at which A and B are satisfied are t' and t'' respectively; hence $t' = t''$. In case (b), we have A and PB satisfied at some time earlier than t: it must be t', since A is not satisfied at any other time. Hence B is satisfied at a time earlier than t', which means that $t'' < t'$. Reasoning similarly, we conclude that in case (c), $t' < t''$.

We have thus shown that for any times t', t'' preceding an arbitrary time t, at least one of $t' < t'', t' = t''$, and $t'' < t'$ holds, so the frame $(T, <)$ is semi-left-linear. We have not shown that *exactly* one of these relations holds, so the frame may fall short of full left-linearity. For this we require in addition that the precedence relation on T should be asymmetric, and as we have said, there is no way of imposing this condition by means of a tense-logical axiom.

The converse statement, that (LL) is valid on any semi-left-linear frame (and hence on any left-linear one), is simple to demonstrate, and is left as an exercise for the reader (see Exercise 7.5, Question (3)).

In exactly the same way, the schema

$$(RL) \quad FA \land FB \to F(A \land B) \lor F(A \land FB) \lor F(FA \land B).$$

determines that any frame on which it is valid is semi-right-linear. And the combination of (LL) and (RL) determines semi-linearity, apart from the possibility of the frame's being decomposable into more than one disjoint semi-linear component.

Dense and discrete time

We normally think of time instants as forming a *dense* ordering. That is, between any two instants, there is a third. A temporal frame $(T, <)$ is **dense** if it has the propety

> DENSITY
> For all t, t' in T, if $t < t'$ then there is
> a t'' in T such that $t < t''$ and $t'' < t'$.

The opposite of density is *discreteness*, whereby each time (except the first and last if there is a beginning or end to time) is sandwiched between unique **previous** and **next** times. This is an appropriate model if we want to think of time not as made up of durationless instants but as broken down into a discrete series of *intervals*, e.g. hours or days: each day has a unique preceding day ('yesterday'), and a unique next day ('tomorrow').

A temporal frame $(T, <)$ is **discrete** so long as it has the property

> DISCRETENESS
> For every t, u in T,
> (a)   if $t < u$ then there is t' such that
>   (i) $t < t'$, and
>   (ii)   there is no t'' such that
> $t < t''$ and $t'' < t'$
> (b)   if $u < t$ then there is t' such that
>   (i) $t' < t$, and
>   (ii)   there is no t'' such that
> $t' < t''$ and $t'' < t$.

Part (a) of this property may be called RIGHT-DISCRETENESS, and part (b), LEFT-DISCRETENESS.

An example of a temporal frame which is left-discrete but not right-discrete is the frame $(H, <)$, where H consists of the number 0 together with all fractions of the form $1/n$, where n is a positive integer, and $<$ is the normal 'less than' ordering. This is left-discrete, since each fraction $1/n$ has a unique previous fraction, namely $1/(n + 1)$, and 0 is not preceded by anthing; but it is not right-discrete since 0 is followed by something (in fact by everything else in the set H) but is not *immediately* followed by anything, since between 0 and any fraction $1/n$ there always comes the fraction $1/(n + 1)$.

Note that discreteness implies irreflexivity (why?), and that the definition of density only captures our intuitive idea of density on condition that the ordering is irreflexive. For example, the integers ordered by $\leqslant$ may be regarded as a temporal frame: although intuitively it is discrete, it actually satisfies the density property rather than the discreteness one! Arguably, our intuition of discreteness applies rather to the integers ordered by $<$, which *is* discrete by our criterion. In any case, as we saw with linearity, it is expedient, and not unreasonable, to turn a blind eye towards possible failure of irreflexivity.

An appropriate axiom for density is

$$GGA \rightarrow GA.$$

This is the converse of the transitivity axiom $GA \rightarrow GGA$, and just as the latter axiom implies its past-tense version $HA \rightarrow HHA$, so does the density axiom given here imply $HHA \rightarrow HA$. To show this is left as a (rather difficult) exercise for the reader (Exercise 7.5, Question (4)). Equally, density can be expressed by either of the axioms

$$FA \rightarrow FFA$$
$$PA \rightarrow PPA.$$

Unlike density, however, discreteness cannot be characterized by means of any tense-logical axioms.

The reader who is interested in pursuing temporal logic in a computational context is referred to Galton (1987).

Exercise 7.5

(1) Show that the following schemas are universally valid:

    (a) $A \rightarrow HFA$
  *(b) $PGA \rightarrow A$
    (c) $FHA \rightarrow A$

*(2) Show that the rules PG-elim and FH-elim can be derived from the rules HF-intro and GP-intro respectively.

(3) Show that the schema

$$(LL) \quad PA \wedge PB \rightarrow P(A \wedge B) \vee P(A \wedge PB) \vee P(PA \wedge B)$$

is valid in any left-linear frame.

(4) Show that the schema $HHA \rightarrow HA$ can be derived from the axiom $GGA \rightarrow GA$.

Appendix: Solutions to Selected Exercises

We give here solutions to the asterisked questions in the Exercises.

CHAPTER 1

Exercise 1.3

(1)   (a)   Consistent (e.g. if there are 15 people in the room),
     (b)   inconsistent (since no number is both more than 20 and less than 10).

(2)   (a) Logically false, (b) logically true, (f) logically false (John's wife *must* be married!),
     (g) contingent (could be true if John is a bigamist, but it doesn't have to be true).

Exercise 1.4

(1)   (b)   {There are more than twenty people in the room} entails 'There are not fewer than ten people in the room';

       {There are fewer than ten people in the room} entails 'There are not more than twenty people in the room'.

Exercise 1.5

(1)   (a) Valid and sound, (d) invalid, (f) valid; soundness depends on who John and Mary are.
(2)   (a) Suppose $P \vDash Q$ and $Q \vDash R$. Then P cannot be true without Q being true, and Q cannot be true without R being true. Hence P cannot be true without R, i.e. $P \vDash R$.

Exercise 1.6

(a)   Valid schema: A is in B, B is not in C/A is not in C.
(d)   Invalid schema: A is not in B, B is in C/A is not in C. Example: Change 'Scotland' to 'England'.
(f)   Valid schema: Whenever A, B; not B today/not A today.

CHAPTER 2

Exercise 2.1(b)

(1)

A	B	A ⊕ B
T	*T*	*F*
T	*F*	*T*
F	*T*	*T*
F	*F*	*F*

(2) (a) Purely truth-functional.

(c) Suggests temporal sequence ('and later'), but this is not a necessary interpretation, e.g. Mary might be washing up the lunch things while John makes supper. On balance, 'and' is probably best regarded as purely truth-functional here.

(e) Not truth-functional: 'or' here means something like 'also known as'.

(f) Not exactly truth-functional as it stands, but it can be made so by paraphrasing as 'I insist that either you will eat up your vegetables or you will have no pudding.' The 'or' here is most naturally interpreted as exclusive, i.e. it does not admit the possibility of eating up the vegetables and still getting no pudding (which would be most unfair!).

Exercise 2.2

(1) (b)

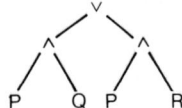

(1) (d)

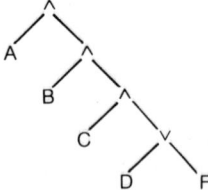

(2) (a) Scope of ∨ : P ∨ (Q ∧ R); scope of ∧ : Q ∧ R
(c) Scope of ∧ : A ∧ B; scope of first ∨ : (A ∧ B) ∨ C; scope of second ∨ : ((A ∧ B) ∨ C) ∨ D.

(3) (a) Depth 0:   (P ∧ Q) ∨ ((R ∧ S) ∨ T)

Depth 1:   P ∧ Q, (R ∧ S) ∨ T

Depth 2:   P, Q, R ∧ S, T

Depth 3:   R, S.

(4) (a) $((A \wedge C) \vee B) \wedge ((B \wedge C) \vee ((A \wedge C) \wedge D))$

(b) $((A \wedge (B \wedge C)) \vee B) \wedge ((B \wedge C) \vee ((A \wedge (B \wedge C)) \wedge D)).$

Exercise 2.3a

(2) 1b, 2b, and 4a remain valid, 3b, 4b, 5a, and 5b become invalid. $A \wedge (B \oplus C)$ logically implies $(A \wedge B) \oplus C$, but not vice versa.

(3) (a) Equivalent to A, i.e. true in each row in which A is true (2 out of 4 rows).

(d) Equivalent to $B \vee C$, i.e. true in each row in which at least one of B and C is true (6 out of 8 rows).

Exercise 2.3b

(2)

$$(A \wedge C) \vee (B \wedge C) \cong (C \wedge A) \vee (C \wedge B) \quad \text{(by 1a)}$$
$$\cong C \wedge (A \vee B) \quad \text{(by 4a)}$$
$$\cong (A \vee B) \wedge C \quad \text{(by 1a)}$$

(4)

$$(A \wedge B) \vee (B \wedge C) \vee (C \wedge A) \cong (A \wedge B) \vee (B \wedge C) \vee (A \wedge C) \quad \text{(by 1a)}$$
$$\cong (A \wedge B) \vee ((B \vee A) \wedge C) \quad \text{(by Question 2)}$$
$$\cong ((A \wedge B) \vee (B \vee A)) \wedge ((A \wedge B) \vee C) \quad \text{(by 4b)}$$
$$\cong (((A \wedge B) \vee B) \vee A) \wedge ((A \wedge B) \vee C) \quad \text{(by 2b)}$$
$$\cong (B \vee A) \wedge ((A \wedge B) \vee C) \quad \text{(by Question 1)}$$
$$\cong (B \vee A) \wedge (A \vee C) \wedge (B \vee C) \quad \text{(by 4c)}$$
$$\cong (A \vee B) \wedge (C \vee A) \wedge (B \vee C) \quad \text{(by 1b)}$$
$$\cong (A \vee B) \wedge (B \vee C) \wedge (C \vee A) \quad \text{(by 1a)}$$

Exercise 2.4a

(1) literal, minterm, maxterm, DNF, CNF; (2) maxterm, DNF, CNF; (4) CNF; (6) none.

Exercise 2.4b

(2) (a) $A \vee (B \wedge C)$; (b) $(A \vee B) \wedge (A \vee C)$

Exercise 2.5

2 (a) Tree diagram:

The truth table has 8 rows; the schema is false in every row except the fourth (i.e., A true, B false, C false).

(c) Tree diagram:

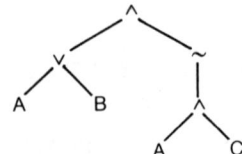

Truth table:

A	B	C	$(A \vee B) \wedge \sim (A \wedge C)$
T	T	T	F
T	T	F	T
T	F	T	F
T	F	F	T
F	T	T	T
F	T	F	T
F	F	T	F
F	F	F	F

(3) (a)

A	B	A|B
T	T	F
T	F	T
F	T	T
F	F	T

(b) Use truth-tables, or, e.g.,

(i) $A|A \cong \sim (A \wedge A)$ (by definition of |)

        $\cong \sim A$ (idempotence of $\wedge$)

(ii) $(A|A)|(B|B) \cong \sim A| \sim B$ (by (i))

            $\cong \sim (\sim A \wedge \sim B)$ (by definition of |)

            $\cong \sim \sim A v \sim \sim B$ (by de Morgan's law)

            $\cong AvB$ (by double negation)

Exercise 2.6

(1) Using truth-tables:

A	B	C	$\sim A \vee B$	$\sim B \vee C$	$\sim A \vee C$
*T	T	T	T	T	T*
T	T	F	T	F	F
T	F	T	F	T	T
T	F	F	F	T	F
*F	T	T	T	T	T*
F	T	F	T	F	T
*F	F	T	T	T	T*
*F	F	F	T	T	T*

(2) (a) Valid, (c) invalid, (e) valid, (g) invalid, (h) valid.

(3) (a) Suppose $A \vee B \vDash C$. If A is true, so is $A \vee B$, hence C is true; therefore $A \vDash C$. Similarly $B \vDash C$. Conversely, suppose $A \vDash C$ and $B \vDash C$. If $A \vee B$ is true then either A is true or B is true. In the former case, C is true (since $A \vDash C$), in the latter case C is true (since $B \vDash C$); so in either case C is true. Hence $A \vee B \vDash C$.
(b) Suppose $A \vDash B$. Then if A is true, B is true, and hence so is $B \vee C$. Similarly, if $A \vDash C$, and A is true, then C is true, and hence so is $B \vee C$. Thus in either case, $B \vee C$ is true, so $A \vDash B \vee C$.
On the other hand, we have $A \vDash B \vee \sim B$ but neither $A \vDash B$ nor $A \vDash \sim B$.

Exercise 2.7

(1) (a) Contingent, (c) tautologous, (d) contradictory.

(3) (b) Suppose (i) A is a tautology and (ii) $A \vDash B$. By (i), A is satisfied by every interpretation; by (ii), B is satisfied by any interpretation that satisfies A, hence by every interpretation. So B is a tautology.

(d) Since $A \vDash A \vee B$, the result is a special case of part (b).

(f) Suppose $A \wedge B$ is a tautology. Then $A \wedge B$ is satisfied by every interpretation. But if an interpretation satisfies $A \wedge B$, it satisfies both A and B. Hence both A and B are satisfied by every interpretation, i.e. A and B are both tautologies.
Suppose, conversely, that A and B are both tautologies. Then A and B are satisfied by every interpretation. But an interpretation that satisfies both A and B also satisfies $A \wedge B$. Hence $A \wedge B$ is satisfied by every interpretation, i.e., $A \wedge B$ is a tautology.

Exercise 2.8

(1) (a)

$$A \leftrightarrow B \cong (A \wedge B) \vee (\sim A \wedge \sim B) \quad \text{(definition of } \leftrightarrow \text{)}$$
$$\cong (A \vee \sim A) \wedge (A \vee \sim B)$$
$$\wedge (B \vee \sim A) \wedge (B \vee \sim B) \quad \text{(distributing } \wedge \text{ over } \vee \text{)}$$
$$\cong (A \vee \sim B) \wedge (B \vee \sim A) \quad \text{(absorption of tautologies)}$$
$$\cong (A \rightarrow B) \wedge (B \rightarrow A) \quad \text{(definition of } \rightarrow \text{)}$$

(or use truth tables)

(f)

$$(A \wedge B) \rightarrow C \cong \sim (A \wedge B) \vee C \quad \text{(definition of } \rightarrow \text{)}$$
$$\cong \sim A \vee \sim B \vee C \quad \text{(de Morgan's law)}$$
$$\cong \sim A \vee (B \rightarrow C) \quad \text{(definition of } \rightarrow \text{)}$$
$$\cong A \rightarrow (B \rightarrow C) \quad \text{(definition of } \rightarrow \text{)}$$

(2) Using truth-table:

(a)

A	B	$A \rightarrow B$	A	B
*T	T	T	T	T*
T	F	F	T	F
F	T	T	F	T
F	F	T	F	F

(d)

A	B	C	A ∨ B	A → C	B → C	C
*T	T	T	T	T	T	T*
T	T	F	T	F	F	F
*T	F	T	T	T	T	T*
T	F	F	T	F	T	F
*F	T	T	T	T	T	T*
F	T	F	T	T	F	F
F	F	T	F	T	T	T
F	F	F	F	T	T	F

(3) (a) Contingent, (d) contingent, (f) tautologous, (h) contradictory, (j) tautologous.

(5) (a) R ∧ ~U → W, U/~W. Invalid.

(b) ~R, F, F → W/~R → W. Valid (note how much more plausible this inference becomes if we modify the conclusion to 'Even if it doesn't rain I will get wet'—this illustrates the subtlety of English usage and its relation to logic).

(g) V ∧ P → C, P, ~C/~V. Valid. Note that if the expression 'this inference' is taken as referring to *this* inference (i.e. the one we are currently discussing), then since it *is* valid, its conclusion is false, and so one of the premisses must be false too. The first and third premisses are clearly true, so the culprit is the second premiss: it has to be false.

Exercise 2.9

(2) (a) (A ∧ ~B) ∨ C, (b) (~A ∧ ~B) ∨ (C ∧ D), (c) **T**.

Exercise 2.10

(1) (a) Well-formed.
(b) Allowable with bracket-dropping but not otherwise.
(e) Ill-formed (we can't have (~ (P_2 ∨ P_0))).
(f) Well-formed.
(j) Allowable with bracket-dropping but not otherwise.
(k) Ill-formed.

(2) (b) ((P ∨ Q) ↔ (Q ∨ P))
(d) ((P ∨ (Q ∧ ~R)) → (P ∨ Q))

(3) (a) Satisfied.   (b) Satisfied.
(d) Falsified.   (f) Satisfied.
(g) Falsified.

(4) (a) S⊨A means that every interpretation which satisfies S also satisfies A, i.e. that no interpretation which satisfies S satisfies ~A, i.e. that no interpretation satisfies S∪{~A}, i.e. that S∪{~A} is unsatisfiable.

(b) Suppose S⊨A → B, and let *I* satisfy S∪{A}. Since *I* satisfies S, and S entails A → B, *I* satisfies A → B. But *I* also satisfies A. Hence *I* satisfies B (since A true and B false makes A→B false). Since I was any interpretation which satisfies $\mathbf{S} \cup \{A\}$, it follows that $\mathbf{S} \cup \{A\} \vDash B$.

Conversely, suppose $\mathbf{S} \cup \{A\} \vDash B$, and let I satisfy $\mathbf{S}$. The only way I can fail to satisfy A→B is by satisfying A while falsifying B. But if I satisfies A then it satisfies $\mathbf{S} \cup \{A\}$, hence also B. So I satisfies A→B, and since I was any interpretation satisfying $\mathbf{S}$, we have $\mathbf{S} \vDash A \to B$ as required.

(5) (a) Unsatisfiable, but not minimally so (since $\{P \to \sim R, P \wedge Q, R\}$ is already unsatisfiable),

(b) minimally unsatisfiable,

(c) satisfiable, hence not minimally unsatisfiable.

If $\mathbf{S}$ is finite and unsatisfiable, keep removing schemas until a satisfiable set is obtained. This is bound to happen, since eventually we reach the empty set, which is satisfiable by default. Put back the last schema removed to obtain an unsatisfiable subset; if this set is not minimally unsatisfiable, repeat the procedure starting with this set, but removing a different schema. This process can't go on for ever, since there are only finitely many schemas to choose from. It will therefore halt eventually at a minimally unsatisfiable set.

If $\mathbf{S}$ is infinite this trick won't work, since we *can* go on removing schemas for ever, and there is no guarantee that any of the sets thereby obtained will be satisfiable. To get the required result, we must show that any unsatisfiable set has a finite unsatisfiable subset (the **compactness theorem** for **Prop**). This is most readily done via a proof system—see Chapter 3.

CHAPTER 3.2

Exercise 3.2

(1) (a)

(1)	P ∧ Q	(premiss)
(2)	R ∧ S	(premiss)
(3)	P	(1, ∧-elim-left)
(4)	S	(2, ∧-elim-right)
(5)	P ∧ S	(3, 4, ∧-intro)

(b)

(1)	P → Q	(premiss)
(2)	P → R	(premiss)
(3)	SUBDERIVATION	
	(3.1) P	(assumption)
	(3.2) Q	(1, 3.1, →-elim)
	(3.3) R	(2, 3.1, →-elim)
	(3.4) Q ∧ R	(3.2, 3.3, ∧-intro)
(4)	P → Q ∧ R	(3, →-intro)

(d)

(1)	SUBDERIVATION	
	(1.1) P ∧ Q	(assumption)
	(1.2) P	(1.1, ∧-elim-left)
	(1.3) ∼P	(premiss)

    (2) ~ (P ∧ Q)   (1, ~-intro)

(g)

    (1) P   (premiss)
    (2) Q ∨ R   (premiss)
    (3) SUBDERIVATION
        (3.1) Q   (assumption)
        (3.2) P ∧ Q   (1, 3.1, ∧-intro)
        (3.3) (P ∧ Q) ∨ (P ∧ R)   (3.2, ∨-intro-right)
    (4) SUBDERIVATION
        (4.1) R   (assumption)
        (4.2) P ∧ R   (1, 4.1, ∧-intro)
        (4.3) (P ∧ Q) ∨ (P ∧ R)   (4.2, ∨-intro-left)
    (5) (P ∧ Q) ∨ (P ∧ R)   (2, 3, 4, ∨-elim)

(i)

    (1) P   (premiss)
    (2) P → Q   (premiss)
    (3) SUBDERIVATION
        (3.1) ~R   (assumption)
        (3.2) Q   (1, 2, →-elim)
        (3.3) ~Q   (premiss)
    (4) ~~R   (3, ~-intro)
    (5) R   (4, ~-elim)

(This example illustrates the principle that from a contradictory set of premisses, anything follows.)

(k)

    (1) (P ∨ Q) ∨ R   (premiss)
    (2) SUBDERIVATION
        (2.1) P ∨ Q   (assumption)
        (2.2) SUBDERIVATION
            (2.2.3) P   (assumption)
            (2.2.4) P ∨ (Q ∨ R)   (2.2.3, ∨-intro-right)
        (2.3) SUBDERIVATION
            (2.3.1) Q   (assumption)
            (2.3.2) Q ∨ R   (2.3.1, ∨-intro-right)
            (2.3.3) P ∨ (Q ∨ R)   (2.3.2, ∨-intro-left)
        (2.4) P ∨ (Q ∨ R)   (2.1, 2.2, 2.3, ∨-elim)
    (3) SUBDERIVATION
        (3.1) R   (assumption)
        (3.2) Q ∨ R   (3.1, ∨-intro-left)
        (3.3) P ∨ (Q ∨ R)   (3.2, ∨-intro-left)
    (4) P ∨ (Q ∨ R)   (1, 2, 3, ∨-elim)

(2) (a)

    (1) SUBDERIVATION
        (1.1) P ∧ ~P   (assumption)
        (1.2) P   (1.1, ∧-elim-left)
        (1.3) ~P   (1.1, ∧-elim-right)
    (2) ~ (P ∧ ~P)   (1, ~-intro)

(c)

(1)   SUBDERIVATION
 (1.1) R   (assumption)
 (1.2) SUBDERIVATION
  (1.2.1) P   (assumption)
  (1.2.2) SUBDERIVATION
   (1.2.2.1) Q   (assumption)
   (1.2.2.2) R   (1, rep)
  (1.2.3) Q→R   (1.2.2, →-intro)
 (1.3) P→(Q→R)   (1.2, →-intro)
(2) R→(P→(Q→R))   (1, →-intro)

(e)

(1) P ∨ ~P   (LEM)
(2) SUBDERIVATION
 (2.1) P   (assumption)
 (2.2) SUBDERIVATION
  (2.2.1) Q   (assumption)
  (2.2.2) P   (2.1, rep)
 (2.3) Q→P   (2.2, →-intro)
 (2.4) (P→Q) ∨ (Q→P) (2.3, ∨-intro-left)
(3) SUBDERIVATION
 (3.1) ~P   (assumption)
 (3.2) SUBDERIVATION
  (3.2.1) P   (assumption)
  (3.2.3) SUBDERIVATION
   (3.2.3.1) ~Q   (assumption)
   (3.2.3.2) P   (3.2.1, rep)
   (3.2.3.3) ~P   (3.1, rep)
  (3.2.3) ~~Q   (3.2.3, ~-intro)
  (3.2.5) Q   (3.2.4, ~-elim)
 (3.3) P→Q   (3.2, →-intro)
 (3.4) (P→Q) ∨ (Q→P) (3.3., ∨-intro-right)
(4) (P→Q) ∨ (Q→P) (1, 2, 3, ∨-elim)

Exercise 3.3

(1) (a)

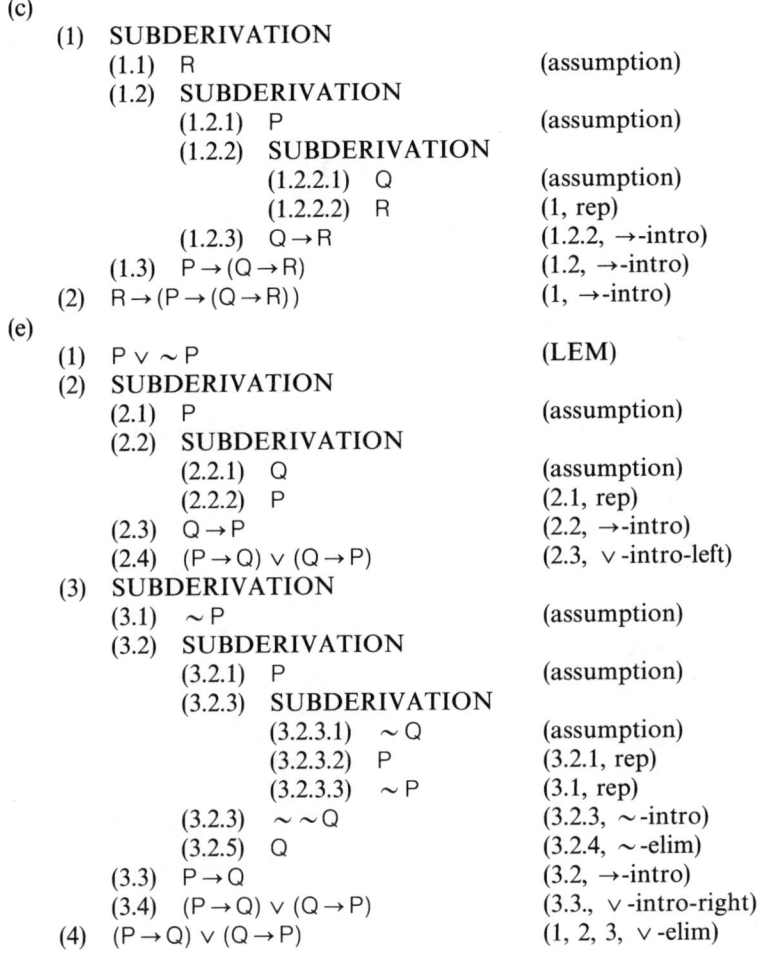

The inference is valid.

(b)

```
                          × (1)   P ∨ Q
                          × (2)   R ∨ S
                          × (3)   ∼ (P ∨ S)
                          ─────────────────── 3
                               (4)   ∼ P
                               (5)   ∼ S
                  ──────────────────────────────── 1
        (6)   P                    (7)   Q
        ══════════ 4         ───────────────── 2
                               (8)  R     (9)  S
                                          ══════════ 5
```

The branch ending at (8) remains open, so the inference is invalid.

(c)

```
                      × (1)   P → Q
                      × (2)   P → R
                      × (3)   ∼ (P → Q ∧ R)
                      ───────────────────────── 3
                         (4)  P
                      × (5)   ∼ (Q ∧ R)
        ──────────────────────────────────────── 1
      (6)   ∼ P                        (7)   Q
      ══════════ 4              ──────────────────────────────── 2
                         (8)   ∼ P                    (9)  R
                         ══════════ 4           ────────────────── 5
                                           (10)   ∼ Q       (11)   ∼ R
                                           ══════════ 7    ══════════ 9
```

The inference is valid.

(f)

```
                          × (1)   P → Q
                            (2)   P
                            (3)   ∼ Q
                            (4)   ∼ R
                      ──────────────────────── 1
                   (5)   ∼ P              (6)   Q
                   ══════════ 2           ══════════ 3
```

The inference is valid.

(i)

```
                          × (1)   P ∨ Q
                          × (2)   R ∨ S
                          × (3)   ∼ (P → S)
                          ─────────────────── 3
                               (4)  P
                               (5)   ∼ S
                  ──────────────────────────────── 1
```

$$\begin{array}{c} \underline{(6) \quad P} \\ \overline{(8) \quad R \qquad (9) \quad S} \\ \end{array} \; 2$$

$$\begin{array}{c} \underline{(7) \quad Q} \\ \overline{(10) \quad R \qquad (11) \quad S} \\ \end{array} \; 2$$

The inference is invalid.

Exercise 3.4

(1) (b) Proof that $A \rightarrow B \vDash \sim (A \wedge \sim B)$:

(1)	$A \rightarrow B$	(premiss)
(2)	**SUBDERIVATION**	
	(2.1) $A \wedge \sim B$	(assumption)
	(2.2) A	(2.1, $\wedge$-elim-left)
	(2.3) B	(1, 2.2, $\rightarrow$-elim)
	(2.4) $\sim B$	(2.1, $\wedge$-elim-right)
(3)	$\sim (A \wedge \sim B)$	(2, $\sim$-intro)

Proof that $\sim (A \wedge \sim B) \vDash A \rightarrow B$:

(1)	$\sim (A \wedge \sim B)$	(premiss)
(2)	**SUBDERIVATION**	
	(2.1) A	(assumption)
	(2.2) **SUBDERIVATION**	
	(2.2.1) $\sim B$	(assumption)
	(2.2.2) $A \wedge \sim B$	(2.1, 2.2.1, $\wedge$-intro)
	(2.2.3) $\sim (A \wedge \sim B)$	(1, rep)
	(2.3) $\sim \sim B$	(2.2, $\sim$-intro)
	(2.4) B	(2.3, $\sim$-elim)
(3)	$A \rightarrow B$	(2, $\rightarrow$-intro)

(2) ($\rightarrow$-elim):

$$\begin{array}{|l} P \\ \sim (P \wedge \sim Q) \\ \hline Q \end{array}$$

Proof:

(1)	P	(premiss)
(2)	**SUBDERIVATION**	
	(2.1) $\sim Q$	(assumption)
	(2.2) $P \wedge \sim Q$	(1, 2.1, $\wedge$-intro)
	(2.3) $\sim (P \wedge \sim Q)$	(premiss)
(3)	$\sim \sim Q$	(2, $\sim$-intro)
(4)	Q	(3, $\sim$-elim)

($\rightarrow$-intro):

$$\begin{array}{|l} \begin{array}{|l} P \\ \hline Q \end{array} \\ \hline \sim (P \wedge \sim Q) \end{array}$$

Proof:

(1) SUBDERIVATION
     (1.1) $P \wedge \sim Q$ (assumption)
     (1.2) P   (1.1, $\wedge$-elim-left)
     (1.3) Q   (rule $P \vdash Q$)
     (1.4) $\sim Q$ (1.1, $\wedge$-elim-right)
(2) $\sim (P \wedge \sim Q)$ (1, $\sim$-intro)

Exercise 3.6

(1) (a) $\{P, R, S\}$, (b) $\{\sim Q, R\}$, (d) P, (e) $\{Q, \sim R, S, T\}$, (f) **T**
(2) Clauses (a), (c), (e), (g), and (h) are Horn clauses.
(3) (a) Unsatisfiable:

$$\sim S \xrightarrow{\sim R \vee S} \sim R \xrightarrow{\sim Q \vee R} \sim Q \xrightarrow{P \vee Q} P \xrightarrow{\sim P \vee S} S \xrightarrow{\sim S} \textbf{F}$$

   (b) Satisfiable:

Let $\mathbf{S} = \{P \vee Q \vee R, \sim S, \sim Q \vee S, \sim P \vee S\}$. Then

$$\mathbf{R(S)} = \mathbf{S} \cup \{P \vee R \vee S, Q \vee R \vee S, \sim Q, \sim P\}$$
$$\mathbf{R^2(S)} = \mathbf{R(S)} \cup \{P \vee R, Q \vee R, R \vee S\}$$
$$\mathbf{R^3(S)} = \mathbf{R^2(S)} \cup \{R\}.$$

No further resolutions are possible, so for all $n > 3$, $\mathbf{R^n(S)} = \mathbf{R^3(S)}$, and since this set does not contain **F**, the original set **S** must be satisfiable.

(4) A suitable sequence is

$$\{\sim P, S\} \xrightarrow{\{\sim S, T\}} \{\sim P, T\} \xrightarrow{\{R, \sim T\}} \{\sim P, R\} \xrightarrow{\{\sim P, \sim R\}} \sim P \xrightarrow{\{P, Q\}} Q \xrightarrow{\sim Q} \textbf{F}$$

(5) In clausal form, we must check the set

$$\{P \vee Q, \sim P \vee R \vee S, \sim R \vee T, \sim R \vee U, \sim U \vee S \vee \sim T, \sim S, \sim Q\}$$

for satisfiability.
   A refutation sequence is

$$
\begin{array}{l}
\sim R \vee T \\
\quad | \; \sim U \vee S \vee \sim T \\
\sim R \vee \sim U \vee S \\
\quad | \; \sim R \vee U \\
\sim R \vee S \\
\quad | \; \sim P \vee R \vee S \\
\sim P \vee S \\
\quad | \; P \vee Q
\end{array}
$$

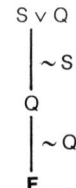

Hence the set of clauses is unsatisfiable, so the inference is valid. (A refutation *tree* would be acceptable here instead of a refutation sequence.)

CHAPTER 4

Exercise 4.1

(1) Schema (d). A, B, C, represent the properties of being a rectangle, square, or oblong, respectively; D represents the property of having unequal diagonals. The inference is valid.

(2) Schema (b). a, b, and c represent Julia, David, and Rosemary, respectively; R represents the relation expressed by 'is taller than'. The inference is valid, but not by virtue of this schema (cf. Section 6.1—the inference is not valid in the predicate calculus, but it is valid in the first-order theory of transitivity).

(5) Schema (c). A represents the property of going (on whatever occasion is being talked about); a and b represent John and Mary respectively. The inference is valid (since if John goes but Mary doesn't, it is not the case that either everybody goes or nobody goes).

Exercise 4.2

(1)

	Referring expression(s)	Predicate
(a)	'Antony'	'...met Carol in Leeds'
	'Carol'	'Antony met...in Leeds'
	'Leeds'	'Antony met Carol in...'
	'Antony', 'Carol'	'...met ____ in Leeds'
	'Antony', 'Leeds'	'...met Carol in ____'
	'Carol', 'Leeds'	'Antony met...in ____'
	'Antony, 'Carol', 'Leeds'	'...met ____ in ---'
(b)	'Charles'	'...lives in the biggest city in England'
	'England'	'Charles lives in the biggest city in...'
	'the biggest city in England'	'Charles lives in...'
	'Charles', 'England'	'...lives in the biggest city in ____'

'Charles', 'the biggest city in England'	'...lives in ____'
(g) 'The author of "The Merchant of Venice"' '"The Merchant of Venice"'	'...was an Englishman' 'The author of...was an Englishman'

(*Note*: it would not be correct to analyse example (g) into a referring expression 'Venice' and a predicate 'The author of "The Merchant of..." was an Englishman'. The statement in (g) does not ascribe any property to Venice!)

(2) (a) 2 is even (true)
    (d) 1 in greater than 5 (false)
    (h) Either 6 is a multiple of 2 or 6 is prime (true)
    (j) 9 is even if and only if 9 is prime (true).

(3) (b) $P(g)$
    (c) $M(f, c)$
    (e) $P(c) \lor G(b, d)$
    (i) $M(h, b) \to P(h) \lor G(h, b)$.

Exercise 4.3

(1) We need only translate the schemas (a)–(d) of which the inferences are interpretations, for example:

$$\text{(a)} \quad \frac{\exists x (A(x) \land B(x))}{\sim \exists x (C(x) \land B(x))}$$
$$\sim \exists x (A(x) \land C(x))$$

$$\text{(c)} \quad \frac{\forall x \sim A(x) \lor \forall x A(x)}{A(a) \to \sim A(b)}$$
$$\sim A(a)$$

(3) (a) (i) $\sim \exists x (P(x) \land \sim \exists y (Q(y) \land R(x, y)))$
        (ii) $\forall x (P(x) \to \sim \forall y (Q(y) \to \sim R(x, y)))$
    (b) (i) $\exists x \sim \exists y (F(x, y) \lor G(x, y))$
        (ii) $\sim \forall x \sim \forall y (\sim F(x, y) \land \sim G(x, y))$

(4) (a) Use m for 'Mary', $P(\_)$ for '____ is a professor', and $A(\_\_\_\_, ...)$ for '____ admires...':

$$\forall x (P(x) \to A(m, x)).$$

    (c) $A(m, m)$ (schematic letters as in (a))
    (e) Use $L(\_)$ for '____ is a lecture', $S(\_)$ for '____ is a student', and $A(\_\_\_\_, ...)$ for '____ attended...'. Four equally acceptable schemas:

$$\sim \exists x (L(x) \land \forall y (S(y) \to A(y, x)))$$
$$\forall x \sim (L(x) \land \forall y (S(y) \to A(y, x)))$$
$$\forall x (L(x) \to \sim \forall y (S(y) \to A(y, x)))$$
$$\forall x (L(x) \to \exists y (S(y) \land \sim A(y, x)))$$

    (f) With schematic letters as in (e),

$$\sim \exists x (L(x) \land \exists y (S(y) \land A(y, x)))$$

(plus variations along the lines of those given for (e)).

    (h) Use $L(\_)$ for '_ is a language', e for 'English', $M(\_\_\_\_, ...)$ for '____ has more words than...':

$$\sim \exists x(L(x) \wedge M(x, e)).$$

(i)   Use d for 'Daniel' and A(___,...) for '___ admires...':

$$\forall x(A(x, d) \rightarrow A(d, x)).$$

(l)   Use E(_) for '_ is an even number', P(_) for '_ is prime', G(___,...) for '___ is greater than...', and a for 'two'. Two equally acceptable schemas are

$$\sim \exists x(E(x) \wedge G(x, a) \wedge P(x))$$
$$\forall x(E(x) \wedge G(x, a) \rightarrow \sim P(x))$$

(n)   Use C(_) for '_ is a cake', M(___,...) for '___ makes...', H(_) for '___ will be hungry'. Two acceptable schemas are

$$\exists x \exists y(C(y) \wedge M(x, y)) \rightarrow \sim \exists x H(x)$$
$$\forall x \forall y \forall z(C(y) \wedge M(x, y) \rightarrow \sim H(z))$$

(o)   Use C(_) and M(___,...) as in (n), E(___,...) for '___ will eat...', and a for 'John'. Two acceptable schemas are

$$\forall x \forall y(C(y) \wedge M(x, y) \rightarrow E(a, y))$$
$$\forall y(C(y) \wedge \exists x M(x, y) \rightarrow E(a, y))$$

(s)   Use A(_) for '_ is an adult', C(_) for '_ is a child', S(___,...) for '___ has seen...':

$$\exists x(A(x) \wedge \sim \exists y(C(y) \wedge \forall z(S(x, z) \rightarrow S(y, z))))$$

(That is, there is at least one adult who can truthfully say 'No child has seen everything that I've seen'. Alternative readings are, perhaps, possible.)

Exercise 4.4

(1) (b)   'The biggest city in England' can be decomposed into the function expression 'The biggest city in...' and the proper name 'England'. Schema: $P(a, f(b))$.

(d)   'Ben's mother' is 'The mother of...' applied to 'Ben'; 'Frances's brother' is 'The brother of...' applied to 'Frances'. Schema: $R(f(a), g(b))$.

(2) (a)   The empty set is not a member of itself (i.e., $\emptyset \notin \emptyset$).

(c)   Every set is equal to the union of itself with the empty set (i.e. for any set X, $X = X \cup \emptyset$).

(e)   Set union is commutative, i.e. for any sets X and Y, $X \cup Y = Y \cup X$.

(3) (b) $\sim \exists x M(x, a)$ *or* $\forall x \sim M(x, a)$.

(d) $\forall x \forall y \forall z(M(x, y) \wedge M(x, z) \rightarrow M(x, i(y, z)))$.

Exercise 4.5

(1) (a)   correct

(b)   incorrect (no quantifier binding x)

(c)   incorrect (brackets not needed round G(x, a) and y not binding anything)

(f)   correct

(g)   incorrect (function symbols used as predicate letters)

(j)   incorrect (function symbol used as variable).

(2)   simple schemas: $P(a), P(b), Q(b, c)$

Predicate: $Q(b, \_)$

Schema: $\forall y Q(b, y)$

Schemas: $P(b) \wedge \forall y Q(b, y), P(a) \wedge \forall y Q(b, y)$

Predicate: $P(\_) \wedge \forall y Q(\_, y)$

Schema: $\exists x (P(x) \wedge \forall y Q(x, y))$

Schema: $\exists x (P(x) \wedge \forall y Q(x, y)) \vee (P(a) \wedge \forall y Q(b, y))$.

(3)   (a)   Unary: $P(\_\_\_, b), P(a, \_\_\_)$

Binary: $P(\_\_\_, \ldots)$

(b)   Unary: $P(\_\_\_) \to R(a, b), P(a) \to R(\_\_\_, b), P(\_\_\_) \to R(\_\_\_, b),$
$P(a) \to R(a, \_\_\_)$

Binary: $P(\_\_\_) \to R(a, \ldots), P(\_\_\_) \to R(\_\_\_, \ldots)$

Ternary: $P(\_\_\_) \to R(\ldots, ---)$

(c)   Unary: $P(\_\_\_) \to \exists x R(a, x), P(a) \to \exists x R(\_\_\_, x),$
$P(\_\_\_) \to \exists x R(\_\_\_, x)$

Binary: $P(\_\_\_) \to \exists x R(\ldots, x)$

(4)   (a)   Unary: $f(\_\_\_, b), f(a, \_\_\_), f(\_\_\_, \_\_\_)$

Binary: $f(\_\_\_, \ldots)$

(d)   Unary: $g(f(\_\_\_), h(f(b))), g(f(a), h(f(\_\_\_))), g(f(\_\_\_), h(f(\_\_\_)))$

Binary: $g(f(\_\_\_), h(f(\ldots)))$

Exercise 4.6

(1)   (a)   'a' is an English word (true).

(e)   'he' is an English word (true).

(g)   'athe' is an English word (false).

(k)   'ehe' is an English Word (false).

(l)   're' is a substring of 'the' (false).

(o)   Every string is a substring of itself (true).

(q)   Some string is a substring of the string obtained by applying to it the rule 'remove the first letter if there is more than one letter in the string' (true—because of one-letter strings).

(u)   There is a pair of English words such that if you concatenate them in either order the result is an English word (true—e.g. 'book' and 'case').

(2)   (a)   0 is even (true).

(e)   The square of 0 is even (true).

(g)   The sum of 0 and 1 is even (false).

(k)   The sum of the squares of 3 and 1 is even (true).

(l)   3 divides 1 (false).

(o)   Every natural number divides itself (true).

(q)   Some natural number divides its own square (true).

(u)   There is a pair of even numbers such that on adding them in either order an even number results (true).

(The oddity of (u) and (v) arises from the fact that in this interpretation $g(x, y)$ is always equal to $g(y, x)$)

(3)   (b) $\sim \forall x Q(f(x), x)$

(d) $\forall x(\sim P(x) \rightarrow \sim P(f(x)))$
(f) $\forall x \forall y (P(g(x,y)) \rightarrow P(g(f(x),f(y))))$

Exercise 4.7

(1) (a)   Satisfiable; e.g., put $D = \{0,1\}, I(P) = \{(0,1),(1,0)\}$.
    (b)   Unsatisfiable. By the first schema, something is P to everything, hence to itself, contradicting the second schema.
(2) (a)   Logically true, (b) neither, (c) logically false.
(3) (a)   The intersection of **M**(A) and **M**(B)
    (b)   The set-difference of **M**(A) and **M**(B), i.e. the set of elements of the former set which are not elements of the latter.

CHAPTER 5

Exercise 5.1

(1) (1) $\exists x \sim F(x)$ (premiss)
    (2)   SUBDERIVATION
          (2.1) $\sim F(a)$ (assumption)
          (2.2)   SUBDERIVATION
                  (2.2.1) $\forall x F(x)$ (assumption)
                  (2.2.2) $F(a)$ (2.2.1, $\forall$-elim: x/a)
                  (2.2.3) $\sim F(a)$ (2.1, rep)
          (2.3) $\sim \forall x F(x)$ (2.2, $\sim$-intro)
    (3) $\sim \forall x F(x)$ (1, 2, $\exists$-elim)
(2) (a) $\forall$-intro not possible at (2.2) (a in undischarged assumption)
    (c) $\exists$-elim not possible at 4 (a in premiss).
(3) (a)   (1) $\forall x(P(x) \rightarrow Q(x))$ (premiss)
          (2) $P(a)$ (premiss)
          (3) $P(a) \rightarrow Q(a)$ (1, $\forall$-elim: x/a)
          (4) $Q(a)$ (2, 3, $\rightarrow$-elim)
          (5) $\exists x Q(x)$ (4, $\exists$-intro: x/a)
    (b)   (1) $\exists x(P(x) \wedge Q(x))$ (premiss)
          (2) $\forall x(Q(x) \rightarrow R(x))$ (premiss)
          (3)   SUBDERIVATION
                (3.1) $P(a) \wedge Q(a)$ (assumption)
                (3.2) $P(a)$ (3.1, $\wedge$-elim-left)
                (3.3) $Q(a)$ (3.1, $\wedge$-elim-right)
                (3.4) $Q(a) \rightarrow R(a)$ (2, $\forall$-elim: a/x)
                (3.5) $R(a)$ (3.3, 3.4, $\rightarrow$-elim)
                (3.6) $P(a) \wedge R(a)$ (3.2, 3.5, $\wedge$-intro)
                (3.7) $\exists x(P(x) \wedge R(x))$ (3.6, $\exists$-intro: x/a)
          (4) $\exists x(P(x) \wedge R(x))$ (1, 3, $\exists$-elim: a/x)
    (d)   (1) $\exists x P(x) \rightarrow Q(a)$ (premiss)
          (2)   SUBDERIVATION
                (2.1) $P(b)$ (assumption)
                (2.2) $\exists x P(x)$ (2.1, $\exists$-intro: x/b)
                (2.3) $Q(a)$ (1, 2.2, $\rightarrow$-elim)

(3) $P(b) \rightarrow Q(a)$ $(2, \rightarrow\text{-intro})$
(4) $\forall x(P(x) \rightarrow Q(a))$ $(3, \forall\text{-intro: } x/b)$

(g) (1) $\forall x \exists y(P(x) \rightarrow R(x, y))$ (premiss)
 (2) **SUBDERIVATION**
  (2.1) $P(a)$ (assumption)
  (2.2) $\exists y(P(a) \rightarrow R(a, y))$ $(1, \forall\text{-elim: } x/a)$
  (2.3) **SUBDERIVATION**
   (2.3.1) $P(a) \rightarrow R(a, b)$ (assumption)
   (2.3.2) $R(a, b)$ $(2.1, 2.3.1, \rightarrow\text{-elim})$
   (2.3.3) $\exists y R(a, y)$ $(2.3.2, \exists\text{-intro: } y/b)$
  (2.4) $\exists y R(a, y)$ $(2.2, 2.3, \exists\text{-elim: } y/b)$
 (3) $P(a) \rightarrow \exists y R(a, y)$ $(2, \rightarrow\text{-intro})$
 (4) $\forall x(P(x) \rightarrow \exists y R(a, y))$ $(3, \forall\text{-intro: } x/a)$

(h) (1) $\forall x(P(x) \rightarrow \exists y R(x, y))$ (premiss)
 (2) $P(a) \rightarrow \exists y R(a, y)$ $(1, \forall\text{-elim: } x/a)$
 (3) $P(a) \vee \sim P(a)$ (LEM)
 (4) **SUBDERIVATION**
  (4.1) $P(a)$ (assumption)
  (4.2) $\exists y R(a, y)$ $(2, 4.1, \rightarrow\text{-elim})$
  (4.3) **SUBDERIVATION**
   (4.3.1) $R(a, b)$ (assumption)
   (4.3.2) **SUBDERIVATION**
    (4.3.2.1) $P(a)$ (assumption)
    (4.3.2.2) $R(a, b)$ $(4.3.1, \text{rep})$
   (4.3.3) $P(a) \rightarrow R(a, b)$ $(4.3.2, \rightarrow\text{-intro})$
   (4.3.4) $\exists y(P(a) \rightarrow R(a, y))$ $(4.3.3, \exists\text{-intro: } y/b)$
  (4.4) $\exists y(P(a) \rightarrow R(a, y))$ $(4.2, 4.3, \exists\text{-elim: } y/b)$
 (5) **SUBDERIVATION**
  (5.1) $\sim P(a)$ (assumption)
  (5.2) **SUBDERIVATION**
   (5.2.1) $P(a)$ (assumption)
   (5.2.2) **SUBDERIVATION**
    (5.2.2.1) $\sim R(a, b)$ (assumption)
    (5.2.2.2) $\sim P(a)$ $(5.1, \text{rep})$
    (5.2.2.3) $P(a)$ $(5.2.1, \text{rep})$
   (5.2.3) $\sim \sim R(a, b)$ $(5.2.2, \sim\text{-intro})$
   (5.2.4) $R(a, b)$ $(5.2.3, \sim\text{-elim})$
  (5.3) $P(a) \rightarrow R(a, b)$ $(5.2, \rightarrow\text{-intro})$
  (5.4) $\exists y(P(a) \rightarrow R(a, y))$ $(5.3, \exists\text{-intro: } y/b)$
 (6) $\exists y(P(a) \rightarrow R(a, y))$ $(3, 4, 5, \vee\text{-elim})$
 (7) $\forall x \exists y(P(x) \rightarrow R(x, y))$ $(6, \forall\text{-intro: } x/a)$

(5) (a)(1) **SUBDERIVATION**
  (1.1) $\forall x R(x, a)$ (assumption)
  (1.2) $R(a, a)$ $(1.1, \forall\text{-elim: } x/a)$
 (2) $\forall x R(x, a) \rightarrow R(a, a)$ $(1, \rightarrow\text{-elim})$
 (3) $\forall y(\forall x R(x, y) \rightarrow R(y, y))$ $(2, \forall\text{-intro: } y/a)$
 (d)(1) $\exists z Q(a, z) \vee \sim \exists z Q(a, z)$ (LEM)
 (2) **SUBDERIVATION**
  (2.1) $\exists z Q(a, z)$ (assumption)

(2.2) SUBDERIVATION
      (2.2.1) $Q(a, b)$ (assumption)
      (2.2.2) $Q(a, b) \vee \forall z \sim Q(a, z)$ (2.2.1, $\vee$-intro-right)
      (2.2.3) $\exists y(Q(a, y) \vee \forall z \sim Q(a, z))$ (2.2.2, $\exists$-intro: y/b)
  (2.3) $\exists y(Q(a, y) \vee \forall z \sim Q(a, z))$ (2.2, $\exists$-elim: z/b)

(3) SUBDERIVATION
  (3.1) $\sim \exists z Q(a, z)$ (assumption)
  (3.2) $\forall z \sim Q(a, z)$ (3.1, $\sim$q-swap)
  (3.3) $Q(a, b) \vee \forall z \sim Q(a, z)$ (3.2, $\vee$-intro-left)
  (3.4) $\exists y(Q(a, y) \vee \forall z \sim Q(a, z))$ (3.3, $\exists$-intro: y/b)
(4) $\exists y(Q(a, y) \vee \forall z \sim Q(a, z))$ (1, 2, 3, $\vee$-elim)
(5) $\forall x \exists y(Q(x, y) \vee \forall z \sim Q(x, z))$ (4, $\forall$-intro: x/a)

Exercise 5.2

(a)

(b)

(d)

$\times$ (1) $\exists x P(x) \to Q(a)$
$\times$ (2) $\sim \forall x(P(x) \to Q(a))$
——————————————— 2
$\times$ (3) $\exists x \sim (P(x) \to Q(a))$
——————————————— 3[x/b]
$\times$ (4) $\sim (P(b) \to Q(a))$
——————————————— 4
(5) $P(b)$
(6) $\sim Q(a)$
——————————————————— 1
$\times$ (7) $\sim \exists x P(x)$ (8) $Q(a)$
——————————— 7   ======== 6
(8) $\forall x \sim P(x)$
——————————— 8[b/x]
(9) $\sim P(b)$
========== 5

(g)

(1) $\forall x \exists y (P(x) \to R(x, y))$
$\times$ (2) $\sim \forall x(P(x) \to \exists y R(x, y))$
——————————————————— 2
$\times$ (3) $\exists x \sim (P(x) \to \exists y R(x, y))$
——————————————————— 3[a/x]
$\times$ (4) $\sim (P(a) \to \exists y R(a, y))$
——————————————————— 4
(5) $P(a)$
$\times$ (6) $\sim \exists y R(a, y)$
——————————————————— 1[a/x]
$\times$ (7) $\exists y (P(a) \to R(a, y))$
——————————————————— 7[b/y]
$\times$ (8) $P(a) \to R(a, b)$
——————————————————————— 8
(9) $\sim P(a)$ (10) $R(a, b)$
======== 5   ——————————— 6
           (11) $\forall y \sim R(a, y)$
           ——————————— 11[b/y]
           (12) $\sim R(a, b)$
           ============ 10

(h)

(1) $\forall x(P(x) \to \exists y R(x, y))$
$\times$ (2) $\sim \forall x \exists y (P(x) \to R(x, y))$
——————————————————— 2
$\times$ (3) $\exists x \sim \exists y (P(x) \to R(x, y))$
——————————————————— 3[a/x]
$\times$ (4) $\sim \exists y (P(a) \to R(a, y))$
——————————————————— 4
(5) $\forall y \sim (P(a) \to R(a, y))$
——————————————————— 1[a/x]
$\times$ (6) $P(a) \to \exists y R(a, y)$
————————————————————————— 6

(7) $\sim P(a)$
—————————————— 5[a/y]
(9) $\sim (P(a) \to R(a, a))$
—————————————— 9
(10) $P(a)$
(11) $\sim R(a, a)$
══════════ 7, 10

× (8) $\exists y R(a, y)$
—————————————— 8[b/x]
(12) $R(a, b)$
—————————————— 5[b/y]
× (13) $\sim (P(a) \to R(a, b))$
—————————————— 13
(14) $P(a)$
(15) $\sim R(a, b)$
══════════ 12

3   (a)

(1) $\forall x \exists y R(x, y)$
× (2) $\sim \exists x \forall y R(x, y)$
—————————————— 2
(3) $\forall x \sim \forall y R(x, y)$

The tree is complete but still open, so the inference has been successfully invalidated. However, no model has been constructed for schemas 1 and 2.

(c)

(1) $\forall x(P(x) \to \exists y Q(x, y))$
× (2) $\exists y \sim Q(a, y)$
× (3) $\sim \sim P(a)$
—————————————— 3
(4) $P(a)$
—————————————— 1[x/a]
× (5) $P(a) \to \exists y Q(a, y)$
—————————————— 5

(6) $\sim P(a)$
══════════ 4

× (7) $\exists y Q(a, y)$
—————————————— 2
(8) $\sim Q(a, b)$
—————————————— 7
(9) $Q(a, c)$
—————————————— 1
(10) $P(b) \to \exists y Q(b, y)$
⋮

The tree can be continued indefinitely, since each time schema 1 is activated, a new existentially quantified schema is produced, which allows the introduction of a new constant; this in turn allows further activations of schema 1. At no stage is a contradiction generated.

Exercise 5.3

(1)   (a)   Unifiable; mgu $= \{x/g(a)\}$
      (b)   Not unifiable
      (c)   Not unifiable
      (f)   Unifiable; mgu $= \{x/f(a, y), z/f(a, f(a, y))\}$
      (h)   Not unifiable
(2)   (a)   The set of clauses to refute is:

          (1) $\{\sim P(x), Q(x)\}$
          (2) $\{\sim Q(y), R(y)\}$

(3) $\{P(a)\}$
(4) $\{\sim R(a)\}$

Refutation sequence:

$$\{P(a)\} \xrightarrow{1,\{x/a\}} \{Q(a)\} \xrightarrow{2,\{y/a\}} \{R(a)\} \xrightarrow{4} \mathbf{F}$$

(b) The set of clauses to refute is:

(1) $\{P(a, w)\}$
(2) $\{\sim P(x, y), Q(x, y)\}$
(3) $\{\sim Q(z, f(z))\}$

Refutation sequence:

$$\{\sim Q(z, f(z))\} \xrightarrow{2,\{x/z, y/f(z)\}} \{\sim P(z, f(z))\} \xrightarrow{1,\{w/f(a), z/a\}} \mathbf{F}$$

(d) The set of clauses to refute is:

(1) $\{\sim R(x, y), \sim R(y, z), R(x, z)\}$
(2) $\{\sim R(u, v), R(v, u)\}$
(3) $\{R(w, f(w))\}$
(4) $\{\sim R(a, a)\}$

Refutation sequence:

$$\{\sim R(a, a)\}$$

$$\Big| \quad 1, \{x/a, z/a\}$$

$$\{\sim R(a, y), \sim R(y, a)\}$$

$$\Big| \quad 3, \{w/a, y/f(a)\}$$

$$\{\sim R(f(a), a)\}$$

$$\Big| \quad 2, \{u/a, v/f(a)\}$$

$$\{\sim R(a, f(a))\}$$

$$\Big| \quad 3, \{w/a\}$$

$$\mathbf{F}$$

(3) (b) Set of clauses to refute is:

(1) $\{P(a)\}$
(2) $\{Q(a)\}$
(3) $\{\sim Q(x), R(x)\}$
(4) $\{\sim P(x), \sim R(x)\}$

Refutation sequence:

$$\{\sim P(x), \sim R(x)\} \xrightarrow{1,\{x/a\}} \{\sim R(a)\} \xrightarrow{3,\{x/a\}} \{\sim Q(a)\} \xrightarrow{2} \mathbf{F}$$

(d) Set of clauses to refute is:

(1) $\{\sim P(x), Q(a)\}$
(2) $\{P(b)\}$
(3) $\{\sim Q(a)\}$

Refutation sequence:

$$\{\sim Q(a)\} \xrightarrow{\ 1\ } \{\sim P(x)\} \xrightarrow{\ 2,\{x/b\}\ } \mathbf{F}$$

CHAPTER 6

Exercise 6.1

(1) (a) Transitive only
    (b) Reflexive and transitive
    (d) Reflexive and symmetric
    (f) Symmetric only
    (g) Reflexive, symmetric, and transitive
    (h) None.

(2) (a) Valid, (b) invalid, (d) valid, (g) valid, (h) invalid.

(3) (a) Suppose $\sim \forall x \forall y (R(x,y) \to \sim R(y,x))$ is satisfied in some interpretation. This is equivalent to $\exists x \exists y (R(x,y) \wedge R(y,x))$, so we can introduce constants a and b such that $R(a,b) \wedge R(b,a)$ is satisfied. If axiom (T) holds, then $R(a,a)$ must be satisfied, so axiom (I) does not hold. Hence if both axioms are satisfied in an interpretation, so must $\forall x \forall y (R(x,y) \to R(y,x))$ be.

(Of course, a formal derivation of the schema from the axioms would also be a good answer.)

Exercise 6.2

(1) (a) (1) $a = f(b)$ (premiss)
        (2) $b = f(c)$ (premiss)
        (3) $a = f(f(c))$ (2, 1, =-elim)
        (4) $\forall x (f(f(x)) = x)$ (premiss)
        (5) $f(f(c)) = c$ (4, $\forall$-elim: x/c)
        (6) $a = c$ (3, 5, =-trans)

    (b) (1) $\exists x \forall y (x = y)$ (premiss)
        (2)   SUBDERIVATION
            (2.1) $\forall y (a = y)$ (assumption)
            (2.2) $a = b$ (2.1, $\forall$-elim: y/b)
            (2.3) $a = c$ (2.1, $\forall$-elim: y/c)
            (2.4) $b = c$ (2.2, 2.3, =-elim)
            (2.5) $\forall x \forall y (x = y)$ (2.4, $\forall$-intro: x/b, y/c)
        (3) $\forall x \forall y (x = y)$ (1, 2, $\exists$-elim: x/a)

(This result shows us that if one thing is identical to everything, then all things are identical, i.e. there is only one thing.)

    (d) (1) $\forall x \forall y \exists z (f(x) = z \wedge g(y) = z)$ (premiss)
        (2) $\exists z (f(g(a)) = z \wedge g(a) = z)$ (1, $\forall$-elim: x/g(a), y/a)
        (3)   SUBDERIVATION
            (3.1) $f(g(a)) = b \wedge g(a) = b$ (assumption)
            (3.2) $f(g(a)) = b$ (3.1, $\wedge$-elim-left)

		(3.3) $g(a) = b$	(3.1, $\wedge$-elim-right)
		(3.4) $f(g(a)) = g(a)$	(3.3, 3.2, $=$-elim-right)
		(3.5) $\exists x(f(x) = x)$	(3.4, $\exists$-intro: x/g(a))
	(4)	$\exists x(f(x) = x)$	(2, 3, $\exists$-elim: z/b)

(2) (c) (1) **SUBDERIVATION**

(1.1)	$F(a)$	(assumption)
(1.2)	$a = a$	($=$-intro)
(1.3)	$a = a \wedge F(a)$	(1.2, 1.1, $\wedge$-intro)
(1.4)	$\exists y(y = a \wedge F(y))$	(1.3, $\exists$-intro: y/a)

(2) **SUBDERIVATION**

(2.1)	$\exists y(y = a \wedge F(y))$	(assumption)

    (2.2) **SUBDERIVATION**

(2.2.1)	$b = a \wedge F(b)$	(assumption)
(2.2.2)	$b = a$	(2.2.1, $\wedge$-elim-left)
(2.2.3)	$F(b)$	(2.2.1, $\wedge$-elim-right)
(2.2.4)	$F(a)$	(2.2.2, 2.2.3, $=$-elim)

(2.3)	$F(a)$	(2.1, 2.2, $\exists$-elim: y/b)
(3)	$F(a) \leftrightarrow \exists y(y = a \wedge F(y))$	(1, 2, $\leftrightarrow$-intro)
(4)	$\forall x(F(x) \leftrightarrow \exists y(y = x \wedge F(y)))$	(3, $\forall$-intro: x/a)

(3) (a) A suitable set of axioms is

(F1) $\forall x(L(x) \rightarrow S(x))$
(F2) $\forall x(S(x) \rightarrow S(n(x)))$
(F3) $\forall x(S(x) \wedge S(y) \rightarrow S(k(x, y)))$
(F4) $\forall x(S(x) \wedge S(y) \rightarrow S(a(x, y)))$
(F5) $\forall x(S(x) \wedge S(y) \rightarrow S(c(x, y)))$
(F6) $\forall x(S(x) \wedge S(y) \rightarrow S(e(x, y)))$
(F7) $\forall x(S(x) \rightarrow L(x) \vee$
$$\exists y(S(y) \wedge x = n(y))$$
$$\exists y \exists z(S(y) \wedge S(z) \wedge$$
$$x = k(y, z) \vee x = a(y, z) \vee x = c(y, z) \vee x = e(y, z)))$$

(Axiom F7 is required to ensure that nothing is a schema except as a consequence of the other axioms.)

(b) So long as we assume the number of schematic letters is finite, we can get what we want as follows:

  (i) there is an interpretation which falsifies every schematic letter;

  (ii) for each schematic letter, there is an interpretation which satisfies that letter and no others;

  (iii) for any two interpretations, there is an interpretation which satisfies a schematic letter if and only if it is satisfied by at least one of the two given interpretations.

These conditions are expressed by the following axioms:

$$(I1) \quad \exists x(I(x) \wedge \forall y(L(y) \rightarrow \sim T(y, x)))$$
$$(I2) \quad \forall x(L(x) \rightarrow \exists y(I(y) \wedge \forall z(L(z) \rightarrow (T(z, x) \leftrightarrow z = y))))$$
$$(I3) \quad \forall x \forall y(I(x) \wedge I(y) \rightarrow \exists z(I(z) \wedge \forall w(L(w) \rightarrow (T(w, z) \leftrightarrow T(w, x) \vee T(w, y)))))$$

Exercise 6.3a

(1) Assume we can derive $\Phi(s(a))$ from $\Phi(a)$. Then we have:

  (1) $\Phi(0)$ (premiss)

(2) SUBDERIVATION
    (2.1) $\Phi(a)$ (assumption)
         $\vdots$
    (2.2) $\Phi(s(a))$ (2.1, assumed derivation)
(3) $\Phi(a) \rightarrow \Phi(s(a))$ (2, $\rightarrow$-intro)
(4) $\forall x(\Phi(x) \rightarrow \Phi(s(x)))$ (3, $\forall$-intro: x/a)
(5) $\Phi(0) \wedge \forall x(\Phi(x) \rightarrow \Phi(s(x)))$ (1, 4, $\wedge$-intro)
(6) $(\Phi(0) \wedge \forall x(\Phi(x) \rightarrow \Phi(s(x)))) \rightarrow \forall x \Phi(x)$ (S3)
(7) $\forall x \Phi(x)$ (5, 6, $\rightarrow$-elim)

Exercise 6.3b

(1) Proof of (MT1):

    (1) $0^*0 = 0$ (M1: x/0)
    (2) SUBDERIVATION
        (2.1) $0^*a = 0$ (assumption)
        (2.2) $0^*s(a) = 0^*a + 0$ (M2: x/0, y/a)
        (2.3) $0^*s(a) = 0 + 0$ (2.1, 2.2, =-elim)
        (2.4) $0 + 0 = 0$ (A1: x/0)
        (2.5) $0^*s(a) = 0$ (2.3, 2.4, =-trans)
    (3) $\forall x(0^*x = 0)$ (1, 2, ind: $\Phi/0^*\_ = 0$)

Proof of (MT4):

    (1) $0^*(b^*c) = 0$ (MT1: x/b*c)
    (2) $0^*b = 0$ (MT1: x/b)
    (3) $0^*c = 0$ (MT1: x/c)
    (4) $(0^*b)^*c = 0$ (2, 3, =elim-right)
    (5) $0^*(b^*c) = (0^*b)^*c$ (4, 1, =-elim-right)
    (6) $\forall y \forall z(0^*(y^*z) = (0^*y)^*z)$ (5, $\forall$-intro: y/b, z/c)
    (7) SUBDERIVATION
        (7.1) $\forall y \forall z(a^*(y^*z) = (a^*y)^*z)$ (assumption)
        (7.2) $a^*(b^*c) = (a^*b)^*c$ (7.1, $\forall$-elim: y/b, z/c)
        (7.3) $s(a)^*(b^*c) = a^*(b^*c) + b^*c$ (MT2: x/a, y/b*c)
        (7.4) $s(a)^*(b^*c) = (a^*b)^*c + b^*c$ (7.2, 7.3, =-elim)
        (7.5) $(a^*b)^*c + b^*c = (a^*b + b)^*c$ (MT6: x/a*b, y/b, z/c; =-symm)
        (7.6) $s(a)^*b = a^*b + b$ (MT2: x/a, y/b)
        (7.7) $(a^*b)^*c + b^*c = (s(a)^*b)^*c$ (7.6, 7.5, =-elim-right)
        (7.8) $s(a)^*(b^*c) = (s(a)^*b)^*c$ (7.4, 7.7, =-trans)
        (7.9) $\forall y \forall z(s(a)^*(y^*z) = (s(a)^*y)^*z)$ (7.8, $\forall$-intro: y/b, z/c)
    (8) $\forall x \forall y \forall z(x^*(y^*z) = (x^*y)^*z)$ (6, 7, ind: $\Phi/\forall y \forall z(\_^*(y^*z) = (\_^*y)^*z)$)

(2) (a) (1) $a^*s(0) = a^*0 + a$ (M2)
            (2) $a^*0 = 0$ (M1)
            (3) $a^*s(0) = 0 + a$ (2, 1, =-elim)
            (4) $0 + a = a$ (AT1)
            (5) $a^*s(0) = a$ (3, 4, =-trans)
            (6) $\forall x(x^*s(0) = x)$ (5, $\forall$-intro: x/a)

    (d) (1) $\forall x(\sim s(x) = 0)$ (S1)

(2) **SUBDERIVATION**

    (2.1) $0 = s(0)$ (assumption)

    (2.2) $s(0) = 0$ (2.1, =-symm)

    (2.3) $\sim s(0) = 0$ (1, ∀-elim: x/0)

(3) $\sim 0 = s(0)$ (2, $\sim$-intro)

(4) **SUBDERIVATION**

    (4.1) $\sim a = s(a)$ (assumption)

    (4.2) **SUBDERIVATION**

        (4.2.1) $s(a) = s(s(a))$ (assumption)

        (4.2.2) $a = s(a)$ (4.2.1, s-elim)

        (4.2.3) $\sim a = s(a)$ (4.1, rep)

    (4.3) $\sim s(a) = s(s(a))$ (4.2, $\sim$-intro)

(5) $\forall x(\sim x = s(x))$ (3, 4, ind: $\Phi/\sim\_ = s(\_)$)

(h) (1) **SUBDERIVATION**

    (1.1) $0^*a = 0$ (assumption)

    (1.2) $0 = 0$ (=-intro)

    (1.3) $0 = 0 \lor a = 0$ (1.2, $\lor$-intro-right)

(2) $0^*a = 0 \rightarrow 0 = 0 \lor a = 0$ (1, $\rightarrow$-intro)

(3) $\forall y(0^*y = 0 \rightarrow 0 = 0 \lor y = 0)$ (2, ∀-intro: y/a)

(4) **SUBDERIVATION**

    (4.1) $\forall y(a^*y = 0 \rightarrow a = 0 \lor y = 0)$ (assumption)

    (4.2) **SUBDERIVATION**

        (4.2.1) $s(a)^*b = 0$ (assumption)

        (4.2.2) $s(a)^*b = a^*b + b$ (MT2: x/a, y/b)

        (4.2.3) $a^*b + b = 0$ (4.2.2, 4.2.1, =-elim)

        (4.2.4) $a^*b + b = 0 \rightarrow b = 0$ (question 2(e): x/a*b, y/b)

        (4.2.5) $b = 0$ (4.2.3, 4.2.4, $\rightarrow$-elim)

        (4.2.6) $s(a) = 0 \lor b = 0$ (4.2.5, $\lor$-intro-left)

    (4.3) $s(a)^*b = 0 \rightarrow s(a) = 0 \lor b = 0$ (4.2, $\rightarrow$-elim)

    (4.4) $\forall y(s(a)^*y = 0 \rightarrow s(a) = 0 \lor y = 0)$ (4.3, ∀-intro: y/b)

(5) $\forall x\forall y(x^*y = 0 \rightarrow x = 0 \lor y = 0)$

                (3, 4, ind: $\Phi/\forall y(\_^*y = 0 \rightarrow \_ = 0 \lor y = 0)$)

CHAPTER 7

Exercise 7.1a

(1) (a) W_1, (b) none, (d) W_2, W_5–W_{10}, (e) W_2, W_4–W_{10}, (f) all, (j) all except W_8.

(2) (b) (i) $\Diamond P \land \sim \Diamond Q$, (ii) $\Diamond Q \land \sim \Diamond\Diamond Q$,

    (d) W_2 and W_{10}.

(3) The only changes from the original table are:

	P	$\Diamond P$	$\Box P$	Q	$\Diamond Q$	$\Box Q$	R	$\Diamond R$	$\Box R$
W_2					T	T	T			
W_5	T	T	T							
W_9								T	T	T
W_{10}					T	T	T			

Exercise 7.2

(1) (a) Let (W, R, I) be any interpretation, and let w be any state in W. Suppose $\vDash_{W,R,I} \Box(A \wedge B)[w]$, and let w' be any state accessible from w. Then $\vDash_{W,R,I} A \wedge B[w']$ and hence $\vDash_{W,R,I} A[w']$ and $\vDash_{W,R,I} B[w']$. Since w' is any state accessible from w, this means that $\vDash_{W,R,I} \Box A[w]$ and $\vDash_{W,R,I} \Box B[w]$, and hence $\vDash_{W,R,I} \Box A \wedge \Box B[w]$. It follows that $\Box(A \wedge B) \rightarrow \Box A \wedge \Box B$ is universally valid, as required.

(2) (a) We shall construct an interpretation which falsifies $\Box A \rightarrow A$ at some state. Let $W = \{w, w'\}$, and $R = \{(w, w')\}$. Let $I(A) = \{w'\}$. Then $\vDash_{W,R,I} A[w']$, and hence $\vDash_{W,R,I} \Box A[w]$. On the other hand, $\vDash_{W,R,I} \sim A[w]$, and hence $\vDash_{W,R,I} \sim (\Box A \rightarrow A)[w]$. So $\Box A \rightarrow A$ is not universally valid.

Exercise 7.3

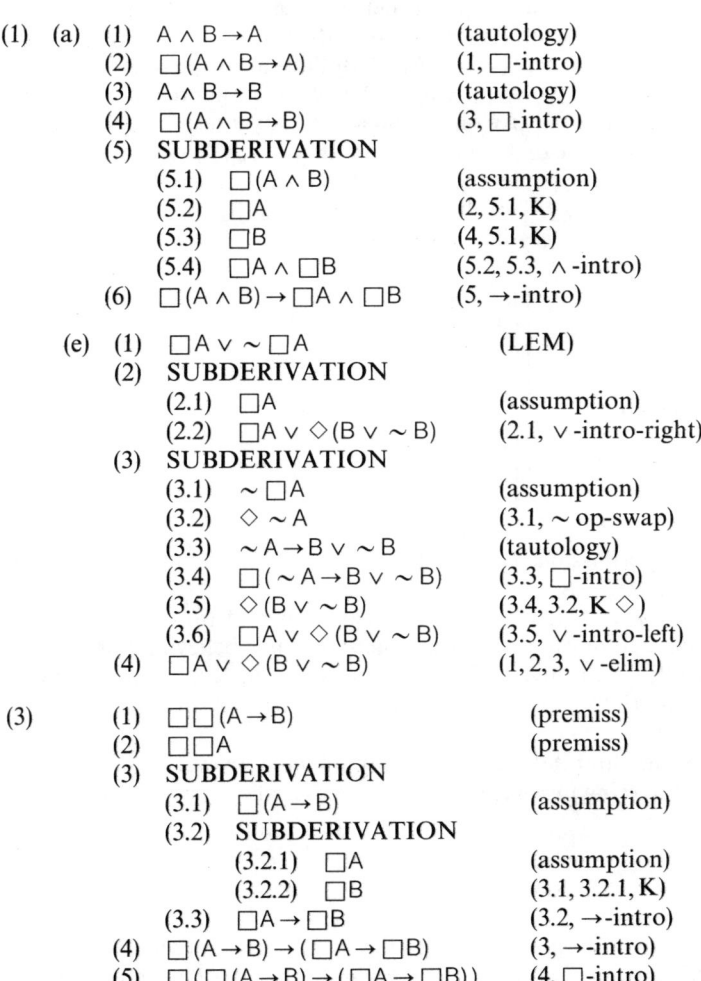

(1) (a) (1) $A \wedge B \rightarrow A$ (tautology)
      (2) $\Box(A \wedge B \rightarrow A)$ (1, $\Box$-intro)
      (3) $A \wedge B \rightarrow B$ (tautology)
      (4) $\Box(A \wedge B \rightarrow B)$ (3, $\Box$-intro)
      (5)   **SUBDERIVATION**
           (5.1) $\Box(A \wedge B)$ (assumption)
           (5.2) $\Box A$ (2, 5.1, K)
           (5.3) $\Box B$ (4, 5.1, K)
           (5.4) $\Box A \wedge \Box B$ (5.2, 5.3, $\wedge$-intro)
      (6) $\Box(A \wedge B) \rightarrow \Box A \wedge \Box B$ (5, $\rightarrow$-intro)

    (e) (1) $\Box A \vee \sim \Box A$ (LEM)
      (2)   **SUBDERIVATION**
           (2.1) $\Box A$ (assumption)
           (2.2) $\Box A \vee \Diamond(B \vee \sim B)$ (2.1, $\vee$-intro-right)
      (3)   **SUBDERIVATION**
           (3.1) $\sim \Box A$ (assumption)
           (3.2) $\Diamond \sim A$ (3.1, $\sim$ op-swap)
           (3.3) $\sim A \rightarrow B \vee \sim B$ (tautology)
           (3.4) $\Box(\sim A \rightarrow B \vee \sim B)$ (3.3, $\Box$-intro)
           (3.5) $\Diamond(B \vee \sim B)$ (3.4, 3.2, K $\Diamond$)
           (3.6) $\Box A \vee \Diamond(B \vee \sim B)$ (3.5, $\vee$-intro-left)
      (4) $\Box A \vee \Diamond(B \vee \sim B)$ (1, 2, 3, $\vee$-elim)

  (3)   (1) $\Box\Box(A \rightarrow B)$ (premiss)
      (2) $\Box\Box A$ (premiss)
      (3)   **SUBDERIVATION**
           (3.1) $\Box(A \rightarrow B)$ (assumption)
           (3.2)   **SUBDERIVATION**
               (3.2.1) $\Box A$ (assumption)
               (3.2.2) $\Box B$ (3.1, 3.2.1, K)
           (3.3) $\Box A \rightarrow \Box B$ (3.2, $\rightarrow$-intro)
      (4) $\Box(A \rightarrow B) \rightarrow (\Box A \rightarrow \Box B)$ (3, $\rightarrow$-intro)
      (5) $\Box(\Box(A \rightarrow B) \rightarrow (\Box A \rightarrow \Box B))$ (4, $\Box$-intro)

(6) $\Box(\Box A \to \Box B)$	(5, 1, K)
(7) $\Box\Box B$	(6, 2, K)

Exercise 7.4

(1) (a) (1) $\Diamond\Diamond A$ (premiss)

     (2) SUBDERIVATION

(2.1) $\sim\Diamond A$	(assumption)
(2.2) $\Box\sim A$	(2.1, $\sim$ op-swap)
(2.3) $\Box\Box\sim A$	(2.2, trans-$\Box$)
(2.4) $\sim\Diamond\Diamond A$	(2.3, op-swap)
(2.5) $\Diamond\Diamond A$	(1, rep)
(3) $\sim\sim\Diamond A$	(2, $\sim$-intro)
(4) $\Diamond A$	(3, $\sim$-elim)

(2) Suppose first that $\Box A \to A$ is valid in (W, R). Let w be any state in W. Define an interpretation function I such that $I(A) = \{w' \in W \mid wRw'\}$. Then $\vDash_{W,R,I} \Box A[w]$. We also have $\vDash_{W,R,I} \Box A \to A[w]$, and hence $\vDash_{W,R,I} A[w]$. But this means that $w \in I(A)$, hence that wRw. But w was any state in W, so R is a reflexive relation on W.

    Suppose conversely that R is reflexive on W, and let I be any interpretation function on (W, R). Let w be any member of W, and suppose that $\vDash_{W,R,I} \Box A[w]$. This means that $\vDash_{W,R,I} A[w']$ for any w' accessible from w in (W, R). Since R is reflexive, w itself is accessible from w, hence $\vDash_{W,R,I} A[w]$. It follows that $\vDash_{W,R,I} \Box A \to A[w]$, and since I and w were arbitrary, $\Box A \to A$ is valid in (W, R).

(3) (a) (1) SUBDERIVATION

(1.1) $\Box A$	(assumption)
(1.2) $\Diamond\Box A$	(1.1, refl-$\Diamond$)
(2) $\Box A \to \Diamond\Box A$	(1, $\to$-intro)
(3) $\Box(\Box A \to \Diamond\Box A)$	(2, $\Box$-intro)
(4) SUBDERIVATION	
(4.1) $\Box A$	(assumption)
(4.2) $\Box\Box A$	(4.1, trans-$\Box$)
(4.3) $\Box\Diamond\Box A$	(3, 4.2, K)
(5) $\Box A \to \Box\Diamond\Box A$	(4, $\to$-intro)

(6) (a) $\Diamond A$ is read 'It is permissible that A'. Schemas: (i) invalid, (ii) valid, (iii)–(v) probably do not make sense.

    (d) $\Diamond A$ is read 'John does not disbelieve A'. Schemas: (i) invalid, (ii) valid, (iii) doubtfully valid (invalid if John can disbelieve something while believing that he does not), (iv) doubtfully valid (invalid if John can falsely believe that he believes something), (v) doubtfully valid (invalid if John can believe something without believing that he does).

    Note carefully the doubtful status of many of these answers: great caution must be exercised in trying to apply modal logic to everyday ideas such as belief.

Exercise 7.5

(1) (b) Let $(T, <, I)$ be any interpretation, and let t be any time in T. Suppose $\vDash_{T,<,I} PGA[t]$. This means that for some $t' < t, \vDash_{T,<,I} GA[t']$. This in turn means

that for any $t'' > t'$, $\vDash_{T, <, I} A[t'']$. But $t > t'$, hence $\vDash_{T, <, I} A[t]$. Since $(T, <, I)$ and t were arbitrary, $PGA \rightarrow A$ is universally valid.

(2) Assume HF-intro. Then we have

(1) PGA   (assumption)
(2) SUBDERIVATION
    (2.1) $\sim A$ (assumption)
    (2.2) HF $\sim A$ (2.1, HF-intro)
    (2.3) $\sim PGA$ (2.2, op-swap)
    (2.4) PGA   (1, rep)
(3) $\sim \sim A$ (2, $\sim$-intro)
(4) A   (3, $\sim$-elim)

The derivation of FH-elim from GP-intro is precisely analogous.

Bibliography

M. Bergmann, J. Moor, and J. Nelson (1980), *The Logic Book* (New York: Random House).

E. W. Beth (1955), 'Semantic entailment and formal derivability', *Mededelingen van de Koninklijke Nederlandse Akademie van Wetenschappen, Afdeling Letterkunde,* N.R. **18** (13) (Amsterdam); reprinted in J. Hintikka (ed.), *The Philosophy of Mathematics* (Oxford University Press, 1969).

A. Church (1936), 'An unsolvable problem of elementary number theory', *The American Journal of Mathematics,* **58**, 345–63, reprinted in M. Davis (ed.), *The Undecidable* (Raven Press, New York, 1965).

S. Cook (1971), 'The complexity of theorem proving procedures', *Proc. Third Annual ACM Symposium on the Theory of Computing* (New York: ACM).

R. W. Floyd (1967), 'Assigning meaning to programs', *Proceedings of Symposia in Applied Mathematics* (American Mathematical Society), **19**, 19–32.

J. H. Gallier (1987), *Logic for Computer Science: Foundations of Automatic Theorem Proving* (New York: John Wiley and Sons).

A. Galton (ed.) (1987), *Temporal Logics and their Applications* (London: Academic Press).

K. Gödel (1930), 'Die Vollständigkeit der Axiome des logischen Funktionenkalküls', *Monatshefte für Mathematik und Physik,* **37**, 349–60; English translation, 'The completeness of the axioms of the functional calculus of logic', in J. van Heijenoort (ed.), *From Frege to Gödel: A Source Book in Mathematical Logic, 1879–1931* (Harvard University Press, 1967).

K. Gödel (1931), 'Über formal unentscheidtbare Sätze der *Principia Mathematica* und verwandter Systeme I', *Monatshefte für Mathematik und Physik,* **38**, 173–198; English translation 'On formally undecidable propositions of the *Principia Mathematica* and related systems, I' in M. Davis (ed.), *The Undecidable* (Raven Press, New York, 1965).

L. Henkin (1949), 'The completeness of the first-order functional calculus', *Journal of Symbolic Logic,* **14**, 159–66; reprinted in J. Hintikka (ed.), *The Philosophy of Mathematics* (Oxford University Press, 1969).

C. A. R. Hoare (1969), 'An axiomatic basis for computer programming', *Communications of the ACM,* **12**, 576–83.

D. Hofstadter (1979), *Gödel, Escher, Bach: an Eternal Golden Braid* (Brighton: Harvester Press).

C. Hogger (1984), *Introduction to Logic Programming* (London: Academic Press).

G. Hughes and M. Cresswell (1968), *An Introduction to Modal Logic* (London: Methuen).

R. Jeffrey (1965), *Formal Logic: Its Scope and Limits* (New York: McGraw-Hill).

J. W. Lloyd (1984), *Foundations of Logic Programming* (Berlin: Springer).

Z. Manna and A. Pnueli (1981), 'Verification of concurrent programs: the temporal framework', in R. S. Boyer and J. S. Moore (eds.), *The Correctness Problem in Computer Science* (London: Academic Press).

E. Nagel and J. Newman (1958), *Gödel's Proof* (London: Routledge and Kegan Paul).

B. Russell (1905), 'On denoting', *Mind,* **14**, 479–93, reprinted in H. Feigl and W. Sellars (eds.), *Readings in Philosophical Analysis* (New York: Appleton-Century-Crofts, 1949).

R. Smullyan (1968), *First Order Logic* (Berlin: Springer).

H. Wilf (1988), *Algorithms and Complexity* (Prentice-Hall).

Index